Joint Ventures in Construction 2:
Contract, Governance, Performance and Risk

ice

Institution of Civil Engineers

publishing

Joint Ventures in Construction 2: Contract, Governance, Performance and Risk

Edited by **Kiyoshi Kobayashi, Khairuddin Abdul Rashid, Masamitsu Onishi and Sharina Farihah Hasan**

Published by ICE Publishing, 40 Marsh Wall, London E14 9TP.

Full details of ICE Publishing sales representatives and distributors can be found at:

www.icevirtuallibrary.com/info/printbooksales

First edition published 2012

www.icevirtuallibrary.com

A catalogue record for this book is available from the British Library

ISBN: 978-0-7277-5783-8

ICE Publishing is a division of Thomas Telford Ltd, a wholly-owned subsidiary of the Institution of Civil Engineers (ICE).

Printed and bound in Great Britain by CPI Group (UK) Ltd, Croydon, CR0 4YY

FSC
www.fsc.org
MIX
Paper from responsible sources
FSC® C013604

CONTENTS

ACKNOWLEDGEMENTS

We would like to thank all contributors for their consent to have their works incorporated as chapters in this book. The chapters were based on selected papers that were presented during past series of The International Conference on Multi-national Joint Venture for Construction Works that were held annually since 2005. This book represents the Conferences from 2008 to 2011.

We would also like to thank Ms. Carolyn Mizutani for her effort in assisting the Editors in proof-reading the manuscript and Mr. Saifull Bahri Shamshul Anwar for formatting the book.

We would like to express our heart-felt gratitude to the institutional support provided by many organizations and institution and specifically to the assistance provided by Kyoto University of Japan and the International Islamic University Malaysia.

Last but not least, we are indebted to the following individuals for their continuous support and encouragement;

- Professor Dr. Yuzo Matsuoka, Director of Kyoto University's GCOE (Global Centre of Excellence). This book represents the effort being promoted by the GCOE under the 'Global Centre for Education and Research on Human Security Engineering for Asian Megacities'
- Dr Yuzo Ohnishi, Professor Emeritus, Executive Member and Vice President of Kyoto University
- Professor Dr. Hiroyasu Ohtsu, Vice Dean of Graduate School of Engineering, Kyoto University, and
- Professor Dato' Sri Dr. Zaleha Kamaruddin, Rector, International Islamic University Malaysia

Kiyoshi Kobayashi, Khairuddin Abdul Rashid,
Masamitsu Onishi and Sharina Farihah Hasan
May, 2012

DEDICATION

This book is dedicated to the victims of the March 11th 2011 earthquake and tsunami tragedy, Tohoku region, Japan. May the victims rest in peace and their families be blessed with strength and fortitude.

THE EDITORS

Kiyoshi KOBAYASHI, BSc., MSc., Dr Eng., is Professor of infrastructure economics at Kyoto University. He serves on the editorial boards of the American Society of Civil Engineers, the Annals of Regional Science, the Journal of Applied Regional Science and the Japan Journal of Civil Engineering. He was also the Editor in chief for the Journal of Japan Society of Civil Engineers and Vice Editor in chief, Journal of Infrastructure Systems. He is the author and co-editor of 53 books and more than 370 technical papers.

KHAIRUDDIN Abdul Rashid, Dip., BSc., MSc., PhD., MRISM, BQSM, is Dean of the Kulliyyah of Architecture and Environmental Design, Professor at the Department of Quantity Surveying, and Head of the Procurement and Project Delivery System Research Unit, International Islamic University Malaysia (IIUM). He is Co-founder and Co-editor of the Journal of Quantity Surveying and Construction Business and Editor in chief of the Journal of Architecture, Planning and Construction Management and reviewer of several international refereed journals on construction in developing countries, construction management and economics. He has authored, co-authored, edited and co-edited 12 books and more than 155 technical papers on quantity surveying, procurement, contract including shari'ah compliant and PPP/PFI.

Masamitsu ONISHI, BSc., MSc., Dr. Eng., is Assistant Professor at the Department of Urban Management, Graduate School of Engineering, Kyoto University, Japan. His doctoral degree was awarded by Kyoto University. He has published 40 academic papers which concern dispute resolution in construction projects, public private partnership, project financing. Dr. Onishi is a board member of Research Committee of Private Finance Initiative in Japan Society of Civil Engineering (JSCE). He is a recipient of JSCE young researcher award 2006 and best paper award 2008.

SHARINA FARIHAH Hasan, Dip., BSc., MSc. (Eng.), PhD. (Eng.), is Assistant Professor at the Department of Quantity Surveying, Kulliyyah of Architecture and Environmental Design and Associate Member of the Procurement and Project Delivery System Research Unit, International Islamic University Malaysia. She is an Editorial Board member of the Journal of Architecture, Planning and Construction Management. She has presented and published works on quantity surveying, pre-qualification of contractors and labour market and productivity. Dr. Sharina has authored and co-authored more than 20 papers, in books chapters and proceedings.

CONTRIBUTORS' BIOGRAPHIES

(Listed according to Chapter)

KHAIRUDDIN Abdul Rashid, Dip., BSc., MSc., PhD., MRISM, BQSM, is Dean of the Kulliyyah of Architecture and Environmental Design, Professor at the Department of Quantity Surveying, and Head of the Procurement and Project Delivery System Research Unit, International Islamic University Malaysia (IIUM). He is Co-founder and Co-editor of the Journal of Quantity Surveying and Construction Business and Editor in chief of the Journal of Architecture, Planning and Construction Management and reviewer of several international refereed journals on construction in developing countries, construction management and economics. He has authored, co-authored, edited and co-edited 12 books and more than 155 technical papers on quantity surveying, procurement, contract including shari'ah compliant and PPP/PFI (Chapter 1 & 11).

Graham M. WINCH, BA., MA., MSc., PhD., MAPM, MCIOB, is Professor of Project Management in Manchester Business School and Jubilee Professor, Chalmers University. He has published over 40 refereed journal articles, complemented by numerous book chapters, conference papers, and research reports. He has held numerous EPSRC awards. He was PI for *ProRIde: Project Risk Identification* and the MBS contribution to *Network Enabled Capability through Innovative Systems Engineering* in collaboration with BAE Systems and a number of UK universities (Chapter 2).

Apichart PRASITSOM, BEng., MEng., is researcher/PhD. student at the Department of Civil Engineering, Faculty of Engineering, Chulalongkorn University, Bangkok, Thailand. His research interests are construction joint ventures, risk management, performance management, strategic management and building code (Chapter 3).

Veerasak LIKIHITRUANGSILP, BSc., MSc., PhD., is Assistant Professor at the Department of Civil Engineering, Faculty of Engineering, Chulalongkorn University, Bangkok, Thailand. He has published and presents works in the areas of civil engineering, construction law and construction joint ventures (Chapter 3).

Katsuji ISHIHARA, BSc., MSc., MBA, PhD., is Professor at the Graduate School of Management, Kyoto University, Japan. He is also a senior executive general manager and principal consultant at the Nikken Sekkei Research Institute, Japan. His experience in practice include a advisory team leader in many PFI/PPP projects in Japan (Chapter 4 & 16).

Sufang WENG, BSc., MSc., is a full-time CSR consultant in Taiwan. She studied Private Finance Initiatives and CSR (Corporate Social Responsibility) topics in the Graduate School of Management, Kyoto University after she had worked in LED industry in Taiwan for 6 years (Chapter 4).

AZMAN Mohd Noor, BA., MA., PhD., is Associate Professor at the Department of Fiqh and Usul al Fiqh, Kulliyyah of Islamic Revealed Knowledge and Human Sciences, International Islamic University Malaysia. In 2006, he was appointed as a Shariah advisor at Al-Rajhi Bank, Malaysia and continues to hold the position until present. He has published papers in local and international journals and books (Chapter 5).

SAIDATOLAKMA Mohd Yunus, BA., MA., is PhD. student in Islamic Banking and Finance at II*i*BF, International Islamic University Malaysia (IIUM). She has a Bachelor of Islamic Revealed Knowledge (specializing in Fiqh and Usul al-Fiqh) from IIUM and an MA in Islamic Revealed Knowledge (specializing in Fiqh) from the same university (Chapter 5).

Gautam RAY, BSc., MSc., LL.B., MA., Ph.D., is Professor at the Graduate School of Management, Kyoto University. He is a civil servant of the rank of Additional Secretary to the Government of India and presently serving a foreign service deputation tenure in Japan. He was in charge of a World Bank Project in India for capacity development in tax administration. He was also a World Bank Scholar at Boston University between 1995 and 1997 and empanelled as a fiscal expert in the International Monetary Fund, Washington D.C. in 1993 (Chapter 6).

Kerry LONDON, BSc.(Arch), BArch., MBldg., PhD., is Professor of Construction Management at the RMIT University. She has published two key texts in supply chain theory and practice, Construction Supply Chain Economics in the Spons Research Series and Construction Supply Chain Management Handbook by Taylor and Frances in the US. Her research interests are construction supply chain theory and practice, design management and internationalization (Chapter 7).

Jessica SIVA, BAppSc.(Arch), BArch(Hons), MPhil., is PhD scholarship holder at RMIT University, Australia in the School of Property, Construction and Project Management. She has over 40 research publications in the areas of design and client management, international collaborative practice, government supply chain management, social, cultural and intellectual capital, sustainable urban development decision-making policy, process and practice and mega projects (Chapter 7).

Tsung-Chieh TSAI, PhD., is Associate Professor of Graduate Institute of Construction and Property Management at the National Yunlin University of Science and Technology. He is the author and co-editor of 5 books and more than 69 technical papers (Chapter 8).

Hung-Chi CHEN, PhD., is Secretary at the Directorate Room in Chiayi County Government, Taiwan. Chen's areas of specialization are Regional and City Development, Land Utilizing, Natural Disasters Recovering Management (Chapter 8).

Min-Lan YANG, PhD., is Assistant manager in JU-MING construction Co., Ltd. Taiwan. Yang's areas of specialization include Construction Management, Engineering Management and Design Management (Chapter 8).

AINUL JARIA Maidin, LL.B., Master in Comparative Law, PhD., is Associate Professor and Head of the Research Management Unit, Ahmad Ibrahim Kulliyyah of Laws, International Islamic University Malaysia. She is a Non-Practising Advocate and Solicitor of High Court of Malaya, Consultant for Project to Review the National Land Code 1965 for Department of Director General of Land and Mines, etc. She has published in local and international journals in areas relating to land tenure, development control and environmental protection (Chapter 9).

Masamitsu ONISHI, BSc., MSc., Dr. Eng., is Assistant Professor at the Department of Urban Management, Graduate School of Engineering, Kyoto University, Japan. His doctoral degree was awarded by Kyoto University. He has published 40 academic papers which concern dispute resolution in construction projects, public private partnership, project financing. Dr. Onishi is a board member of Research Committee of Private Finance Initiative in Japan Society of Civil Engineering (JSCE). He is a recipient of JSCE young researcher award 2006 and best paper award 2008 (Chapter 10, 15 & 17).

Toshihiko OMOTO, BSc., MSc., Dr. Eng., FICE, FCIArb, is former Professor and currently a Visiting Professor at Kyoto University teaching contract administration and dispute resolution. He is a Fellow of the Chartered Institute of Arbitrators, UK (FCIArb), and a FIDIC President's Approved Adjudicator. His experiences as Neutral include a Dispute Board Member in China, Madagascar and Japan, an arbitrator in Nepal and a mediator in Japan. He is serving as an Assessment Panel for Adjudicators of FIDIC (APA) (Chapter 10).

Kiyoshi KOBAYASHI, BSc., MSc., Dr. Eng., is Professor of infrastructure economics at Kyoto University. He serves on the editorial boards of the American Society of Civil Engineers, the Annals of Regional Science, the Journal of Applied Regional Science and the Japan Journal of Civil Engineering. He was also the Editor in chief for the Journal of Japan Society of Civil Engineers and Vice Editor in chief, Journal of Infrastructure Systems. He is the author and co-editor of 53 books and more than 370 technical papers (Chapter 10, 12, 15, 17 & 18).

Lei SHI, BSc., MSc., Dr. Eng., is Associate Professor at Department of Construction Management, Dalian University of Technology, Dalian, China. Dr. Shi earned the doctoral degree in Urban Management from Kyoto University (Japan) and master degree in Architecture and Building Science from Tohoku University (Japan), as well as the Bachelor's degree in Architecture from Dalian University of Technology. His research interests include Construction Contract, Construction Institutional Design, Infrastructure Management and Risk Management (Chapter 12 & 17).

Taisuke MIYAO, LLM, is Procurement Specialist of Asian Development Bank. Mr. Miyao previously worked as Director of Procurement Supervision Division of Japan International Cooperation Agency which is the government aid agency in Japan. He has numerous experiences in procurement reform activities including commenting on draft procurement laws, and consultant recruitment and bidding for infrastructure projects in developing countries. Admitted in the bar association of the State of New York (Chapter 12).

Myungsik DO, BSc., MSc., Dr. Eng., is Associate Professor in the Department of Urban Engineering, Hanbat National University, Korea. His research interests are transportation planning, SOC project evaluation and asset management of infrastructure facilities. He serves on the editorial boards of the Journal of the Korean Society of Civil Engineers and Journal of Korean Society of Transportation. He is a recipient of the 2001 KSCE Young Researcher Award (Chapter 13).

Sang Hyuk LEE, BSc., MSc., PhD., is post-doctoral researcher in the Department of Urban Engineering, Hanbat National University, Korea. His research interests are in transportation asset management especially pavement management system (PMS) with risk and uncertainty management (Chapter 13).

Joseph H.L. CHAN, BSc.(Hons), MSc., PhD., MHKIS, MRICS, is Postdoctoral Research Fellow at the Department of Civil Engineering of The University of Hong Kong. He is a quantity surveyor by profession. His research interests cover construction procurement management, carbon labelling schemes, and construction safety. He has published a number of research papers on the identification, analysis and allocation of key risk factors associated with target cost contracts in leading academic journals and international conference proceedings. (Chapter 14).

Daniel W.M. CHAN, BEng, (Hons), PhD., MAPM, MHKICM, MASCE, is Associate Professor in Construction Project Management at the Department of Building and Real Estate of The Hong Kong Polytechnic University. His current research interests include construction time performance, construction procurement systems, relational and collaborative contracting, partnering and alliancing, target cost contracting, public private partnership and construction safety management. He has published 12 refereed research monographs, 1 scholarly textbook, 2 book chapters, 71 journal articles, and 93 conference papers. Daniel is Member of the Editorial Board for The Open Construction and Building Technology Journal (TOCBTJ) since 2008. (Chapter 14).

Patrick T.I. LAM, Dip.Finance, MSc., PhD., AP(HK), MHKIS, MHKIE, MHKICM, MSISV, MSIArb, MRICS, MCIOB, RPE, RPS, CCE, is Associate Professor in Construction Economics and Project Financing at the Department of Building and Real Estate of The Hong Kong Polytechnic University. He is a quantity surveyor by profession. His present research interests encompass procurement systems, project financing, public private partnership, construction contracts and specifications, together with quality and buildability issues (Chapter 14).

Yasuhiro SHOJI, BEng., MEng., MA, P.E.Jp., is a General Manager of Utsunomiya National Highway Office, Kanto Regional Development Bureau, Ministry of Land, Infrastructure, Transport and Tourism(MLIT). He has worked for the ministry as a researcher and a practitioner for more than twenty years, specializing in road planning, project evaluation, disaster risk management, and private-private-partnership(PPP). He serves on the editorial board of the Infrastructure Planning Review, Japan Society of Civil Engineers(JSCE) (Chapter 15).

SHARINA FARIHAH Hasan, Dip., BSc., MSc. (Eng.), PhD. (Eng.), is Assistant Professor at the Department of Quantity Surveying, Kulliyyah of Architecture and Environmental Design and Associate Member of the Procurement and Project Delivery System Research Unit, International Islamic University Malaysia. She is an Editorial Board member of the Journal of Architecture, Planning and Construction Management. She has presented and published works on quantity surveying, pre-qualification of contractors and labour market and productivity. Dr. Sharina has authored and co-authored more than 20 papers, in book chapters and proceedings (Chapter 18).

Kakuya MATSUSHIMA, BSc., MSc., Dr. Eng., MJSCE, is Associate Professor at the Department of Urban Management, Graduate School of Engineering, Kyoto University Japan. He serves on the editorial board of Infrastructure Planning Review, Japan Society of Civil Engineers (JSCE). He is the author and co-editor of 2 books and more than 100 technical reports including more than 30 refereed journal articles. His major is transportation economics and infrastructure planning (Chapter 18).

LIST OF TABLES

LIST OF FIGURES

LIST OF APPENDICES

(Listed according to Chapter)

LIST OF ABBREVIATIONS

AC	Appeal Cases
ADR	Alternative Dispute Resolution
AEC	Architectural, Engineering and Construction
AHP	Analytic Hierarchy Process
AMI	Australian-Malaysian Institute
APEC	Asia-Pacific Economic Cooperation
BdB	Building down Barriers
BLR	Building Law Reports
BOO	Building-Operation-Owned
BOOT	Building-Operation-Owned-Transfer
BOP	Bottom of Pyramid
BOT	Build-Operate-Transfer
BTL	Build Transfer Lease
BTO	Build Transfer Operate
C&OA	Construction and Operation Agreement
CBD	Central Business District
CGE	Computable General Equilibrium
CIDB	Construction Industry Development Board Malaysia
CJV	Construction Joint Venture
CLB	Company Law Board
CLJ	Construction Law Journal
CPA	Construction and Planning Agency
CPIPL	Chatterjee Petrochem (India) Pvt.Ltd
CPMC	Chatterjee Petrochem (Mauritius) Company Co.
CPS	Concept Phase Submission
CSC	Contractors' Service Centre
DAB	Dispute Adjudication Board
DB	Design and Build
DBO	Design-Build-Operate
DIPP	Department of Industrial Policy and Promotion
DoD	Department of Defense
DRB	Dispute Review Board
DSC	Domestic Subcontractor
D-Supplier	Domestic Supplier
EOU	Export Oriented Units
ESCO	Energy Service Companies
EV	Equivalent Variation
FDI	Foreign Direct Investment
FIDIC	Fédération Internationale des Ingénieurs-Conseils
GCW	Government Contract Works
GDP	Gross Domestic Product
GEMPACK	General Equilibrium Modeling PACKage
GFI	Global Financial Integrity
GMig2 Model	Global Migration Model Version 2

GMP	Guaranteed Maximum Price
GOI	Government of India
GoWB	Government of West Bengal
GTAP	Global Trade Analysis Project
HC contractor	High Compliant Contractors
HPL	Haldia Petrochemical Company Ltd
HRC	Historic Reference Cost
HSBC	Hongkong and Shanghai Banking Corporation
IBS	Industrialized Building System
ICA	International Committee of Arbitration
ICC	International Chamber of Commerce
IDBI	Industrial Development Bank of India
ILS	Integrated Logistic Support
IMF	International Monetary Fund
IOC	Indian Oil Corporation
IST	Integrated Supply Teams
JV	Joint Venture
JVA	Joint Venture Agreement
LC contractor	Low Compliant Contractors
LCC	Life Cycle Costing
LTH	Lembaga Tabung Haji
M&E	Mechanical & Electrical
MABC	Malaysian-Australian Business Council
MAIWP	Majlis Agama Islam Wilayah Persekutuan
MLJ	Malayan Law Journal
MNC	Multinational Corporations
MOU	Memorandum of Understanding
MRG	Minimum Revenue Guarantee
MUL	Maruti Udyog Ltd.
NEC2	New Engineering Contract Version 2
NEC3	New Engineering Contract Version 3
NETIS	New Technology Information System
NPV	Net Present Value
NSC	Nominated Subcontractor
N-Supplier	Nominated Supplier
OECD	Organization for Economic and Cultural Development
OIC	Organization of Islamic Cooperation
P.A	Participant's Account
PAM	Pertubuhan Arkitek Malaysia
P.b.u.h	Peace be upon him
PFI	Private Finance Initiative
PPI	Private Partnership in Infrastructure
PPP	Public Private Partnership
PWD	Public Works Department
RBI	Reserve Bank of India
REHDA	Real Estate and Housing Developers' Association
RIBA	Royal Institute of British Architects
ROW	Rest of World
S.O	Superintending Officer

SCI	Social, Cultural and Intellectual
Sdn Bhd	Sendirian Berhad (Private Limited)
SMC	Suzuki Motor Corporation
SOC	Social Overhead Capital
SPC	Special Purpose Company
SWOT	Strengths, Weaknesses, Opportunities and Threats
TCC	Target Cost Contract
TCG	The Chatterjee Group of companies
THSRC	Taiwan High Speed Rail Corp.
UK	United Kingdom
UNCTAD	United Nations Conference on Trade and Development
UNIDO	United Nations Industrial Development Organization
VFM	Value for Money
WBIDC	West Bengal Industrial Development Corporation
WTO	World Trade Organization

INTRODUCTION

Kiyoshi KOBAYASHI, KHAIRUDDIN Abdul Rashid,
Masamitsu ONISHI and SHARINA FARIHAH Hasan

Since the publication of *Joint Ventures in Construction* (Kobayashi, Khairuddin, Ofori and Ogunlana, 2009) debates on the subject of international multi-national joint ventures for construction works continue. Academics, researchers, practitioners and policy-makers argued and counter-argued, proposed and counter-proposed theories and view-points in their endeavor to make project coalitions in international multi-national joint ventures for construction works more effective and efficient.

Such debates are to be expected. This is due to reasons including the various and continuous changes taking place in the world's economy and politics, in technology, in the way in which construction businesses are being conducted, projects implemented and in the way the public, key stakeholders and the governments, initiate, manage and satisfy or otherwise of their respective expectations arising from the projects they were involved.

In *Joint Ventures in Construction* Kobayashi et.al (2009) reviewed issues of cultures and ethics, specific considerations of co-operation and relationships, legal, contractual and governmental aspects relating to international multi-national joint ventures for construction works, and the broader issues surrounding investment, productivity and subcontracting associated with construction works including those on joint ventures.

In furthering effort to provide a platform for the process of consultation on the subject of international multi-national joint ventures for construction works to continue the Editors present this book entitled JOINT VENTURES IN CONSTRUCTION 2: CONTRACT, GOVERNANCE, PERFORMANCE AND RISK.

Typically, most books tend to present discussions on construction joint ventures in the context of a national project environment and that the discussions almost always focus on joint ventures between constructors. In this book the Editors focus their thoughts on international multi-national joint ventures for construction works, and specifically on the key stakeholders, i.e. the governments, constructors, consultants and others that are involved in international joint venture projects' coalitions.

In an international multi-national joint venture for construction works the project's environment is almost always characterized by multi-dimensional and multi-cultural relationships. Hence, issues such as aims and objectives of the key stakeholders that may or may not necessarily be consistent with each other, and with that of the project's aims and objectives are considered common.

In addition, the respective stakeholder's aims and objectives may be in line with each other but dissimilar in terms of levels or they might be at opposing ends of the project spectrum and probably even at odds with the bigger interests of society and nation.

Furthermore, the project's environment may be complicated by other issues and interests that are either directly or indirectly related to the project due to the nature of construction related businesses and the fact that the market is dominated, more than before, by private sector-led economic model (Privatization, Private Finance Initiative and Public Private Partnership), monetarism and supply-side economics; and that more and more joint venture projects involve multi-nationals and multi-cultural contractual relationships[i].

In this Introduction the Editors refer to the aims and objectives of the key stakeholders, in relation to a project involving international and multi-national partners, as 'expectations' of the stakeholders.

In most standard construction textbooks the key stakeholders' expectations are managed in a typical project management, contract administration and the likes approaches and techniques. In almost all the cases however, the focus would be on achieving the project's aims and objectives that concerns time, cost and quality. But the outcome of such approaches and techniques, while on the one hand they might bring positive results in terms of meeting the project's aims and objectives but on the other hand they seldom address the expectations of the key stakeholders in the most efficient and effective manner.

Therefore, it is not be unexpected that research and media reports continue to highlight the continuous and upward trends in disputes arising from dis-satisfied stakeholders in local and in international multi-national construction joint ventures[ii]. The situation, in recent times, is exacerbated by continuous and upward increases in material and fuel prices, political and economic instabilities, or due to nature's intervention such as tsunami and earthquake in some parts of the world.

Given the above arguments, herein we would like to propose an alternative style of managing international multi-national joint ventures for construction works that in our opinion could provide a more efficient and effective way in managing the various kinds and levels of expectation of the key stakeholders. We refer to this alternative style as **'rationalization of expectations'**.

Expectation of key stakeholders, in the context of international multi-national joint ventures for construction works, may be organized by basing the expectations on their background such as their organizational objectives and common business customs in their originated countries particularly in a multi-national context. For example, if a government is involved then the expectations of the government into the joint venture would typically include social welfare benefits, protection of the environment, politically correct development and of project outcome, the provision of employment during the construction phases and aspects such as competitive advantage, retention of capital outflow, the use of indigenous labor and the likes.

In the case of the constructors, their expectations would typically relate to profit maximization, speedy project approval, minimum or little statutory control on the construction site and objective project outcome, notwithstanding other social and welfare obligations.

Somewhere in between the government and the constructors, lie the consultants. Their expectations may come into play if their presence in the joint venture is included. For example, consultants, being private companies, may share much, if not all, of the expectations of the constructors but being a party that would normally be aligned with the client (government) for they may be under the employment of the client, they may share some of the expectations of the government. See illustration in Figure I.1.

In addition, there may be other partners in the joint venture such as funder, sub-contractors and material and equipment suppliers. These parties also have their own expectations.

Figure I.1: Stakeholders' expectations in construction joint ventures

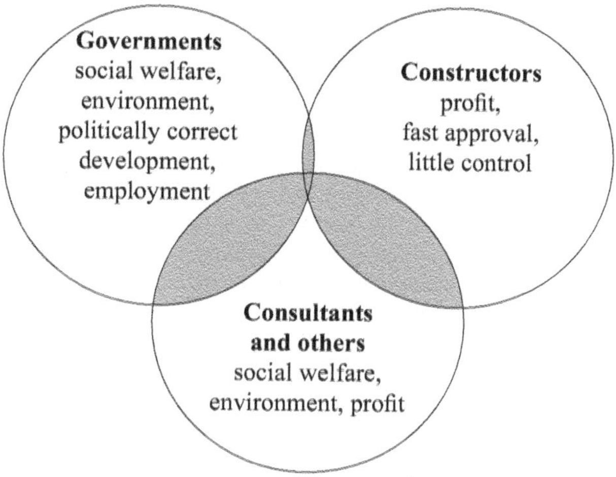

Figure I.2: Desired and non-desired zones of expectations

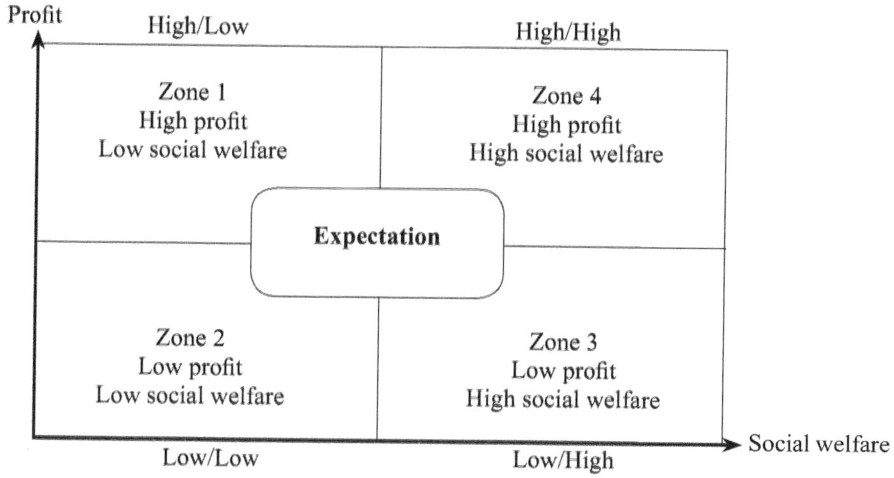

Expectations are matters that the expectations proponents would like to achieve and they would do what it takes, within the ambit of the laws and other social and cultural norms, to achieve them. Referring to Figure I.2, it is Zone 4 that all constructors aspire to be in and they would do what it takes in order to avoid Zones 1, 2 and 3, but especially Zone 2. For governments, typically they would aspire to be in Zone 3 for if they end up in the other zones they would be open for public criticisms and other negative implications.

Thus, complications would arise due to the various expectations, especially when these expectations are multi-dimensional in nature, i.e. owned by different stakeholders, expressed at various levels, at different times and that some might change over time during the project duration. Inconsistency of expectations among stakeholders may arise due to the difference in social norms in a multi-national environment. Failure to achieve/satisfy some, if not all of the expectations, by any of the stakeholders may lead to conflict. Theoretically, no stakeholder accepts failure and the consequential negative implications.

In the study of conflict management, within the context of a construction contract, most standard form of contract propose various methods of identification of conflicts and disputes and dispute resolution techniques. However, almost always that these standard forms of contracts focus on identification and resolution of disputes arising from and related to the main players in the contract only, i.e. the Client and the Contractor. But in international multi-national joint ventures for construction works the potential for conflict to happen extend beyond the contracting partners. Conflict among the stakeholders who are partners in the joint venture, if not efficiently and effectively managed, may eventually lead to dispute in the contract and could cause various negative implications to the project and its outcome. Consequently there is an urgent need for a more efficient and effective style of managing the above mention expectations.

The Editors' idea on the concept of rationalization of expectations focuses on taking steps to identify, acknowledge, and distribute various interests of individuals and organization working within a combined environment and project. Therefore, the concept may, on the one hand be seen as an approach in seeking a compromise between opposing parties akin to seeking a win-win conclusion to an ongoing dispute, but on the other hand it may also be related to managing the outcome of expectation not realized, the principal reason being to minimize potential negative implications that might arise to the affected stakeholders' and to the project.

Thus, it is within the context of the various expectations and the need for these expectations to be rationalized, as applicable, in typical international multi-national joint ventures for construction works that the Editors are making their case. The Editors propose the following steps to be taken:

(1) Identify the key stakeholders in the joint venture;
(2) For each stakeholder, identify their expectations;
(3) List down the expectations of each stakeholder and categorize these expectations possibly into three categories of low, medium and high expectation. The intention is to prioritize expectations that have the highest potential for conflict to arise if they are not realized, and expectations that have the highest positive impact to the stakeholders' interests and that of the project's objectives if they are being realized. These expectations therefore, require attention and rationalization;
(4) For those expectations that have been identified to be on the high category, identify possible ways for the expectations to be shared by all stakeholders, failing which the expectations would then be identified as the ones that have the highest requirement for them to be rationalized; and
(5) The focus of the process of rationalization of expectations would involve;
    (a) making stakeholders initially share the expectations, or
    (b) if not, accept there would be expectations that might not be realized for the sake of achieving the greater objectives of the key stakeholders' collective expectations and of the project's objectives.
    (c) Techniques such as those applied in project governance, alternative dispute resolution (for example, mediation) and construction partnering could be employed to seek agreement of the various stakeholders. The Editors' idea on the concept of rationalization of expectations of stakeholders in international multi-national joint ventures for construction works is still at the embryonic stage but they make no claim that the idea is having a pioneer in status. They acknowledge further and in-depth research and consultation are required before the idea could be moved to the next level. To this end the Editors invite readers to provide ideas and constructive criticisms.

This book JOINT VENTURES IN CONSTRUCTION 2: CONTRACT, GOVERNANCE, PERFORMANCE AND RISK discusses some of the points related to expectations of stakeholders in international multi-national joint ventures for construction works and on how these expectations could be rationalized and managed. Researchers from Asia and the Pacific region namely Malaysia, Japan, Thailand, China, Hong Kong, Taiwan, Korea, India, Australia and UK presented their research works.

Consequently, this book presented in three parts and consisting of eighteen chapters, is considered unique in the sense that it represents the workings of international multi-national joint ventures for construction works as seen from both national and international perspectives with special emphasis on;

- Part one, consisting of five chapters (Chapter 1 to Chapter 5), reviews issues on contract and contract administrative structures in joint ventures for construction works;
- Part two, consisting of seven chapters (Chapter 6 to Chapter 12), examines specific considerations on aspects related to governance in joint ventures for construction works; and
- Part three, consisting of six chapters (Chapter 13 to Chapter 18), reviews issues related to performance and risk in joint ventures for construction works.

Endnotes

(i) EC Harris, Construction disputes on the rise, in the *Construction and Contract News*, p60-61, available online at http://www.building.hk/forum/2011_0729const_donr.pdf, accessed 1st May 2012.

(ii) *Ibid*, during 2010 almost 31% of joint ventures for construction work resulted in disputes. Two key reasons given for the disputes are related to the contract administrator's 'lack of understanding of contract procedure' and 'a partiality to the employer's interests.'

SECTION 1:
CONTRACT

Joint Ventures in Construction 2
ISBN 978-0-7277-5783-8

doi: 10.1680/jvc2.57838.001

ice
Institution of Civil Engineers
publishing

Chapter 1

Terms and Conditions Agreed on by Parties in Construction Work Joint Ventures

Khairuddin Abdul Rashid, International Islamic University Malaysia

Introduction

In the context of construction work, when two or more parties agree to become partners in a joint venture to execute a project, they may either (i) combine their resources and co-operate during the bidding and the construction processes, jointly sharing the risks and profits of the joint venture according to the resources each has brought to the joint venture; or (ii) enter a bid together but confine their responsibilities, profits, and losses only to the part of the work they undertake individually. Of these two types of joint venture, the former arrangement is known as 'integrated' and the latter as 'separate' [1].

In addition, joint ventures may be in the form of either (i) unincorporated, or (ii) incorporated. Briefly, an unincorporated joint venture does not involve any structural changes, and the parties retain their own independence, basing their co-operation entirely upon the contract. Such a joint venture may be viewed as a project-based venture, which is terminated once the construction project is complete and all obligations among the parties settled. Each party is liable with regard to its own losses and share of profits. An incorporated joint venture involves structural changes and is akin to a legal partnership involving shareholders' agreement or a corporate joint venture [2].

In a joint venture for construction work, the joint venture agreement is of paramount importance, as it spells out aspects including equity, joint-control, liability for losses, risks, profit sharing, etc., relating to each party to the joint venture.

However, unlike in construction projects where standard forms of contract are almost always readily available, parties to a joint venture for construction work, usually contractors, rely on a joint venture agreement that is normally custom-made, or on an in-house contract drafted by the parties themselves. Up to now, some joint ventures may have been established on an ad-hoc or possibly incomplete basis, or even entirely orally. The key question therefore is what are the terms and conditions parties agree on in a joint venture?

There is a dearth of literature related to joint venture for construction work addressing the subject of joint venture agreements. For example, in past conferences on construction joint ventures organized by Kyoto University in Japan, International Islamic University Malaysia, and their partners[i], the aspects of joint venture most frequently addressed include critical success factors, the role of government, public-private-partnership, and culture, but there has been no examination of the terms and conditions of the joint venture agreement itself.

Objective and Methodology

A study was carried out to seek answers to key question regarding joint venture agreements for construction work: i.e., what are the terms and conditions parties agree on in a joint venture?

The focus of the study is on joint venture agreements for construction work projects procured through conventional methods of procurement. Hence, joint venture agreements for construction projects procured through modern methods of procurement, such as public private partnership, privatization, and the private finance initiative are outside the scope of the current study.

In the current study, joint venture agreements for construction work were examined in order to:

- Identify common terms and conditions of joint venture agreements; and
- Develop a model of a joint venture agreement, i.e., of the terms and conditions considered appropriate for use in joint ventures for construction work.

The principal methodology used for the study is critical examination of three joint venture agreements for construction work projects made available to the researcher [ii]. A summary of the said joint venture agreements is given in Table 1.1.

Table 1.1: Joint venture agreements examined in the study

	Project	Brief description of the joint venture
1	Road construction (Selangor, Malaysia, 2002)	Unincorporated joint venture between two parties. The joint venture includes submission of a project proposal, execution, completion, implementation, performance and maintenance of the work.
2	Sewage treatment plant (Selangor, Malaysia, 1994)	A Memorandum of Understanding (MOU) for an unincorporated joint venture between two parties. The proposed joint venture includes submitting a project proposal, designing and constructing the work on a turnkey basis.
3	Joint venture agreement for infrastructure project	A template of a joint venture agreement for unincorporated joint venture to tender and for the performance of the project.

Critical Examination of Joint Venture Agreements

Joint venture agreement No. 1: AB – BP JV

About the joint venture

The joint venture is between two contracting firms, for the purpose of this chapter referred to as *AB Sdn. Bhd.* and *BP Sdn. Bhd. AB Sdn. Bhd.* was invited by the Employer to submit a proposal to design, finance, and construct a roadwork project. *AB Sdn. Bhd.* requested *BP Sdn. Bhd.* to prepare and submit the proposal for and on its behalf and in its name to the Employer, with the intention and at the request of *BP Sdn. Bhd.* that if the proposal was accepted by the Employer, the roadwork project would be awarded by *AB Sdn. Bhd.* to an unincorporated joint venture to be formed between *AB Sdn. Bhd.* and *BP Sdn. Bhd.* on a sub-contract basis.

The proposal was accepted by the Employer and *AB Sdn. Bhd.* was awarded the project. Subsequently, an unincorporated joint venture between *AB Sdn. Bhd.* and *BP Sdn. Bhd.* under the name and style of *AB – BP JV* for the purpose of designing, financing, constructing, completing, performing, and maintaining the roadwork project was formed. As initially agreed between *AB Sdn. Bhd.* and *BP Sdn. Bhd.*, the former sub-lets the entire roadwork contract to *AB – BP JV*; i.e., *AB – BP JV* acts as a sub-contractor to *AB Sdn. Bhd.* Figure 1.1 illustrates the style of the joint venture and the relationship between the joint venture and the Employer.

Figure 1.1: *AB – BP JV*

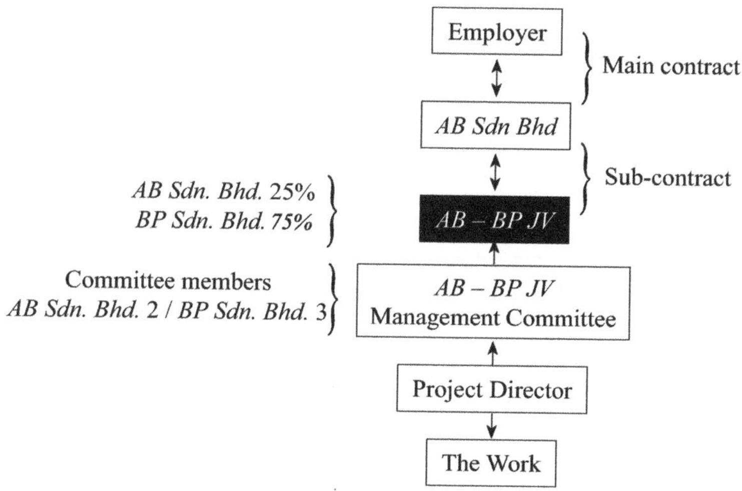

About the joint venture agreement

The *AB – BP JV* joint venture agreement is in three parts: (i) articles of agreement; (ii) conditions of contract; and (iii) schedules to the conditions of contract. Salient provisions of the agreement are given in Table 1.2.

Table 1.2: Salient provisions of the AB – BP JV agreement

	Heading / sub-heading	**Brief description of the contract provision**
Part 1	The agreement	The agreement is between *AB Sdn. Bhd.* and *BP Sdn. Bhd.*
	Articles A – D	The articles set out the background of the joint venture and outline the parties' desire to enter into an unincorporated joint venture under the name and style of *AB – BP JV* and for *AB – BP JV* to act as sub-contractor to *AB Sdn. Bhd.*
Part 2	Conditions of contract clauses 1 – 26	

	Heading / sub-heading	Brief description of the contract provision
	Clause 1 – purpose of the joint venture	*BP Sdn. Bhd.* shall prepare and submit a proposal for and on behalf of and in the name of *AB Sdn. Bhd.* to the Employer, and if the proposal is accepted by the Employer the contract shall be awarded by *AB Sdn. Bhd.* to an unincorporated joint venture to be formed between *AB Sdn. Bhd.* and *BP Sdn. Bhd.* on a sub-contract basis.
	Clause 2 – name and address	Name and address of the joint venture.
	Clause 3 – the joint venture	Definition of the joint venture, i.e. that the parties shall act together in the joint venture, that the joint venture is a contractual affiliation for the performance of the work, that the joint venture is a subcontractor to the main contract work, and that there is no intention of creating a legal entity.
	Clause 4 – exclusivity	The parties shall cooperate exclusively among themselves.
	Clause 5 – share of participation	The parties agree that the financial interest in the joint venture will be as follows: *AB Sdn. Bhd.* 25 %; *BP Sdn. Bhd.* 75 %. The rights, interests, liabilities, obligations, and all profits or losses are to be shared or borne by the parties in accordance with the said proportions.
	Clause 6 – duration of the agreement	The joint venture agreement shall be from the date the agreement is signed and shall be automatically terminated upon the first occurrence of any one of the following events: (i) The Employer awards the project to a party other than *AB Sdn. Bhd.*; (ii) The Employer discontinues the project prior to the award; (iii) In the case of award of the contract to *AB Sdn. Bhd.* by the Employer, upon full and final settlement of all contractual accounts and claims.
	Clause 7 – management committee	A Management Committee shall be set up, comprised of 5 members: 2 members from *AB Sdn. Bhd.* and 3 from *BP Sdn. Bhd.* The Management Committee shall consider any matters related to the Work. Decisions of the Management Committee shall require the majority vote of those present at the meeting. Expenses shall be borne by the respective party.

	Heading / sub-heading	Brief description of the contract provision
	Clause 8 – project director	*AB Sdn. Bhd.* shall appoint a Project Director. The Project Director oversees the carrying out of the work in accordance with the sub-contract agreement between *AB Sdn. Bhd.* and *AB – BP JV.* He reports to the Management Committee.
	Clause 9 – matters relating to the work	The parties agree that *AB – BP JV* shall act as a sub-contractor to *AB Sdn. Bhd.* The contract sum shall be the contract value of the contract awarded by the Employer to *AB Sdn. Bhd.* less the Guaranteed Profit to *AB Sdn. Bhd. BP Sdn. Bhd.* is the lead party of the joint venture.
	Clause 10 and 11 – accounts	Details of the joint venture account include opening a joint venture checking account, nominating persons to act on behalf of each party, establishing procedures to receive and make payments, making credit arrangements, etc.
	Clauses 16 – good faith	The parties acknowledge that it is not practical to make provision for every contingency that may arise in the course of the performance of the work; therefore, they shall negotiate in good faith at any unforeseen occurrence and amend the joint venture agreement accordingly.
	Clause 23 – governing law	The laws of Malaysia.
	Clause 24 – arbitration	Any disputes between the parties shall be settled amicably, failing which the dispute shall be settled by decision of a single arbitrator to be agreed upon between the parties, with the award of the arbitrator being final and binding.
Part 3	Schedules to the conditions of contract	Correspondence between the parties, guarantees, and indemnity

Key observations on the joint venture agreement are as follows:

- The joint venture is of the 'integrated' type and is 'unincorporated.'
- The joint venture acts as a sub-contractor to the Main Contractor in a contract signed between the Employer and the Main Contractor.
- The Main Contractor is a party to the joint venture.
- The establishment of the joint venture was initiated during the tender stage. The parties defined their roles and functions beginning with the bid submission, with the intent of carrying out the work on a joint venture basis in the event their bid was accepted.

Joint venture agreement No. 2: *PECB – MHES JV*

About the joint venture

The document examined is not a joint venture agreement but a Memorandum of Understanding

(MOU) between two contracting firms, for the purpose of this chapter referred to as *PEC Bhd.* and *MHES Sdn. Bhd.*

The MOU between *PEC Bhd.* and *MHES Sdn. Bhd.* relates the parties' desire to form a formal joint venture agreement in the event that a proposal submitted by *PEC Bhd.* for and on behalf of the proposed joint venture is accepted by the Client. The proposal concerned the design and construction of a new sewage treatment plant on a Turnkey basis. The formal joint venture agreement, under the name and style of *PECB – MHES JV* for the purpose of designing and constructing the new sewage treatment plant, shall be formed within 4 weeks of award of the project by the Client to the joint venture. Figure 1.2 illustrates the style of the proposed joint venture and the relationship between the joint venture and the Client.

Figure 1.2: *PECB – MHES JV*

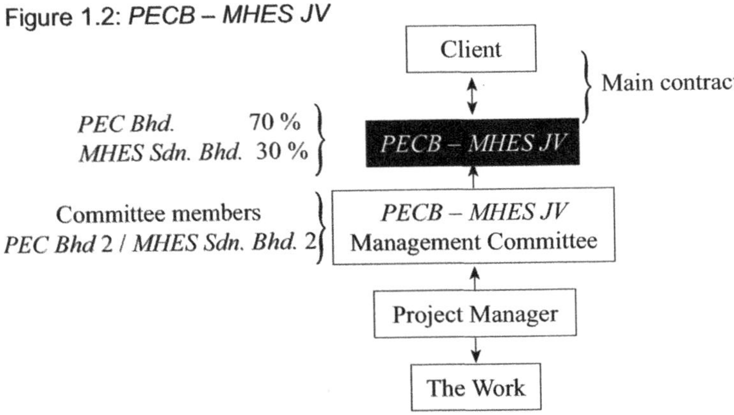

About the joint venture agreement

The MOU for the *PECB – MHES JV* joint venture is in two parts: (i) articles of agreement; and (ii) conditions of contract. Salient provisions of the MOU are given in Table 1.3.

Table 1.3: Salient provisions of the *PECB – MHES JV* agreement

	Heading / sub-heading	Brief description of the contract provision
Part 1	The agreement	The agreement, in the form of an MOU, is between *PEC Bhd.* and *MHES Sdn. Bhd.* The MOU expresses the parties' desire to form a formal joint venture agreement under the name and style of *PECB – MHES JV* within 4 weeks of the proposed joint venture securing an award from the Client.
Part 2	Conditions of contract clauses 1 – 7	
	Clause 1 – establishment of joint venture	Parties agree, as soon as possible after the MOU is executed and upon securing of a contract by the Client, to take steps to enter into a formal joint venture agreement for the purpose of performing the said Work.

	Heading / sub-heading	Brief description of the contract provision
	Clause 2 – equity of joint venture	The equity of the joint venture shall be as follows: *PEC Bhd.* 70 %; *MHES Sdn. Bhd.* 30 %.
	Clause 3 – undertaking	The parties shall cooperate exclusively among themselves.
	Clause 4 – management and implementation of the work	A Management Committee shall be formed to administer the affairs of the joint venture. The Management Committee shall have 4 members, 2 from each party. Its Chair shall be appointed from among the representatives of *PEC Bhd.* Decisions of the Management Committee shall require a majority vote of those present at the meeting. Expenses shall be borne by the respective party. The division of responsibility of the parties to the Work is as follows: *MHES Sdn. Bhd.* shall be responsible for the design, construction, and commissioning of the new sewage treatment plant; *PEC Bhd.* shall be responsible for the design, construction, and commissioning of earthworks and infrastructure. The execution and carrying out of the Work shall be supervised by a Project Manager to be appointed by the Management Committee.
	Clause 5 – the joint venture account	A bank account shall be operated under the name of the joint venture, with the Managing Director of *PEC Bhd.* and the Managing Director of *MHES Sdn. Bhd.* as the approved signatories.

Key observations on the joint venture agreement are as follows:

- The joint venture is of the 'separated' type and is 'unincorporated'.
- The joint venture acts as the Main Contractor in a contract signed between the Client and the Main Contractor.
- The establishment of the joint venture was initiated during the tender stage by way of an MOU. The parties defined their roles and functions beginning with the bid submission, with the intent of carrying out the work on a joint venture basis in the event their bid was accepted.

Joint venture agreement No. 3: *ABC – XYZ JV*

About the joint venture

The document examined is not a joint venture agreement but a template for the same. It belongs to a contracting firm known to have a series of construction joint ventures in its project portfolio. The template is used as a basis by the firm in negotiations with potential partners prior to establishing a joint venture.

For the purpose of this chapter, the partners to the joint venture are referred to as *ABC Sdn. Bhd.* and *XYZ Sdn. Bhd.* The subject of the joint venture is any construction project. The joint venture template indicates that the proposed joint venture covers participation of the parties in tender exercises and if successful, executes the project for the Employer. Figure 1.3 illustrates the style of the proposed joint venture and the relationship between the joint venture and the Employer.

Figure 1.3: *ABC – XYZ JV*

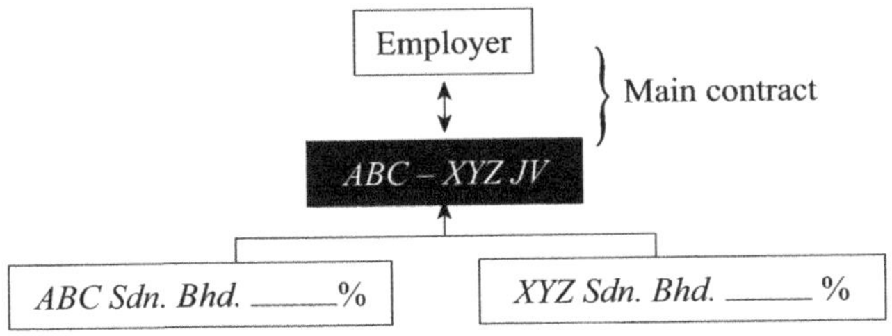

About the joint venture agreement

The agreement template for *ABC – XYZ JV* joint venture is in two parts: (i) articles of agreement; and (ii) conditions of contract. Salient provisions of the agreement template are given in Table 1.4.

Table 1.4: Salient provisions of the ABC – XYZ JV agreement

	Heading / sub-heading	**Brief description of the contract provision**
Part 1	The agreement	The agreement is between *ABC Sdn. Bhd.* and *XYZ Sdn. Bhd.* The agreement expresses the parties' desire to form a formal joint venture agreement under the name and style of *ABC – XYZ JV* in the event that their tender is successful and the joint venture is awarded the project by the Employer.
Part 2	Conditions of contract clauses 1 – 13	
	Clause 1 – name, address and purpose of the joint venture	Parties agree to form a joint venture to be known as *ABC – XYZ JV* for the sole purpose of participating in the tender exercise for the project and if successful, to execute the project for the Employer.
	Clause 2 – liabilities of the parties	Each party of the joint venture shall be jointly and severally liable to the Employer and notwithstanding this, each party agrees and covenants to indemnify the other against all loss, damages, claims, proceedings and expenses howsoever suffered by or brought against the other by reason of its non-compliance with any of the provisions of the agreement or by reason of any of its acts of omission, negligence, or breach. The parties agree that the agreement is not a partnership or any other form of entity having a separate legal personality.

	Heading / sub-heading	Brief description of the contract provision
	Clause 3 – extent of participation	For all purposes of the submission of the tender and the performance of the project, the parties shall share all costs, benefits and/or liabilities incurred by the joint venture in the following proportions (percentage to be inserted upon the conclusion of negotiations): *ABC Sdn. Bhd.*: ______ %; *XYZ Sdn. Bhd.*: ______ %.
	Clause 4 – leadership	The agreement names a person from *ABC Sdn. Bhd.* to represent the joint venture and names two others, one from each party, to act as signatories to sign all documents for the tender and the contract.
	Clause 5 – expenses	Each party shall bear all costs incurred by itself.
	Clause 7 – exclusivity	The parties shall cooperate exclusively among themselves insofar as the project is concerned.
	Clause 8 – validity	The joint venture agreement shall be from the date the agreement is signed and shall be automatically terminated upon the first occurrence of any one of the following events: (i) The employer cancels the project; (ii) The Employer rejects the tender and awards the project to others; (iii) The contract for the project has been fully performed and there are no further liabilities between the joint venture and the Employer as well as between the parties of the joint venture.
	Clause 9 – force *majeure*	The obligations of any party shall be suspended and it shall not be deemed in default or liable for damages if such party is prevented from complying therewith by acts of God, wars, insurrections, riots, strikes, lockouts, acts of governmental agencies, etc.
	Clause 10 – arbitration	Disputes if any shall be referred to the decision of a single arbitrator to be agreed upon between the parties. No party shall take any dispute to any court until after an award has been made.
	Clause 12 – extent of the agreement	The agreement shall be binding on the successors-in-title and permitted assigns of the parties thereto.
	Clause 13 – governing law	The laws of Malaysia.

Key observations on the joint venture agreement are as follows:

- The proposed joint venture is of the 'separated' type and is 'unincorporated.'
- The joint venture acts as the Main Contractor in a contract signed between the Employer and the Main Contractor.

- The establishment of the joint venture is to be initiated during the tender stage. The parties agree to participate jointly in the tender exercises and to execute the Work jointly if successful in the tender.

Synthesis of the terms and conditions of joint venture agreements identified

The articles of agreement and terms and conditions present in the three joint venture agreements described herein were listed and examined. Table 1.5 provides the list of the said articles of agreement and terms and conditions (in the form of clause headings).

Table 1.5: Articles of agreement, terms and conditions of the joint venture agreements

AB – BP JV	*PECB – MHES JV*	*ABC – XYZ JV*
Articles of agreement 1. Details of the parties	Articles of agreement 1. Details of the parties	Articles of agreement 1. Details of the parties
Terms and conditions 2. Style and purpose of the joint venture 3. Name and address of the joint venture 4. Exclusivity 5. Share of participation 6. Duration of the agreement 7. Management of the joint venture 8. Project management 9. Accounts 10. Insurance 11. Severability 12. Covenant for further assurance 13. Waiver 14. Good faith 15. Indemnity 16. Confidentiality 17. Assignment 18. Compliance with laws 19. Event of default 20. Cost and expenses 21. Governing law 22. Dispute resolution 23. Notices 24. Interpretation and miscellaneous	Terms and conditions 2. Style and purpose of the joint venture 3. Share of participation 4. Undertaking 5. Management of the joint venture 6. Project management 7. Accounts	Terms and conditions 2. Style and purpose of the joint venture 3. Name and address of the joint venture 4. Each party of the joint venture shall be jointly and severally liable to the Employer 5. Share of participation 6. Management of the joint venture 7. Expenses 8. Confidentiality 9. Exclusivity 10. Validity and termination 11. Force majeure 12. Dispute resolution 13. Notices 14. Extent of the agreement 15. Governing law

It is considered appropriate for the articles of agreement, terms and conditions to be synthesized. The synthesis aims to produce a list of articles of agreement, terms and conditions considered representative of joint venture agreements to be used by parties when they form joint ventures for construction work. The synthesis focuses on those articles of agreement, terms and conditions that may be considered omnipresent. In the synthesis, articles of agreement, terms and conditions that are present in at least two of the three joint venture agreements are considered omnipresent.

The synthesized articles of agreement, terms and conditions were presented to three Quantity Surveyors who are currently involved in executing construction work in joint venture projects. Their views were sought on the suitability of the synthesized articles of agreement, terms and conditions to represent a basic joint venture agreement that they would be willing to agree on. The outcome of the processes of synthesis and validation is a list of nineteen articles of agreement, terms and conditions that are considered to be representative of basic joint venture agreements for construction work projects. All nineteen articles of agreement, terms and conditions are classified into five key categories, as illustrated in Figure 1.4. Brief explanatory notes for each article of agreement, as well as for terms and conditions of the joint venture agreement are given in Table 1.6.

Figure 1.4: Categories of provisions in a basic joint venture agreement

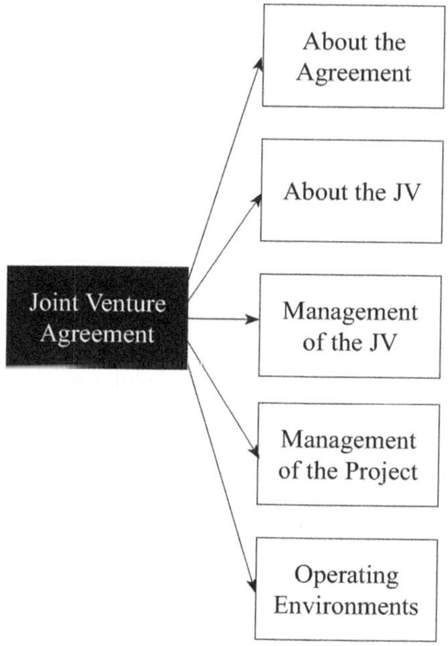

Table 1.6: Brief explanatory notes for each article of agreement, terms and conditions of the joint venture agreement

Articles of agreement, terms and conditions	Brief notes
Articles of agreement 1. About the agreement	The formal preamble, recitals, and the basic operative terms of the agreement between the parties forming the joint venture: date of the agreement, identity of the parties and their addresses, background leading to the formation of the agreement, brief description of the work to be executed, documents forming the agreement, and attestation clauses to be completed by the respective parties.
Terms and conditions *About the JV* 2. Style and purpose of the joint venture	Name and style of the joint venture (incorporated or unincorporated) and the purpose of the joint venture.
3. Name and address of the joint venture	Name and address of the joint venture.
4. Liabilities of parties	Each party of the joint venture shall be jointly and severally liable to the Employer and each party agrees to indemnify the other against all loss, damages, etc.
5. Exclusivity	The parties shall cooperate exclusively among themselves throughout the joint venture.
6. Share of participation	The equity of each party in the joint venture, in proportions; thus *First Party* _____%; *Second Party* _____%. In addition, specific responsibilities, liabilities, etc., of each party shall be identified and specified.
7. Duration of the agreement	Commencement and expiration / termination of the joint venture agreement.
Management of the JV 8. Management of the joint venture	Typically, a joint venture shall be managed by a committee. Members of the committee are appointed by the respective joint venture partner so as to represent the proportion stated in 6 above.
Management of the project 9. Project management	Typically, the carrying out of the work shall be delegated to a Project Manager. The Project Manager is appointed by and reports to the Management Committee stated in 8 above.

Operating environments 10. Accounts	Typically, parties to a joint venture agree to operate a joint account under the name of the joint venture. Each party shall assign a person to be the approved signatories. Expenses / incomes shall be recognized under this bank account.
11. Good faith	The parties shall negotiate in good faith for any unforeseen occurrence and amend the joint venture agreement accordingly.
12. Force *majeure*	The obligations of any party shall be suspended and it shall not be deemed in default or liable for damages if such party is prevented from complying therewith by acts of God, wars, insurrections, riots, strikes, lockouts, acts of governmental agencies, etc.
13. Confidentiality	All matters relating to the joint venture shall be regarded by the parties as being confidential and shall not be disclosed to any external party without prior written consent of the other member of the joint venture.
14. Assignment	Neither party may assign or cause the joint venture to assign all or any of the rights, interests, benefits and obligations of the respective party without prior written consent of the other member of the joint venture.
15. Event of default	In the event that any party shall go into liquidation or shall become insolvent or enter into any deed or arrangement with its creditors, the said party shall be obliged to sell their interest in the joint venture to the remaining party at such price to be determined by an independent auditor and the costs of such appointment shall be borne by the parties in equal shares.
16. Governing law	Typically, the laws of Malaysia will be the governing law for joint venture agreements formed in Malaysia.
17. Dispute resolution	Typically, the alternative dispute resolution technique named in a joint venture agreement is Arbitration.

18. Notices	Notices between the parties shall be made in writing. The agreed modes and address for delivery of notices shall be identified and specified.
19. Interpretation and miscellaneous	Miscellaneous provisions, e.g. meanings of words (singular-plural, genders), etc.

Challenges and problems in forming a joint venture

The following are some of the challenges and problems faced by parties when they initiated and negotiated joint venture agreements for construction work projects[3]:

(1) Choosing a partner or partners
(2) Deciding whether to form an integrated or non-integrated style of JV
(3) Deciding on the roles and responsibilities of the parties
(4) Deciding on extent of participation – such as how the actual work should be physically split so as to reflect the share of participation by each party (e.g. in the construction of a bridge across a river on a 50-50 share basis, how would the project be divided: would it be split in the middle, and who should connect the two parts?)
(5) Leader of the JV
(6) Level of acceptance of risk by parties
(7) Mutual trust
(8) Establishing the Management Committee
(9) Establishing the site team – i.e. the project manager and his team
(10) Adopting and assimilating a common system and work culture
(11) Establishing a financial management system, including the opening of a joint account
(12) Non-performance of a party to a JV – such as to require special protection by way of indemnity or parental guarantee or having a provision in the JV agreement for a party to take control of the defaulting party's scope of work, to mitigate loss and reputation
(13) Apportionment of common costs among the parties, such as maintenance costs and utility bills
(14) Sharing of costs of preliminaries in the event of a time overrun due to one of the parties or the Employer
(15) Lacking teamwork among the members of the JV

Conclusion

This chapter reported on a study seeking answers to a key question regarding joint venture agreements for construction work projects: i.e., what are the terms and conditions parties in a joint venture agree on? The findings of the study, regarding the common terms and conditions of joint venture agreements, suggested that the key provisions of the agreement shall include:

- The formal preamble, recitals, and the basic operative terms of the agreement between the parties forming the joint venture: date of the agreement, identity of the parties and their addresses, background leading to the formation of the agreement, brief descriptions of the work to be executed, attestation, etc;
- Information about the joint venture: i.e., name and style of the joint venture, its purpose, equity / share of participation, etc.;
- Management of the joint venture;
- Management of the project; and

- The operating environment of the joint venture, encompassing such matters as accounts, assignment, default, termination/liquidation, dispute resolution, governing law, etc.

A list comprising nineteen articles of agreement, terms and conditions that are considered appropriate and representative of basic joint venture agreements for construction work projects has been developed and validated.

In addition, the study has identified key challenges and problems faced by parties when they initiated and negotiated joint venture agreements for construction projects.

A comprehensive, well-thought-out, and well-written joint venture agreement goes a long way in helping to reduce the liability of the joint venture and of the partners to the joint venture. Consequently, it is necessary for parties aspiring to form a construction work joint venture to acquire the necessary knowledge and skills so that they are able to arrive at a more holistic and effective joint venture agreement, i.e. an agreement that is able to fulfill their needs and requirements, as well as those of the project.

In this chapter, a model of a joint venture agreement and the key challenges and problems related to the formation of joint ventures for construction work have been identified and described. Parties may use these data as a basis for their negotiations.

References

[1] Onishi, M., Zhang, S., and Kobayashi, K. (2005) The Structure of Joint Venture Agreement and Social Efficiency. In Kobayashi, K. and Khairuddin Abdul Rashid (Eds) (2005) *Towards New Concepts and Paradigms in Multi-national Joint Venture Contracting.* Penang: Kyoto University, Japan and International Islamic University Malaysia.

[2] Likhitruangsilp, V. and Mekkriengkrai, S. (2007) Civil Liability of Construction Joint Ventures under Thai Legal System. In Kobayashi, K., Ogunlana, S., Onishi, M., and Sharina Farihah Hasan (Eds) (2007) *International Joint Ventures: Reaching Strategic Goals.* Bangkok: Asian Institute of Technology, Thailand and Kyoto University, Japan.

[3] Ramar, S. (2007) *Joint Venture in Construction Projects.* A special lecture presentation at the International Islamic University Malaysia, Kuala Lumpur, 12th September 2007.

Endnotes

(i) Please refer to the conferences' proceedings, i.e. Kobayashi and Khairuddin, 2005; Kobayashi, Pham, Onishi and Nguyen, 2006; Kobayashi, Ogunlana, Onishi and Hasan, 2007.

(ii) The joint venture agreements were made available to the researcher by his colleagues, who are practicing Quantity Surveyors. Attempts to acquire further joint venture agreements were not successful.

Joint Ventures in Construction 2
ISBN 978-0-7277-5783-8

doi: 10.1680/jvc2.57838.017

Chapter 2

Construction Joint Ventures in the Project Coalition: A Conceptual Framework

Graham M. Winch, The University of Manchester

Introduction

Much of the research and best practice in construction procurement focuses on the relationship between the client and its first tier supplier or suppliers. Important as this relationship is for the success of the project process, it is not the whole story in governing the project coalition, defined as the entire network of firms mobilised by the client to achieve the project mission. This chapter will focus on developing a conceptual framework for understanding the various forms of collaboration *between* firms within that project coalition, rather than their relationship with the client. Thus we are not concerned with the business-to-business relationships between the client as principal and its suppliers of construction resources as agents, but with how the suppliers as coalitions of agents ensure that they mobilise the resources required to meet their commitments to the client in various kinds of joint venture. It will thereby advance a concept of the construction project coalition as a "nexus of treaties" [1] chartered by the client.

The chapter is structured as follows. First, a more formal definition of the issues will be proposed in terms of the problem of the vertical and horizontal governance of transactions. Then the argument will move on to focus on the mobilization and motivation of external resources, leaving internal resources for another day. Most of the governance issues here are very similar to those in the relationship between the client and the principal contractor, so attention will be directed towards those features of governance specific to the sub-contract relationship. Then we will move to a case example of an innovative approach to the construction supply chain before making some concluding comments.

Horizontal and Vertical Governance

Clients may decide to use existing resources that are available to them in-house, or they may choose to procure them from one or more additional firms – what is known as the make/buy decision – when mobilizing the resources required to achieve the project mission. In the first case, the process of mobilization and motivation is essentially an operational one, with the management of human resources at its core. In the second, further firm-to-firm relationships are formed; in turn these sub-contractor firms then have to make the same make/buy decision. Thus a chain of firms linked through a series of contracts can develop behind the *first-tier* suppliers in direct contract with the client. This chain of firms linked through a series of sub-contracts has become known as the *supply chain*.

In order to make the argument here clearer, it is useful to distinguish between vertical and horizontal transaction governance in the construction project process. Vertical governance focuses on the shifting set of transactions between the client and its first-tier suppliers. Depending on the level of integration of the project coalition structure chosen, the number of vertical transactions can vary from one (for a completely integrated project coalition) to more than 50 (for a construction-management type mediated project coalition, ranging from the acoustics engineer to the façade package contractor). Behind each one of these vertical transactions there may be one or more tiers of horizontal transactions. Where all resources mobilized are in-house to the supplier, no further external transaction needs to be governed; where an extensive supply chain develops, there may be a large number of horizontal transactions which need to be co-ordinated. These relationships are illustrated in Figure 2.1, which captures the different types of contracts that are deployed within the construction project coalition. The external ones in the horizontal dimension are usually called the supply chain; the vertical ones can usefully be defined as the *project chain*. The relative importance for the client of the supply chain and the project chain will depend on the procurement route chosen. In integrated procurement routes such as turnkey, the supply chain issues predominate, while in mediated procurement routes such as construction management project chain issues predominate.

The governance of both vertical and horizontal transactions is subject to the same three contingencies - identified in Figure 2.2 as uncertainty, frequency and asset specificity – that apply to all transactions within the project coalition [2]. However, choosing the most appropriate governance mode for horizontal transactions is usually easier than for principal contracts with the client, because the scope of work for each contract is less, and the impact of each factor is therefore easier to determine. While the provision of resources in-house by the client is unusual, and tends to be restricted to transactions under very high uncertainty, it is much more common in horizontal governance. Indeed, by definition, at some point in the supply chain there must be a switch to in-house resource supply.

Figure 2.1: Horizontal and vertical governance in the project coalition (Source: Winch [3] Figure III.i)

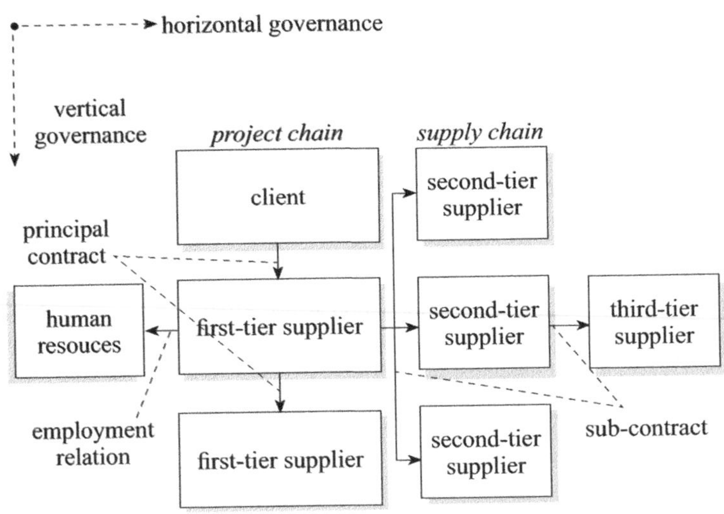

Vertical governance becomes horizontal governance through the process of *integration*. In the classic meaning of the term, (vertical) integration means the integration of companies up and down the supply chain, such as a car assembler purchasing its component suppliers. However, in project-based industries, where projects and firms are arrayed in a matrix, integration means moving up the project chain towards a single point of responsibility in an integrated project coalition. Its inverse is fragmentation into a large number of principal contracts, such as in mediated or unmediated project coalitions (see Winch [3] for definitions of these terms). Integration may also imply integration along the supply chain – there are certainly advocates of this – but actual examples are rare in construction. Cacciatori and Jacobides [4] argue that there has been significant integration on the supply side in UK construction, but they do not give case examples and largely rely on self-defined images of firms for evidence of this development. They also identify competing definitions of what "integration" means between different types of firm in the sector.

Managing the Supply Chain

The supply chain for a given project can be defined as that set of firms engaged in external transactions commencing with a principal contractor, and terminating when external transactions switch to internal ones in the employment relationship. On this definition, a project may have one or more supply chains, depending on the client's choice of number of principal contractors. Supply chains may be switched from project to project complete as in *integrated supply teams* (IST), or more normally they are formed anew for each project. Each firm within the supply chain has to make its own make/buy decision regarding whether to sub-contract further for the resource inputs it requires, or to use its own employees to add value. In principle, the supply chain extends right back to the extraction of the raw materials – iron ore, gypsum, clay, etc. – from the earth, but in practice it is helpful to limit the analysis to the point where the production of materials and components becomes construction-specific. Thus, a quarry for aggregates is part of the construction supply chain; a mine for bauxite is not.

Figure 2.2: The project coalition governance model
(Source: Winch [3] Figure III.i)

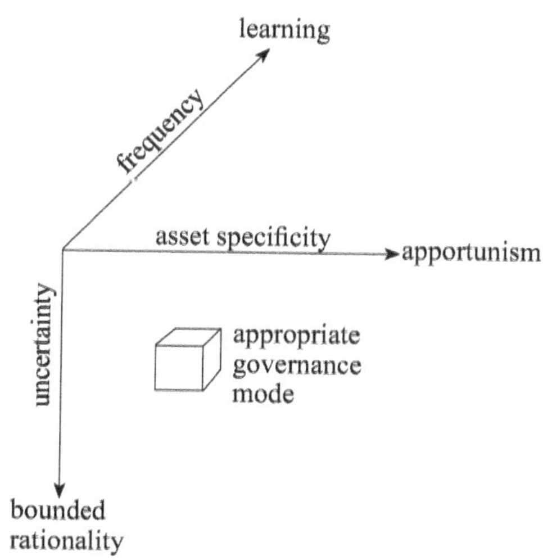

Horizontal governance in construction has always involved market transactions for the supply of specialist skills and services; trades such as roofing, for instance, have traditionally been the province of specialist rather than general contractors. Our concern here is to understand the variety of different external governance modes for different trades. There has long been a debate about why construction firms sub-contract:

- Some argue, in effect, that it is a solution to a transaction cost problem generated to maximise flexibility in coping with the uncertainties in the resources required for a particular project [5].
- Others argue that it is a solution to a production cost problem, driven by specialisation to achieve production efficiencies through economies of scale and the learning curve [6].

Recent developments in supply chain management suggest that both could be right, depending on the trade supplied. The segmentation of the supply chain into those suppliers that are relatively critical to the buying firm's operations and those which are less critical and can be treated as a commodity is central to effective supply chain management [7].

Our concern here is with *relatively* low uncertainty transactions during project execution (as opposed to front-end definition); high uncertainty transactions such as the supply of conceptual design services tend to be governed either internally or professionally. Figure 2.3 identifies the external horizontal governance options under low transaction frequency. Where both of these are low, sequential spot contracts (traditional trade sub-contracts) will be favoured. This is the case for many of the traditional trades where there is a ready supply of firms to do the work and large asset specificities are not generated during project execution – one bricklaying sub-contractor can be relatively easily replaced with another. However, this situation should not be confused with the spot-market contract of classical theory [8]; it remains a relational contract, even if that relationship is pretty distant.

Figure 2.3: Horizontal governance under uncertainty
(Source: Winch [3] Figure 7.3)

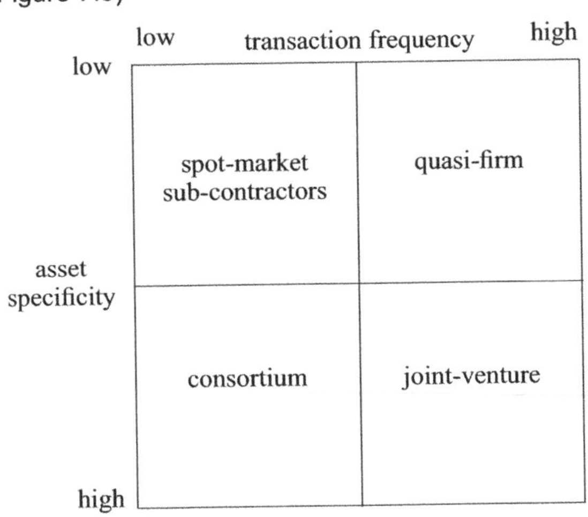

As transaction frequencies rise, other options become available – it becomes possible to enter into repeat relationships with suppliers. This allows the developments of a quasi-firm, where trade sub-contractors are offered repeat contracts by principal contractors conditional on satisfactory performance [9]. This is most favoured where the same resources are required for project after project, such as in housebuilding. Many construction quasi-firms are fairly informal – the same firm is used simply to reduce procurement costs without any strategy for deepening the relationship over time. However, this is changing with the advocacy of partnered ISTs. The 'firm' element of the quasi-firm tells us that important hierarchical elements enter into the relationship, but without any of the obligations on the part of the principal that characterize the employment relationship. Smaller firms within the quasi-firm can find themselves treated in a most cavalier way when it suits the principal contractor. In other cases – such as Japan – the whole construction industry can form itself into a series of tiered quasi-firms around the leading principal contractors. In Shimizu, this tiered 'family' is known as the *kanekikai* which employs a total of nearly 70k people [10].

Where asset specificities are higher – either pre or post-contract – problems emerge of potential opportunistic behaviour by the sub-contractor. There are various ways in which this can happen:

- The sub-contractor owns a proprietary technology required for the project;
- The client specifies the use of a particular supplier – the nominated sub-contractor;
- A particular sub-contractor's work forms a large percentage of the total cost of the works, or is at a key point on the critical path and switching costs arise [11] due to the hold-up problem [12];
- The contract contains a local content or technology transfer clause which has the effect of generating asset specificities for the local partner.

Under such circumstances, a consortium between the sub-contractor and principal contractor will be favoured [13], thereby minimising opportunistic behaviour by the owner of the specific asset at the cost of the principal contractor. By providing for a common incentive framework between the principal and sub-contractor, the consortium arrangement reduces the scope for opportunistic behaviour by the supplier of the key resource, and improves the flow of information between the parties. These issues most frequently arise on building projects around the mechanical and electrical (M&E) services packages, and it is noticeable that it is common for construction firms engaged in principal contracting to own a subsidiary focused on M&E contracting. Where the same resources are required on a programme of projects, then it may be appropriate to form a *joint venture*. It should be noted here that the usage of the term "joint venture" in the construction industry is not in accord with the wider use of the term in economics and management (e.g. Lorange and Roos [14]. In that literature, consortia are, in essence, temporary, while joint ventures usually share equity participation and are often indeterminate in length. The classification used here follows mainstream practice, rather than construction practice.

Managing Consortia and Joint Ventures

The formation of a consortium or joint venture is more complex than that of a traditional sub-contract or a quasi-firm. While in the latter, risks are allocated and managed through the sub-contract, in the former, risks are internalised and shared between the partners without the benefit of hierarchy to resolve problems and disputes. The costs associated with managing the relationships within the alliance are transaction costs, and are typically higher than those incurred within a single organisation. The challenge of selecting and negotiating with potential partners may be called the Jane Austen problem – the problem of finding a partner of appropriate status, wealth and character – in analogy with the continual search by Austen's characters for suitable husbands.

In terms of *status*, the partner firms have to possess complementary resources so that the whole of the alliance is greater than the sum of its parts in terms of value generated to balance the additional transaction costs incurred. This is the central rationale for the formation of a joint venture. Unless the partners to the alliance bring complementary resources, then there is little reason for them to come together and every reason for them to separate should they find out that the dowry was not what was expected. For instance, Medicinq, one of the Prime Supply Chain Partners in the UK National Health Service's original ProCure21 framework (now ProCure21+), formed from three regional builders to enable them to offer coverage over the whole ProCure 21 area; they decided not to bid to be a member of the Scottish equivalent because they had no local partner there.

In terms of *wealth*, the firms have to be equally capable of bearing the risks shared within the alliance. *Integrated consortia* are jointly and severally liable for contract execution, they operate with a common budget and they divide profits and losses at project completion in proportion to the original investment made by each partner. The inability of one partner to meets its obligations shifts that responsibility to the others. For instance, the size of some framework agreements such as UK Ministry of Defence's Regional Prime Contracts valued at up to over £700m over five years means that suppliers prefer to share risks with others through forming joint ventures.

Not all members of alliances are able to share risks equally. For instance, principal contractors on international projects are often required to enter consortia with local suppliers as part of the market entry criteria. Sometimes these local firms lack both assets to back their full participation as well as complementary skills – in effect the team is not very integrated and they are little more than sub-contractors in terms of their relationships with the principal contractor. Another example of this is in framework agreements such as ProCure 21 where the client expects the IST to include suppliers of design and other consultancy services. Such professional service firms are rarely able to share risks on an equal basis with the much better capitalized contractors which is one reason why ISTs are probably closer to a quasi-firm than a joint venture.

In terms of character, the firms have to have working practices that are similar enough to allow them to function effectively together. Firms differ on many dimensions – organizational culture, management practices and strategic intent. Unless the partner firms have the same basic objectives for the joint venture and are able to adapt their existing management systems so that they are compatible with those of their partners, then problems are likely to arise. Organizational cultures are more deep rooted and less amenable to change – they are often the biggest stumbling block to successful joint venture management. Many commentators on joint ventures focus on differences in national cultures as a major management problem in international joint ventures. However, there is no solid evidence that these problems are actually any worse than those caused by differences in organizational cultures between firms of the same nationality. International construction projects work in English and are rooted in an engineering culture that favours fact over opinion. Certainly, on the Channel Fixed Link project, organizational cultures were more of an issue than national cultures [15].

Clustering the Supply Chain

Traditionally in construction, the different members of the project coalition were defined around the manual skills required to work different materials – carpenters and wood, masons and stone, and so on. Although construction technologies have developed enormously over the last 500 years, this basic principle of supply chain organisation has not. As technologies changed, and new ones emerged, they were simply added to the list of specialist trades, thereby creating an increasingly fragmented set of resource bases. This is an important source of inefficiency in both production and transaction cost terms. How, then, might things be organised differently?

A central organizational principle when decision-making under uncertainty is to group together responsibility for those tasks which have the greatest interdependence and hence greatest requirement for information processing between them - Thompson [16] was the first to articulate this principle explicitly. This is because they will have the most intensive information flows as they collectively solve the problems they are assigned. This solution to the problem of interdependency under uncertainty has recently been developed as an alternative basis for the organisation of the resource bases in the supply chain known as supply chain clusters. The concept was developed independently in the UK and Finland (see Gray [17] and Lahdenperä [18]). One of the early implementations of this principle was on the Building down Barriers project presented in below which has had a considerable influence on the UK's Defence Estates organization as a client in recent years and is the inspiration behind the advocacy of integrated supply teams.

The organizational principle of construction clusters is a core team is responsible for the definition of the facility as a technological system, developing a budget and a strategy for its realization. An important part of the realization strategy is the definition of the clusters. The resource bases are clustered around technical sub-systems of the facility such as its structure, envelope or services for the processes of developing the detail design and execution on site. Where appropriate, the cluster leaders can also be members of the core team. The aim is to gain the maximum of information processing within the clusters, so as to seek efficiencies in production costs while clearly defining the interfaces between the clusters, thereby minimizing the amount of information processing required between clusters in pursuit of transaction cost efficiencies. Because the principle of clusters is, in essence, about the supply chain rather than the project chain, it can be implemented within either mediated or integrated project coalitions.

The Building Down Barriers Case

Influenced by its success in reforming defence procurement using the prime contracting concept for weapons and materiel, the UK Ministry of Defence decided in the mid 1990s to develop a new form of project coalition for its constructed facilities. They piloted this new form in the Building down Barriers (BdB) experimental projects. The Army Land Command required new sports facilities at Aldershot and Wattisham, and these two projects were chosen for the experiment with new approaches to procurement. Additional funding from the UK Government allowed a research team from the Tavistock Institute to work with the two project coalitions on an action research basis to facilitate the process and to capture their experience for wider dissemination [19]-[21]. There are a large number of features of this project that are of interest, however, this case will focus on the innovative approach to the management of the supply chain and its mobilization in the reduction of both capital and operating costs.

The BdB project went through three distinct phases:

(1) *Concept (January 1991 to early 1999)*. The first step was to appoint two prime contractors – Laing and AMEC. They then worked with their respective architects and Land Command to work up the design to beyond stage D of the RIBA plan of work. During this phase, decisions were taken as to how the clusters of resource bases were to be organised and their members selected. As appropriate, these cluster leaders and other cluster members participated in the design process. The outcome of this phase was the concept phase submission (CPS), which was evaluated by the Defence Estates Organisation and Army Land Command, together with fixed prices for executing the work from the two prime contractors – £10.8m for Aldershot and £4.2m for Wattisham.

(2) *Detail design and execution (early 1999 to mid 2000)*. In January and February 1999, the construction contracts were signed – AMEC at Aldershot agreed its price on a GMP basis with a gainshare formula, while Laing at Wattisham offered to discount the fixed price submitted with the concept phase submission by 5% in expectation of savings being found during execution. Both buildings were handed over early – Wattisham in February and Aldershot in June 2000.

(3) *Proving (mid 2000 to 2002)*. The prices submitted at the end of phase 1 included the running of the two facilities by the prime contractors for periods of 15 and 24 months respectively from handover.

The expectation was that this new form of project coalition would be able to achieve the 30% cost reduction challenge specified in the Latham Report of 1993, while also improving the whole life cost performance of the facility. The key tool in establishing these savings targets was the historic reference cost (HRC) model for each building. This was an artificial construct generated by pricing from existing records the capital and whole life cost of a facility designed to the normal MoD standards. The HRCs were accepted as being very tight. As Table 2.1 shows, the Latham target was not achieved, but nevertheless important savings were made against a tough benchmark by making capital investments (capex) in a higher quality of specification to achieve greater savings on running and maintenance costs (opex) over the facility life cycle; both projects show roughly a 100% return on the investment in higher quality of specification. Both projects were handed over ahead of time, with excellent safety records and no disputes. User satisfaction with both facilities is high.

All members of the two project coalitions are reported to have achieved good margins between 8% and 24%, and some achieved more due to the effectiveness of the organisation of the works on site. Those firms that did not achieve their budgeted margins attributed it to poor estimating. Client-initiated changes to specification after CPS were minimal, accounting for a £100k increase in the HRCs at Aldershot.

The budget at CPS figure was already the result of significant effort in design development involving formal value engineering, and was fully owned by the prime contractor. For instance, a significant decision was to invest in a relatively expensive combined heat and power unit so as to attack more effectively whole life costs. There is a general agreement that on both projects the fit and finish of the buildings is good compared to what would have been expected through traditional design and build. The differences between the CPS figures and the outturn costs predicted midway through phase 2 in October 1999 were largely due to intensive formal value engineering exercises on both projects. AMEC and Land Command thereby shared savings of £400k. The Laing figure is the discounted fixed price and cannot therefore yield further savings for the client against budget. The savings on Aldershot are intriguingly close to the 5% discount offered by Laing on Wattisham. A significant source of savings during phase 2 on both projects came from collaboration within the groundworks clusters between the structural engineer and the groundworks contractor on the foundation works.

The most difficult clusters to manage were the finishes clusters. These are inherently diverse, and consist of large numbers of very small firms. Many of these firms simply did not have the managerial capacity to engage with the collaborative approach within the clusters. Here, relationships between the trade contractor and the prime contractor tended to me much more "traditional" in terms of the construction industry recipe [3] than with the larger firms working on the main trade packages. The additional managerial effort required by the prime contractors for the finishing trades cut into their margins.

The proving period worked extremely well at Wattisham – Laing put a caretaker on site who handled minor defects and those that he could not handle were swiftly dealt with by Laing or its cluster members. The facility at Aldershot was more complex, and also its use patterns changed from those expected at scheme design. This led to some upgrading of facilities, but also means that the facility is not fully fit for its current purpose.

Table 2.1: The evolution of the budget for Building down Barriers
(Sources: Building down Barriers Evaluation Reports 1-4 (London, Tavistock Institute))
(I am very grateful to Richard Holti for making these figures available to me)

	capex		opex (on NPV basis)	
	HRC	budget	HRC	budget
Phase 1 Concept phase submission				
Aldershot	9.7m	10.7m	16.6m	14.7m
Wattisham	3.8m	4.0m	7.2m	6.7m
Phase 2 midway (October 1999)				
Aldershot	9.8m	10.3m	16.7m	14.3m
Wattisham	3.8m	4.0m	7.1m	6.7m
Completion (final account)				
Aldershot	9.8m	10.3m	16.7m	14.3m
Wattisham	3.8m	4.0m	7.2m	6.2m

A notable feature of the two projects is a sustained attempt to implement the principles of clustering the supply chain as advocated by Gray and Lahdenperä. Around the prime contractor a project core team was formed which included the designers – architects and engineers – and the cluster leaders. While there were slight differences in implementation, the same basic clusters were used on both projects, as illustrated in Table 2.2. The table shows how the design work was divided between the designers working for the prime contractor who were responsible for facility definition and those within the clusters responsible for facility description and execution. It is notable that those cluster members which made the most of the collaborative arrangements and tended to make higher margins were those that had their own design capability – this was particularly important for the early identification of coordination issues with other clusters.

Table 2.2: Organisation for Building down Barriers
(Sources: Building down Barriers Evaluation Reports 1-4: I am very grateful to Richard Holti for making these available to me)

Wattisham Pilot		Aldershot Pilot	
Cluster Leader	Design responsibility	Cluster Leader	Design responsibility
Groundwork: Civil engineering contractor	Scheme and detail design by core team consultant advised by cluster leader	*Civils and groundwork:* Civil engineering contractor	Scheme and detail design by core team consultant advised by cluster leader
Frame and envelope: Steelwork fabricator	Scheme and detail design by core team consultant advised by cluster leader	*Dry envelope:* Steelwork fabricator and roofing and cladding contractor	Scheme by core team consultant; detail by cluster leaders
Swimming pool: Pool contractor	Scheme and detail by cluster leader	*Water treatment:* Water treatment contractor	Scheme and detail by cluster leader
Internal finishes: Architect	Scheme and detail by architect, in consultation with other cluster leaders	*Blockwork and finishes:* Building contractor	Scheme and detail by architect, in consultation with cluster leader
Mechanical and electrical services: M&E contractor	Scheme by core team consultant; detail led by cluster leader	*Mechanical and electrical services:* M&E contractor	Scheme by core team consultant; detail led by cluster leader
		Sports equipment: Sports equipment contractor	Scheme and detail by architect in consultation with cluster leader

The Building down Barriers experiment provides a new rationale for the formation of collaborative arrangements in the construction industry. While on the Wattisham and Aldershot projects, the project coalition was more of a consortium than a joint venture as defined in Figure 2.3 above, the framework agreements developed by Defence Estates and ProCure 21 in the UK were heavily influenced by BdB and offer the opportunity for Integrated Supply Teams to form joint ventures.

Conclusion and Further Work

In this chapter we have moved on from the project chain to investigate more closely the management of the supply chain that delivers construction resources to the project. As always, the basic problem is one of information – how do clients clearly express their requirements and motivate suppliers to meet those requirements under uncertainty? Where certainty is high – for relatively simple facilities, or relatively late in the project life cycle – it is possible to write clear and binding contracts and rely on the market to ensure performance. However, certainty is rarely high enough for this on construction projects. More elaborate arrangements are required that allow contracts to be adjusted during project realization. So far as the supply chain is concerned, the main choice to be made is between sub-contracting for the resources or using in-house resources – in particular employees. The uncertainties inherent in the construction process would encourage most firms to use in-house resources, but the low levels of transaction frequency make this difficult. The response has been to set up tiers of suppliers, with relationships varying from the one-off to the continuing. The horizontal governance framework presented in Figure 2.3 provides additional insight into the issue here, and locates joint ventures as one of four options for horizontal project governance within the project coalition chartered by the client. The Building down Barriers case shows how complementary skills – the status aspect of the Jane Austen problem – can form the basis for consortia and joint ventures within the project coalition.

The approach we have developed here along the dimensions of horizontal and vertical governance within the project coalition presented in Figure 2.1 is, we submit, an important complement to the governance framework for a single transaction presented in Figure 2.2 because it allows the development of a perspective on the project coalition as a nexus of treaties chartered by the client. We still lack good analytical tools for understanding the interactions within this nexus of treaties. Two lines of enquiry that are presently under investigation in the Centre for Infrastructure Development at Manchester Business School are the use of social network analysis [22], [23] and the development of contracting maps to understand procurement strategy within the project chain.

References

[1] Aoki, M. Gustafsson, B. and Williamson, O.E. (1990) *The Firm as a Nexus of Treaties*. Sage, London.

[2] Winch, G.M. (2001) Governing the Project Process: A Conceptual Framework. *Construction Management and Economics* **19**: 799–808.

[3] Winch, G.M. (2010) *Managing Construction Projects: An Information Processing Approach* (2nd ed). Wiley Blackwell, Oxford.

[4] Cacciatori, E. and Jacobides, M.G. (2005) The Dynamic Limits of Specialization: Vertical Integration Reconsidered. *Organization Studies* **26**: 1851-1883.

[5] Ball, M. (1988) *Rebuilding Construction: Economic Change in the British Construction Industry*. Routledge, London.

[6] Eccles, R.G. (1981a) Bureaucratic versus Craft Administration: The Relationship of Market Structure to the Construction Firm. *Administrative Science Quarterly* **26**: 449–469.

[7] Dyer, J.H., Cho, D.S. and Chu, W. (1998) Strategic Supplier Segmentation: The Next 'Best Practice' in Supply Chain Management. *California Management Review* **40**: 57–77.
[8] Macneil, I. (2001) *The Relational Theory of Contract: Selected Works of Ian Macneil.* Sweet and Maxwell, London.
[9] Eccles, R.G. (1981b) The Quasi-Firm in the Construction Industry. *Journal of Economic Behavior and Organization* **2**: 335–357.
[10] Cox, A. and Townsend, M. (1998) *Strategic Procurement in Construction.* Thomas Telford, London.
[11] Winch G. M. (2006) The Governance of Project Coalitions: Towards a Research Agenda. In D. Lowe with R. Leiringer (eds.) *Commercial Management of Complex Projects: Defining the Discipline.* Blackwell, Oxford.
[12] Chang, C-Y. and Ive, G. (2007) The Hold-up Problem in the Management of Construction Projects: A Case Study of the Channel Tunnel. *International Journal of Project Management* **25**: 394-404.
[13] Clark, I. and Ball, D. (1991) Transaction Cost Economics Applied? Consortia within Process Plant Contracting. *International Review of Applied Economics* **5:** 341-357.
[14] Lorange, P. and Roos, J. (1992) *Strategic Alliances: Formation, Implementation and* Evolution. Blackwell, Oxford.
[15] Winch, G.M., Clifton, N. and Millar, C.J.M. (2000) Organisation and Management in an Anglo-French Consortium: the Case of Transmanche Link. *Journal of Management Studies* **37**: 663–685.
[16] Thompson, J.D. (1967) *Organizations in Action: Social Science Bases of Adminstrative Theory.* McGraw-Hill, New York.
[17] Gray, C. (1996) *Value for Money: Helping the UK Afford the Buildings it Likes.* Reading Construction Forum, Reading.
[18] Lahdenperä, P. (1995) *Reorganizing the Building Process: A Holistic Approach.* VTT Publications 258, Espoo.
[19] Holti, R., Nicolini, D. and Smalley, M. (2000) *The Handbook of Supply Chain Management: The Essentials.* CIRIA, London.
[20] Nicolini, D., Tomkins, C., Holti, R., Oldman, A. and Smalley, M. (2000) Can Target Costing be Applied in the Construction Industry – Evidence from Two Case Studies. *British Journal of Management* **11**: 303–24.
[21] Nicolini, D., Holti, R. and Smalley, M. (2001) Integrating Project Activities: The Theory and Practice of Managing the Supply Chain Through Clusters. *Construction Management and Economics* **19**: 37–47.
[22] Pryke, S. D. (2004) Analysing Construction Project Coalitions: Exploring the Application of Social Network Analysis. *Construction Management and Economics* **22**: 787-797.
[23] Pryke, S. D. (2005) Towards a Social Network Theory of Project Governance Construction. *Management and Economics* **23**: 927-939.

Joint Ventures in Construction 2
ISBN 978-0-7277-5783-8

doi: 10.1680/jvc2.57838.029

Chapter 3

Design of Administrative Structures for Construction Joint Ventures

Apichart Prasitsom, Chulalongkorn University
Veerasak Likhitruangsilp, Chulalongkorn University

Introduction

Joint venture (JV) is a domestic or international cooperative enterprise involving two or more legally distinct organizations which join temporarily to undertake a specific project [1], [2]. Construction joint venture (CJV) is a business form widely used for large and complex construction projects because it can enhance contractors' competitiveness by integrating various resources and capabilities from their partners (e.g., capital, material, labor, equipment, technology, know-how, or expertise) as well as allocate risks and liabilities among them.

The scope of CJV is not internationally standardized. According to definitions appearing in various works of literature, common characteristics of JVs can be summarized as follows [3], [4], [5]:

(a) The collaboration is formulated from two or more partners.
(b) A partner can be either a business organization or an individual person.
(c) A JV is formulated to operate only for the duration of a specific project.
(d) The collaboration can be in the form of capital money, resources, know-how, benefit or loss, and risk.
(e) Mutual agreements among partners can be oral or written contracts.
(f) The duration of the collaboration must be clearly specified, with the time frame to be either within specified dates or for the length of time required until every target is achieved.
(g) All partners are jointly and severally liable for obligations to third persons.

Although CJV is a popular business form used in many construction projects worldwide, the performance of most CJVs is still unsatisfactory. Various issues concerning the performance of CJVs are addressed in numerous research works, including performance evaluation, partner selection, trust management, and cultural differences. Administrative structure is widely accepted as a critical factor contributing to the performance and risk of CJV projects [6], [7].

Like the scope of CJV, the definition of CJV has not been standardized. Various temporarily to operate and obligate a particular project by collaborating in various levels is the fundamental concept which can be found in general definitions of the CJV. This may mislead many people that a CJV's administrative structure organizational structure) must always entail collaborative teams that manage every task in the construction project and must use resources shared by every partner. However, there are also other patterns of administrative structure such as the pattern where all tasks are separated into packages and each partner is allowed to manage its own package separately (some consider this a consortium rather than a CJV). This perception can be attributed to the lack of standard definitions of CJV and JV. To clarify this ambiguity, this chapter aims to explore and analyze the significant characteristics of various administrative structures of CJVs in Thailand, and provide guideline for desiring CJV administrative structure

The administrative structure (or organizational structure) represents the relationship among various tasks such as allocation, communication, coordination, and supervision, which have a direct impact on the achievement of organizational objectives. There are two main components of the administrative structure. The first component is tasks individual to each section according to their characteristics. The second is communication and coordination beyond sections [8], [9]. Stipulating CJV administrative structure, a strategic implementation process, is an important step in fulfilling not only all partners' objectives (e.g. profit and learning) but also the CJV's objectives (e.g. time, cost, and quality).

Life Cycle of CJV

Typically, the life cycle of a CJV can be broken down into four major phases: (1) the beginning phase, (2) the formulation phase, (3) the operation phase, and (4) the termination phase [modified from Kotelnikov [10]]. A CJV begins with establishing a strategy and choosing the partners (the beginning phase), which then negotiate, develop, design, and adjust the details of their administrative structure during the bidding stage (the formation phase). The construction works are performed (the operation phase) until they are complete and the CJV is dissolved (the termination phase). Figure 3.1 illustrates the processes in each phase of the CJV's life cycle, which are divided into two levels, namely, the partner level and the CJV level.

Figure 3.1: Life cycle of CJV and main processes in each phase

	I. BEGINNING PHASE	II. FORMATION PHASE	III. OPERATION PHASE	IV. TERMINATION PHASE
Partner Level	Investigate initial details of the construction project Assess organization status Evaluate possible strategies Define objectives for joining the CJV Search, assess, and select partners Sign MOU or pre-bid JV agreement (optional)	Obtain bidding documents Collaborate with other partners Operate own tasks (assess details of the construction project) Sign pre-bid JV agreement Sign JV agreement	Collaborate with other partners Operate own work package (including employee, material, equipment, and capital money) Sign other agreements among partners Follow legal procedures	Collaborate with other partners Operate own work package Sign various agreements among partners to cancel collaboration Follow legal procedures
Construction Joint Venture Level	Obtain bidding documents Communicate with owner and third parties	Obtain bidding documents Prepare bid proposal Plot operation plan, CJV administrative structure, team, employees, duty, and liability Follow legal procedures Communicate with owner and third parties Submit bid	Communicate with owner and third parties Operate collaborative work packages (including employee, material, equipment, and capital money) Follow legal procedures	Adjust property Follow legal procedures Dissolve CJV

As compared with those of other businesses, the CJV's life cycle is unique. For other businesses, firms usually have enough time to develop their business plans and select suitable partners. Once the partner has been selected, it is almost certain that the JV is constituted and can be run. In contrast, construction contractors cannot fully evaluate details of construction projects before selecting partners due to the conditions of the bidding process in most projects. Typically, contractors wishing to bid on a certain project in the form of a CJV must obtain bidding documents on behalf of the CJV only. This means that contractors must choose their partners and constitute CJVs before obtaining bidding documents. The award of a contract determines whether or not the CJVs will continue to perform the work or will be terminated at this point.

Constituting a CJV may involve several contracts and agreements, including the memorandum of understanding (MOU), the pre-bid JV agreement, the JV agreement, and other agreements among partners [11]. Among these legal documents, the JV agreement is considered the most important because it officially formulates a CJV as well as stipulates each partner's rights and liabilities [3]. Several provisions in the JV agreement encompass the CJV administrative structure, including equity share, list of partners, sponsor name (leader partner), authorized representatives, and duties and responsibilities of executive bodies. All of these will directly affect JV project management [12]. Thus, the administrative structure of a CJV must be meticulously designed to maximize the benefits of this business scheme [13].

Research Methodology

This research primarily adopts qualitative methods to examine CJV practices from the viewpoint of the participants and their particular social and institutional contexts [14]. Data were mainly collected by the Delphi method, which is a systematic and interactive research technique for obtaining the judgment of independent experts on a specific topic [15].

The first interview used open-ended questions to identify the characteristics of CJV administrative structures. Open-ended questions allowed respondents to share their experience, knowledge and expertise freely [16]. The questions in the first interview focused on staff's experience regarding the administrative structure of the CJVs in which they had participated or were operating. Fourteen respondents, all executives in seven leading construction firms in Thailand, were interviewed face to face and responded to the questionnaires. The duration of each interview was approximately 45 minutes. Feedback from panelists in the second and third interviews was collected through e-mail questionnaires.

CJV Administrative Structures

After analyzing the results from the Delphi process, we found that CJV administrative structures adopted in the Thai construction industry were quite diverse but can be categorized into four major structures:
(1) Collaborated-operation structure
(2) Separated-operation structure
(3) Mixed-operation structure
(4) Single-partner operation structure.

It should be noted that the CJV administrative structures presented in this chapter represent the practices that were actually implemented by contractors, which were usually different from those officially stipulated in the JV agreements. These practices aimed to enhance the flexibility and efficiency of CJV project management. In such cases, the partners normally arranged undisclosed internal agreements that stipulated their actual rights, duties, and liabilities in the project, which were different from those specified in the official JV agreements. Third parties, however, were not bound by such internal agreements.

Collaborated-operation structure

For the collaborated-operation structure, all tasks in a construction project are handled by a collaborative team consisting of personnel obtained from each partner or from outsourcing. This team is led by a main project manager, who is under the supervision of the executive board of the CJV. No task is allocated to an individual partner. The unique characteristics of the collaborated-operation structure are as follows:

(a) The partners participate in any operation of the construction project via the CJV executive board only.
(b) The CJV executive board and the main project manager are fully responsible for every task.
(c) The main project manager of CJV has no relationship with any partners and is hired to carry out all tasks in the construction project.
(d) All staff members of the CJV, who are nominated by each partner or outsourced, are required to be under the supervision of the main project manager and must take no direct orders from any partners.
(e) The capital money for the entire construction project is directly obtained from the CJV's fund, not from an individual partner.
(f) The CJV's net profit or loss is determined by the overall performance of the construction project and is allotted to each partner based on its proportion of contribution.
(g) The partners are jointly and severally liable for obligations to the project owner and the third person, and share their liability pursuant to their proportions of contribution.

The collaborated-operation structure has not been popular in the Thai construction industry due to its high cost and the complexity of its formation and management. However, according to some JV executives, this structure was used in certain CJV projects because of the difficulty of dividing construction tasks into portions. Figure 3.2 illustrates an example of the collaborated-operation structure of a CJV modified from an actual CJV project.

As can be seen, the MNO Joint Venture comprises of three partners: Contractor M, Contractor N, and Contractor O, with contributions of 34, 33, and 33 percent, respectively. The main project manager, the leader of the project, is not an employee of any partner but is recruited exclusively for the project. All work are carried out by the main project manager, who is under the supervision of an executive board consisting of representatives of each partner. Even though according to the concept of the collaborated-operation structure all partners are jointly responsible for all tasks, tasks are allocated to employees that are specialize. For example, if Contractor O specializes in signaling installation, employees from Contractor O will be assigned signaling tasks for the project. In this particular example, Contractor M and Contractor N collaborated on railway bridge construction because of Contractor M's interest in acquiring the related know-how from Contractor N.

Separated-operation structure

All tasks of CJVs using the separated-operation structure are grouped into work packages, each of which is executed by a certain partner. Typically, no work package will be handled by multiple partners at the same time. Each work package is carried out by a project manager nominated by the partner responsible for that package. All project managers limitedly coordinate with the main project managers and the CJV executive board. The unique characteristics of the separated-operation structure are as follows:

(a) Each partner has full control over its work packages.
(b) The CJV executive board and the main project manager have less authority than those in the collaborated-operation structure.

(c) The main project manager is responsible for such tasks as the overall image of the construction project, communication with the owner and the third person, and coordination among other project managers.
(d) The project manager and personnel of each work package are appointed by a partner handling that work package, and are responsible for their work package only.
(e) Each partner must be responsible for the capital money of its work packages. However, there are some work packages (e.g., general affairs) that are normally funded by the CJV's capital money.
(f) The profit and loss of each partner is evaluated based on the performance of its work packages only.
(g) The partners are jointly and severally liable for obligations to the project owner and the third person for every task, and share their liability pursuant to their proportions of contribution.

Unlike in the collaborated-operation structure, in the separated-operation structure it is not required that the main project manager be outsourced, as he is responsible only for coordination among the partners. In some CJV projects, the project head can be one of the project managers.

Since each work package in the construction project is allocated to a certain partner, this administrative structure is often called a consortium, rather than a JV. However, in Thailand, neither business form has a clear definition. Thus, CJV in this chapter is defined to include any administrative structure in which all partners are jointly liable for all work to the project owner and the third person.

Figure 3.2: Example of the collaborated-operation structure

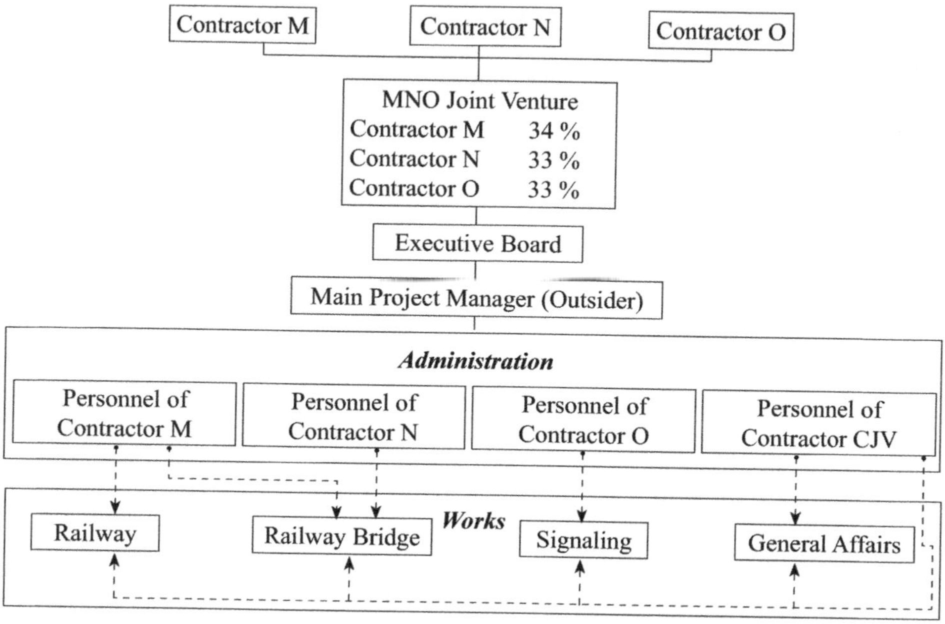

Figures 3.3 and 3.4 display two examples of the separated-operation structure modified from actual CJV projects. Figure 3.3 illustrates a project in which all tasks were divided into three work packages, with Package 1 assigned to Contractor W, Package 2

assigned to Contractor K, and General Affairs handled by the newly-hired JV personnel. In Figure 3.4, the tasks are divided into two work packages: (1) all construction work packages performed by Contractor A and (2) material and equipment procurement handled by Contractor P. For the second package, there was no project manager because Contractor P was responsible only for supplying material equipment and providing consultations for the project. The operation of contractor P is under the supervision of Contractor A's project manager. This particular practice resulted from Contractor A's strategy that aimed to prevent other contractors from cooperating with Contractor P and thus competing in the bidding for this project.

Mixed-operation structure

This CJV administrative structure is a combination of the collaborated-operation structure and the separated-operation structure. That is, some work packages are performed collaboratively by multiple partners, whereas other work packages are assigned to individual partners. This type of structure is usually adopted when the first two structures are inappropriate for the project due to several factors such as task requirements, flexibility of project management, each partner's capacity, resource constraints, and objectives of forming the CJV. Figure 3.5 illustrates an example of the mixed-operation structure, which was modified from an actual CJV project.

Figure 3.3: Example of the separated-operation structure consisting of contractors only

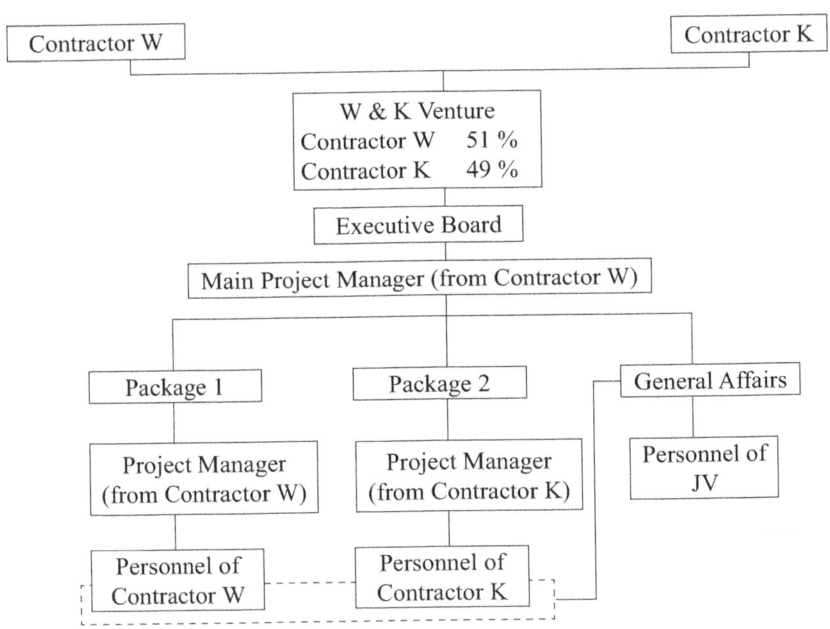

As can be seen, the XYZ Joint Venture is comprised of three contractors. The project's tasks are divided into five work packages. The first three work packages are separately allocated to specific partners, whereas Work Packages 4 and 5 are collaborations between Contractors X and Y, and Contractors X and Z, respectively. In this particular project, the CJV adopted the mixed-operation structure for two main reasons. First, Contractor X was unable to perform

Work Package 4 alone, making it necessary to collaborate with Contractor Y, an experienced underground contractor. Second, Contractor X wished to acquire knowledge of new construction techniques in Work Package 5 from Contractor Z. For such work packages, the collaborated-operation structure was adopted.

Single-partner operation structure

For the single-partner operation structure, one partner performs all tasks in a construction project while the others have no responsibility to execute any task. This structure can be divided into two types based on the characteristics of partners:

(a) Type A

For the first type, all partners are construction firms, one of which has the capacity to perform alone every task in the project but decides to form the CJV out of a desire to decrease competition in bidding, to gain advantage by including a partner who is a domestic firm, or to improve credentials during the bidding process. Once the CJV has won the project, some partners may decide to be idle during the operation. All tasks of the project are then assigned solely to the capable partner. Idle partners, however, obtain a share of profit and are still liable for obligations to the project owner and the third person. Several construction contractors in Thailand have adopted this administrative structure as a business strategy.

Figure 3.4: Example of the separated-operation structure consisting of contractor and supplier

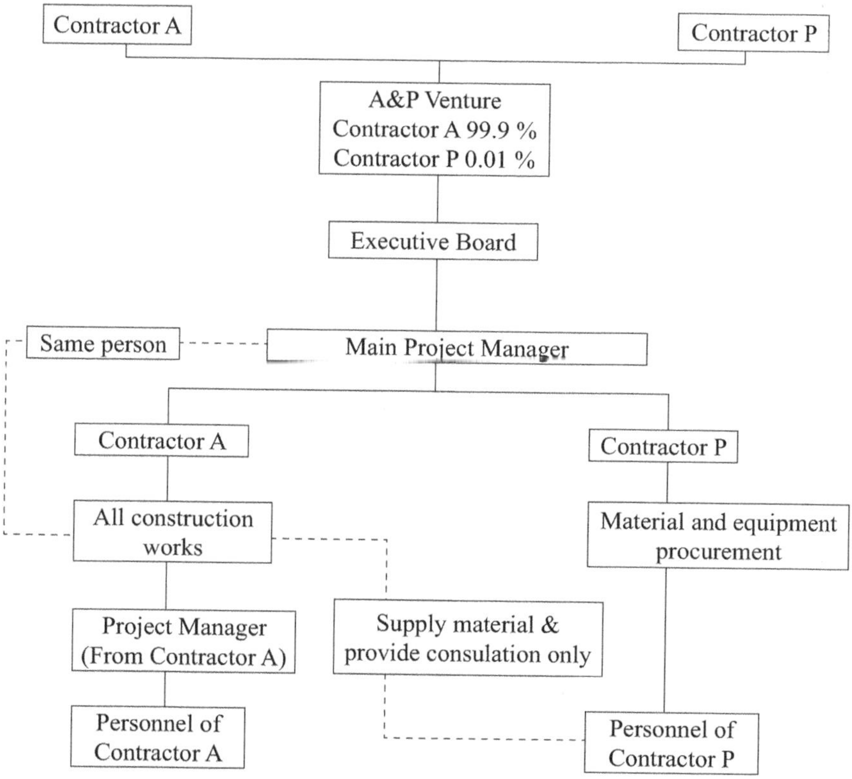

This administrative structure can be found in medium construction projects where the partners are all domestic contractors, and in large construction projects where a foreign contractor is responsible for all work. It was found that this structure was acceptable to owners as long as CJV partners were able to execute all contracted tasks.

Figure 3.6 shows an example of the single-partner operation structure, Type A. Even though Contractor F and Contractor G form an "apparent" F&G Joint Venture, all work is executed only by Contractor F. Contractor G still shares the profit and is jointly liable for the CJV's obligations.

(b) Type B
In this type of arrangement, partners found temporary construction firms. Even though such firms do not have the capacity to execute any tasks in the project, their political connections can facilitate the awarding of a CJV project. Clearly, such malpractice is illegal (e.g., bid rigging) but is presented here only for the sake of completeness.

Figure 3.5: Example of the mixed-operation structure

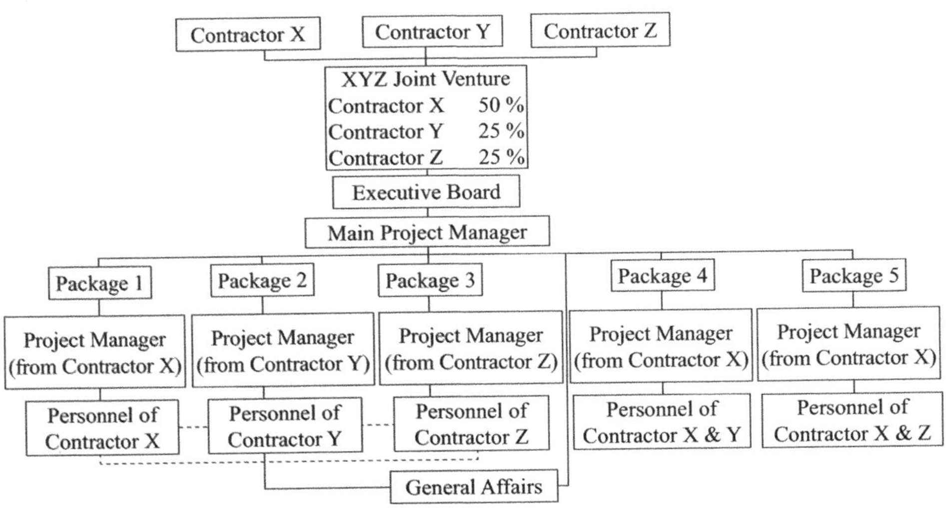

Comparison of CJV Administrative Structures

Once the major administrative structures of CJV used in the Thai construction industry were defined, the characteristics of the collaborated-operation structure and the separated-operation structure, the two basic types were analyzed and compared to identify their advantages and disadvantages, in various issues.

Table 3.1 shows the effects of CJV administrative structures on important management issues. The results consist of three parts: the project level, the operation level, and the partner level [5], [7].

Project level

The level is set to consider the effectiveness of the basic objectives for construction project management which are consisted of project cost, project schedule, and project quality by using the objective indicators.

Figure 3.6: Example of the single-partner operation structure (type A)

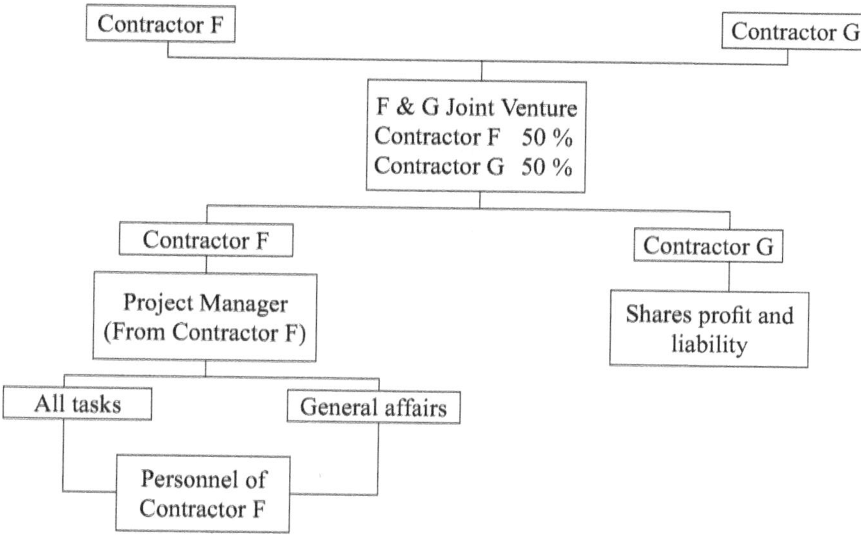

Operation level

This level is considered the effectiveness of management control over the CJV organization. It is measured by the level of effectiveness of operation of CJV organization including formulation, operation, cost, stability, and power of main project manager.

Partner level

The final level focused on the success of a partner's preset objectives are applied by management of each partner. The examples of these objectives are managerial flexibility, sharing risks, sharing resources, financial, know-how transfer enhancing competitiveness, and entering international markets.

As can be seen from Table 3.1, the collaborated-operation structure promotes know-how acquisition as well as facilitates appropriate risk sharing among partners, whereas the separated-operation structure allows flexibility and strong management power for each partner.

Guideline for Designing CJV Administrative Structure

Several factors need to be considered while choosing partners and an appropriate CJV administrative structure for a particular project [7]. Once solid partners have been secured, the leader must clarify the details of the CJV administrative structure by negotiating with all partners. Each partner must deliberate both internal and external factors that may affect not only the effectiveness of management at the project and operation levels but also the success of each partner to achieve its objectives at the partner level [15],[17],[18]. The internal factors of each partner encompass its objectives for joining the CJV, type of business, business status, and resources; whereas, external factors refer to these internal factors in other partners [19], as well as effects from the project and laws. Details of some important factors are as follows.

Table 3.1: Effects of CJV administrative structures on important management issues

Management Issues	Collaborated-Operation Structure	Separated-Operation Structure
Project Level		
Project cost	Lower	Higher (due to setup cost)
Project schedule	Not directly related	Not directly related
Project quality	Not directly related	Not directly related
Operation Level		
Formulation	Slower	Faster
Operation	Slower	Faster
Cost	Higher (due to setup cost)	Lower
Stability	Higher	Lower
Power of main project manager	More	Less
Partner Level		
Managerial flexibility	Less	More
Know-how transfer	Promote	Obstruct
Financial	Share	Individual
Resource	Share	Individual

Current business status

Contractors needs to consider when choosing an appropriate CJV administrative structure is their business status at the moment of joining the CJV, including the number of current projects in hand, capital money status, human resources, and business plans.

Objectives for joining the CJV

Contractors should adopt CJV administrative structures that conforms to their objectives for joining the CJV. For example, the collaborated-operation structure is appropriate for contractors that have limited resources or want to acquire know-how from other partners.

Characteristics of the project

Typically, CJV projects require more construction resources (e.g., capital money, technology, and expertise) than conventional projects do. For a particular CJV project, the allocation of work packages depends primarily upon the partners' capacity, which can be divided into three cases:

(a) If each work package can be executed by a single partner, the separated-operation structure is usually adopted.

(b) If some work packages can be handled by individual partners while some cannot, the CJV usually adopts the mixed-operation structure.

(c) If no work package can be performed by any individual partner, the collaborated-operation structure is the most appropriate.

Type of partner's business

The partners of a CJV can be from different types of business, including construction contractors, engineering firms, consultants, material and equipment suppliers, or financial institutions. This is another important factor to consider when choosing the administrative form because its type of business directly influences a partner's degree of collaboration. For example, a CJV partner that is a construction consultant or an engineering firm would certainly have a degree of collaboration more limited than a partner that is a construction contractor. In such a case, the separated-operation structure would therefore be an appropriate CJV administrative structure.

Conclusion

The CJV administrative structure plays a significant role in the success of CJV project management because it delineates the scope of each partner's tasks and provides a collaborative framework for all partners. This chapter presents four CJV administrative structures commonly used in the Thai construction industry. For the collaborated-operation structure, every work package is executed by a collaborative team of all partners. For the separated-operation structure, each work package is clearly allotted to an individual partner. These two CJV structures are the most fundamental of the mixed-operation structure. Some CJV projects can be executed by a single-operation structure, in which one partner performs all tasks in the project while other partners only share the profit. Each CJV administrative structure entails unique characteristics, advantages, and limitations. Thus, contractors must meticulously design an administrative structure conducive to achieving their objectives for joining the CJV. In addition, it is necessary for contractors to consider other external factors, including the characteristics of the project and other partners' objectives, as well as laws and regulations, to maximize the benefits of adopting this strategy in cooperative business.

References

[1] Geringer, J. M. (1988) *Joint venture partner selection: Strategies for developing countries.* Quorum, New York.

[2] Encyclopedia of Business. (2006) *Reference for Business* [Online]. Available at: http://www.referenceforbusiness.com/encyclopedia/Int-Jun/index.html. [Accessed 27 November 2010].

[3] Pittayaporn, T. (1990) *Legal Status of Joint Venture.* M. Law thesis. Chulalongkorn University, Bangkok.

[4] Likhitruangsilp, V. and Mekkriengkrai, S. (2007) Civil Liability of Construction Joint Ventures under Thai Legal System. *Proceedings of the the 3rd International Conference on Multi-National Joint Venture for Construction Works.* Bangkok, Thailand, 1-2 November 2007.

[5] Prasitsom, A. (2008) *Construction Joint Venture Administration in Thailand.* M. Eng. thesis. Chulalongkorn University, Bangkok.

[6] Fan, L. (1988) Equity joint venture in the Chinese construction industry. *International Journal of Project Management* **6(1)**: 50-58.

[7] Ozorhon, B. Arditi, D. Dikmen, I. and Birgonul, M.T. (2008) Effect of Partner Fit in International Construction Joint Ventures. *Journal of Management in Engineering* **24(1)**: 12-20.

[8] Pugh, D.S. (1997) *Organization Theory: Selected Readings.* Penguin Books, Harmondsworth.

[9] Jacobides, M.G. (2007) The inescapable limits of organizational structure and the unfulfilled role of hierarchy: Lessons from a near war. *Organization Science* **18 (3)**: 455-477.

[10] Kotelnikov, V. (2006) *Joint Venture: Why, What, and How?* [Online]. Available at: http://www.1000ventures.com/ business_guide/jv_main.html [Accessed 25 September 2007].

[11] Likhitruangsilp, V. and Prasitsom, A. (2008) Construction Joint Venture Contracting. *Proceedings of the the COBRA 2008 Research Conference.* Dublin, Ireland, 4-5 September 2008.

[12] Jacob, R. (1999) *Joint ventures information* [Online]. Available at: http://www.cooper.edu/GTK-V/jvent.htm [Accessed 29 September 2007].
[13] Norwood, S.R. and Mansfield, N.R. (1999) Joint venture issues concerning European and Asian construction markets of the 1990's. *International Journal of Project Management* **17(2)**: 89-93.
[14] Kaplan, B. and Maxwell, J.A. (1994) Qualitative Research Methods for Evaluating Computer Information Systems. *Evaluating Health Care Information Systems: Methods and Applications*: 45-68.
[15] Hallowell, M. R. and Gambatese, J.A. (2010) Qualitative Research: Application of the Delphi Method to CEM Research. *Journal of Construction Engineering and Management* **136 (1)**: 99-107.
[16] Ling, F. Y. Y. Pham, V. M. C. and Hoang, T. P. (2009) Strengths, Weaknesses, Opportunities, and Threats for Architectural, Engineering, and Construction Firms: Case Study of Vietnam. *Journal of Construction Engineering and Management* **135 (10)**: 1105-1113.
[17] Fong, C. K. (1985) *Construction Joint Ventures in Singapore*. Butterworth, Singapore.
[18] Luo, Y. (1997) Partner Selection and Venturing Success: The Case of Joint Ventures with Firm in the People's Republic of China. *Organization Science* **8 (6)**: 648-662.
[19] Gale, A. and Luo, J. (2004) Factors affecting construction joint ventures in China. *International Journal of Project Management* **22 (1)**: 33-42.

Joint Ventures in Construction 2
ISBN 978-0-7277-5783-8

doi: 10.1680/jvc2.57838.041

Chapter 4

The Application of the Environmentally Conscious Contract Law to a PFI Project (Japan Case)

Katsuji Ishihara, Kyoto University
Sufang Weng, Kyoto University

Introduction

We now place much greater importance on environmental issues including global warming. The unequivocal evidence of global warming is illustrated by the increase in the average ambient temperatures, the melting of glaciers, and the rising sea level. As the global understanding of climate change and the associated risks that correspond with it continue to develop, people are becoming increasingly demanding about more political actions. Government regulations and contracts have substantial potential for the reduction of global greenhouse gas emissions below current levels.

As environmental consciousness becomes more prevalent, we notice that some pioneers have started to include environmentally consciousness clauses into procurement or construction contracts. Even from the publicly available information, a few cases exist in which the environmental protection responsibility clauses for an operation period are included in the contracts in practice. We found an example: the World Games Main Stadium of 2009 [1] in Kaohsiung, Taiwan.

To be a proactive model of green buildings in Taiwan, the contractor was required to acquire a Green Building Candidate Certificate before beginning construction. They were also required to comply with the Green Building Laws in Taiwan that have nine standards: the diversity of life, the degree of Green, water maintenance in base, energy saving, CO_2 degradation, waste degradation, indoor environmental improvement, water resource improvement, and sewage-handling improvement. The contract obligates the project company to comply with the nine standards during both the construction and operation periods under the periodic audit conducted by the Kaohsiung City Government [2].

Service contracts with Energy Service Companies (ESCO) expect energy cost reductions during a service period. Therefore, it is expected that they continue to make efforts to reduce energy consumption. In addition, we can learn lessons from the procurement contract of the U.S. Department of Defense (DoD) in the 1970s, although they did not require a decrease in environmental impacts. Both are mentioned later in this chapter.

The Private Finance Initiatives (PFI), which is a way of using private finances and skills to deliver capital investment projects, has come to be used both by the central government and local governments, and is widely accepted in Japan. In a conventional procurement of public works, the construction contract is ended simultaneously with the delivery of a building. Therefore, it is difficult for a contractor to take responsibility for the environmental performance in the life of a building after delivery. Based on the above argument, people involved in the PFI have recently begun discussing how to apply Environmentally Conscious Contract Laws into a PFI project. It is considered that this law can be applied well to a comprehensive evaluation of bidders and a proposal system, which are the main characteristics of the PFI. Moreover, since a contractor participates in a project through a project period and payment for the project is made during the period, the applicability is good.

This chapter examines the problem of when is the most appropriate time to apply the Environmentally Conscious Contract Law to a PFI project and proposes suitable application schemes.

Environmentally Conscious Contract Laws

In the case when the government and a public institution purchase goods and service, they use the open bid in which the bidder presenting the lowest price is selected, in principle. This traditional supply system aims at minimizing the expenditures of the government. Consequently, it may be unable to procure goods with a low environmental load that have a high initial price.

In Japan, the Environmentally Conscious Contract Law was enforced for the purpose of contributing to global warming prevention in November 2007. This law, aimed at the national government and the related public sectors, shall use a comprehensive bid evaluation, including not only a price but also the environmental performance. The government shall contract with those who offer the best products and services. This leads to a new society where environmental efforts are rewarded economically. This law stipulates that the head of each ministry shall make public information about the performance of its environmentally conscious contracts and submit a notification to the Minister of the Environment every year.

There has been a "Green Purchase Law" that aims at improving the environment through the measures conducted by the national government before the Environmentally Conscious Contract Law was enforced. This law requests that the national government should take the lead and promote the procurement of products and services which contribute to environmental load reduction. Then the law was aimed at the construction of a sustainable society.

The Green Purchase and Environmentally Conscious Contract Laws do not only lower a buyer's own environmental load but also intend to convert the economy and the entire society into an eco-friendly land by demanding that the supply-side companies offer products with a small environmental load. However, the Green Purchase Law also requires products and services that contribute to environmental load reductions, as summarized in a Table 4.1. In contrast, the Environmentally Conscious Contract Law requires the contract to be relevant to the discharge curtailment of greenhouse gases, etc., itself.

Table 4.1: Green Purchase Law and Environmentally Conscious Contract Law (after the Ministry of the Environment[3])

Law	Green Purchase Law	Environmentally Conscious Contract Law
Characteristics	Specify the environmental performance of goods and services.	Specify the structure of the contract such as the contracting procedures.
Purpose	To promote the procurement of the goods and services that satisfy a required environmental performance.	To specify contract procedures utilizing an environmentally conscious system, such as a comprehensive evaluation and a proposal system. To contract with those who can offer the best environmental performance.
Regulated Products and Contracts	237 items within 18 fields, such as paper, stationery, office automation apparatus, car, uniform, fatigue dress, equipment, disaster prevention stockpile, public work, and service.	Four procurement contract types, such as the purchase of electricity and official vehicles; and service contracts, such as those with ESCO and architects.

The Environmentally Conscious Contract Law intends to reduce the life-cycle environmental load through the procurement of goods and services. Specifically, contracts concerning the following four fields are subject to the law:

(a) Purchase of electricity,
(b) Purchase of official vehicles,
(c) Service contracts with ESCO, and
(d) Service contracts with architects.

CO_2 emissions from office buildings in Japan have been consistently increasing since 1990. Therefore, efforts in the above fourth field, service contracts with architects, are expected. To promote the reduction of emissions, such as greenhouse gases from a building, the government has to ask for a minimum environmental performance for all designs. The government should ask architects for the technical proposal about the curtailment of greenhouse gas emissions to improve the environmental nature of a building. It is appropriate to utilize an architect with outstanding knowledge in environmental design. The architect who has the creativity, knowledge, and experience in environmental design is selected by utilizing an environmentally conscious proposal system. This brings about the application of outstanding knowledge that leads to an improvement in the environmental performance of a building.

Contract Scheme for Reference

Life cycle costing procurement

In 1964, the DoD issued the DoD Directive, an Integrated Logistic Support (ILS). In the ILS, Life Cycle Costing (LCC) contracts are trial conducted on the procurement of tires in airplanes. Subsequently, to promote the adaptability of LCC concepts in military procurement, in 1970, the DoD issued the DoD Guide, LCC-1 "Life Cycle Cost for Equipment," and other DoD Directives to promote the LCC. Presently, in all military affairs, a law has been enacted by the DoD Instruction stating that an LCC analysis should be done. Procurement based on the LCC is similar to an Environmentally Conscious Contract, as both are contracts of parameters based on service life.

In a procurement based on the LCC, compensations for damages on occasions of failure to achieve data prescribed on contracts are claimed as follows:

(a) On the occasion that the functional targets are not fulfilled in the trial tests after delivery, the differences between the target values and real values should be complemented without additional charges. If it is the occasions of the maintenance of the equipment, this rule is applied to the repair of the flaws.

(b) On the occasion that the functional targets are fulfilled, but the operation fees calculated in trial tests after delivery differ from the values in the original proposal/contract, leading to changes in LCC values, the unit price should be changed to make the LCC values calculated in the trial tests the same as the values in the contract.

(c) On the occasion that the LCC calculated by bidders is not fully reliable before the orders are placed, to make sure the LCC is appropriately calculated, the following are considered: to have tests after delivery with penalty clauses and to add guarantee clauses in contracts or to have a bulk contract including maintenance. However, we need to consider the fact that the necessary expenses of tests after delivery and the guarantee might be shifted to the contractors. This is a fact in the cases of the penalty clauses in which the contractor's aim is the complete fulfillment of the contract. It causes problems if the penalty is too strict that even the margin of errors brings about a high penalty. In addition, it is believed that the risk premium is already counted into the bidding price in a bulk contract, including maintenance.

Currently in public procurement in Japan, systems of procurement and bidding contracts conducted by the American DoD have not yet been applied.

Energy Service Company (ESCO)

ESCO means companies guarantee the energy-saving effects by providing a total solution, including equipment reforms and factory/building energy-saving affairs. The characteristic of ESCO is that its income comes from, and is also calculated by, the percentage of the deducted part of the customer's cost savings. After the contract period is over, the total saved cost will be completely the customer's profit.

ESCO, in general, provides the following services and their combinations:

(a) Diagnose the energy situation,
(b) Design and construct energy-saving equipment,
(c) Maintain/operate/manage energy-saving equipment,
(d) Funding,
(e) Guarantee the energy-saving effect, and
(f) Measure and verify the energy-saving effect.

In ESCO contracts, as in the energy-saving technical scheme, there are proposals from bidders. These include:

(1) Predicted and guaranteed deducted sums on energy-saving issues.
(2) Predicted and guaranteed deducted sums on CO_2.

On the occasion that an ESCO failed to achieve the above guarantee, a public sector has the right to provide advice to the ESCO. If the predicted effect is not achieved, an ESCO's reward will be also deducted.

As guaranteeing energy-consuming effects directly relates to environmental responsibility, ESCO contracts are considered an applicable field in Environmentally Conscious Contract Law. Furthermore, ESCO contracts are applicable to existing buildings as it can set standards in the energy use situation. With this information, energy consumption, after ESCO has adapted it, can be reasonably predicted. In equipment preparation and bulk contracts, including the operation phase, ESCO contracts are similar to the PFI. However, newly-set PFI's are very different from ESCO's as it is difficult to set the standards to predict the ready situation.

Environmentally Conscious Efforts in PFI Projects

Representative cases in Japan

In several proactive PFI projects in Japan, environmentally conscious efforts have been incorporated in the bid evaluation and the terms of agreement. However, there is still much room for improvement. In this section, the author will clarify problems with regard to the environmentally conscious efforts by reviewing some PFI projects in Japan.

Environmentally conscious measures used in PFI projects can be classified into the following five types:
(a) Setup of a target of efforts,
(b) Qualitative evaluation in a bid selection process,
(c) Quantitative evaluation in a bid selection process,
(d) Adjustment through energy costs, and
(e) Commitment and penalties for carbon-dioxide emissions, etc.

Almost all of the PFI projects adopt types (a) or (b), whereas few projects adopt type (e).

Choshi Municipal High School, construction and maintenance (Choshi City[4])

Choshi is located in the southernmost end of the Chiba Prefecture, which adjoins Tokyo.Choshi applied the PFI to the design, construction, and maintenance project of a municipal high school, which consists of a classroom building (9,700 m^2) and a gymnasium (4,200 m^2). The tender price of the successful bidder was about 5,991 million yen. The project began in 2008. The project period is 22 years. The unique feature of this project is that it is the first project abiding of the concept of the Environmentally Conscious Contract Law.

The number of full marks in the design proposal it could obtain in total for the proposal evaluation is 75 out of the 200 points. "Resource-savings, energy-savings, and life-cycle-cost-savings" (20 points) are included as an environmentally conscious evaluation items in the design evaluation, as follows:
(1) Is there any proposal about the consideration of the global environment? (5 points)
(2) Is it the equipment plan excellent in the energy-savings and easy-maintenance nature? (5 points)
(3) Are there an appropriate policy and an appropriate view included about the long life of the buildings and durable equipment? (5 points)
(4) Is there any proposal about the life-cycle cost reduction? (5 points)

The amount of electricity (e.g., air-conditioning, lighting, hot-water supply, elevators, etc.), gas (e.g., air-conditioning and hot-water supply), and water services (e.g., air-conditioning and water supply) used are monitored and then computed with the given equation and conversion coefficients (e.g., electricity 0.555 (kg-CO_2/kWh), gas (oil) 3.00 (kg-CO_2/kg), and water 0.425 (kg-CO_2/ m^3) every year for item 2) evaluation of the energy savings nature.

Hirakata Municipal Schools, air-conditioning installation and maintenance (Hirakata City [5])

Hirakata, in the Osaka Prefecture, applied the PFI to the air-conditioning design, installation, and maintenance project of the 74 municipal schools, which consists of 2,001 classrooms. The tender price of a successful bidder was about 5,103 million yen. The project period began in October 2008. The feature of this project is a contract clause that speculates the penalty for failure in energy savings.

In the case that the quantity of the energy used by the air conditioning systems exceeds the level proposed by the successful bidder and that this situation occurs continuously or intermittently, one penalty point is accumulated per classroom each day until the successful corrective measure is confirmed. The service fee for the project company is cut by 100% if the penalty point exceeds 60,001 points. If the accumulated penalty point is less than 600, there is no deduction from the service fee.

Member of the House of Representatives Hall, construction and maintenance (Ministry of Land Infrastructure, Transport and Tourism [6])

The Ministry of Land, Infrastructure, and Transport applied the PFI to the design, construction, and maintenance project of the Members of the House of Representatives Hall, a relatively large building with a floor area of over 209,000 m^2. The project is a BTO and service purchase type project, in which a project company's earning relies solely on the revenue from a public sector. The tender price of a successful bidder was about 110.6 billion yen. The project period began in March 2006 and will end by March 2020 with a 10-year maintenance period. The feature of this project is a contract clause that speculates the penalty for a failure in energy savings.

The evaluation of the bid for this project includes elaborated criteria for environmental proposals. Full marks of the proposal evaluation is 400 out of a total of 1,000 points. 60 points are assigned for the environmental items, which consists of "energy savings (30 point)," "other environmental load reductions (20 point)," and "a reduction in energy costs (10 point)". Qualitative proposals are required in terms of energy savings based on specified conditions and coefficients. The conditions include various assumptions regarding the activities in the building.

Application of environmentally conscious contract laws to PFI projects

The PFI is a comprehensive contract of design, construction, and maintenance/operation. Therefore, environmentally conscious commitments for the operating period could be made and the framework of guaranteeing these commitments should be prepared. Based on this viewpoint, the environmentally conscious contract is highly expected to have obvious effects on the PFI projects. But can it be this simple?

Difficulty in predictions

In the case that a project company takes complete responsibility for maintenance and operations, it is possible for the company to take responsibility for greenhouse gas emissions during an operation period, since the company can manage risks regarding the environment of use and the intensity of use. If a clear reference (such as an existing building of a ESCO business) to predict the situation of use is provided, the project company can also take the risk. However, in general, most PFI projects are newly established and have no existing standards to follow. Accurate information of the use situation is hard to obtain prior bidding; hence, in most cases, the risks in terms of greenhouse gas emissions are excessively transferred to the project company.

In PFI projects, it is difficult to ask a project company for responsibility related to the use of a building because the detailed information on the condition of the building use is rarely shown in a bidding phase. Furthermore, the difficulty is deepened in the cases where a public sector is asked for the design change and other various changes from the bidding conditions. Causation between the changes initiated by a public sector and greenhouse gas emissions becomes complicated in this kind of case.

A method to make up this incompletion in the prediction is to observe the conditions of use immediately after the operation begins and then set the benchmarks based on the discussion between a public sector and a project company after a certain period of operation.

Bidder's actions and bidding evaluation

The contents of the proposals submitted in bidding are based on bidders' good consciousness and strategies. Due to the uncertainty of the quantity of greenhouse gas emissions, Bidder A might predict 50 tons of greenhouse gas emissions, while Bidder B predicts 100 tons, even if both of them expect the same probabilities and an expected value of 100 tons. To solve this problem, we consider an appropriate scheme for a penalty, payment reduction, or implement required corrective measures necessary. Otherwise, bidders may take this advantage of uncertainty and not work hard to achieve the commitments.

In a bidding process, it is suggested that a public sector set the calculation method of greenhouse gas emissions and make a hypothesis of use conditions. Then, the bidders' should develop their proposals based on these methods and conditions. Once all bidders calculate their proposals under the same conditions, we do not have to worry about the accuracy of the condition itself in the bidding evaluation.

Reduction of payments and penalties

One of the simplest methods to reduce greenhouse gas emission is to ask a project company to pay for the energy fees. However, because the sum of the equipment costs and the energy fees may not be positively related to greenhouse gas emissions, the moral hazard of the company might occur.

The reduction on a unitary payment basis could be effective for a Build-Operate-Transfer (BOT) project. On the other hand, the actual situation of the BTO is that the design/construction and maintenance/operation are clearly divided in terms of cash flows, because installment payments of the construction costs of the buildings are confirmed liabilities. Therefore, even if the amount of greenhouse gas emissions during a maintenance/operation period exceeds the agreed value on a contract, it may not be applicable to ask for the payment reduction on unitary payment basis.

Conclusion

A variety of problems, such as abnormal weather, has occurred all around the world due to global warming. Most of us understand the necessity of taking care of global environmental problems. However, not only during the production phase but also during the use phase, greenhouse gas emissions might make up a significant amount of environmental pollution. Buildings and public infrastructures currently in use might be a significant emission source of greenhouse gases. Therefore, it is reasonable to apply the environmental-load deducting framework based on the view of the life cycle to construction activities. Moreover, this is reason that the Environmentally Conscious Contract Law targets buildings.

However, public architecture and public infrastructure in Japan have had respective contracts on their design, construction, and maintenance/operation. It takes time to measure and verify the commitment of a reduction of greenhouse gases in the maintenance and operation phases. This causes a difficulty in applying the concepts of the environmentally conscious contract. In other words, most of the problems in the operation and maintenance phase are actually due to design problems. However, the design contract is closed before the operation period. Therefore, it is difficult for a public sector to prove that the increase in greenhouse gases resulted from flaws by the designers or the construction companies.

The PFI is a comprehensive contract of design, construction, and maintenance/operation. Since the above problems caused by their respective contracts are solved, the fulfillment of the concepts of the environmentally conscious contract could be expected in the PFI. However, this chapter illustrates that many other problems should be solved. Most of them are problems in which a project company cannot control for the risk of an increase in greenhouse gas emissions. As proposed in this chapter, for the purpose of not putting unreasonable risk on the project company, it is appropriate to have verification tests based on pre-agreed standards after the construction phase or to set up benchmarks after a certain period of operation and motivate continuous improvement.

References

[1] The World Games 2009 Kaohsiung (2008). http://www.worldgames2009.tw/wg2009/eng/Venues_connect2.php [Accessed: 2008, September] (for online resources).

[2] Kaohsiung Municipal Government (1995) *Public Report.* http://www.gazettes.com.tw/KS/PDF/K9508310.pdf [Accessed: 2008, September] (for online resources).

[3] Ministry of the Environment (2007) *The Outline of Environmentally-Conscious-Contract and Principles.*

[4] Choshi City (2008) *Draft of Project Contract of Choshi High School PFI Project.* http://www.city.choshi.chiba.jp/edu/education/ky_soumu/pfi/pfi.html.

[5] Hirakata City (2007) *Draft of Project Contract of Hirakata Municipal Schools PFI Project.* http://www.city.hirakata.osaka.jp/freepage/gyousei/kyouiku/sisetu/work/pfi_info2.htm.

[6] Ministry of Land Infrastructure, Transport and Tourism (2005) *Bid Evaluation Criteria of 'Member of the House of Representatives Hall' PFI Project.* http://www.mlit.go.jp/gobuild/pfi/singiin_shu/050531_05_0stdselect.pdf.

Joint Ventures in Construction 2
ISBN 978-0-7277-5783-8

ICE Publishing:
doi: 10.1680/jvc2.57838.049

Chapter 5

The Application of Build, Operate, Transfer (BOT) Contract as a Mode of Financing in Developing *Waqf* Land: Malaysian Experience

Azman Mohd Noor, International Islamic University Malaysia
Saidatolakma Mohd Yunus, International Islamic University Malaysia

Introduction

It has not been denied by any Muslim scholar that after a person's death, he or she will still gain benefits from previous acts of charity (*Sadaqah jāriyah*). The Prophet Muhammad (Peace be upon Him) in a tradition narrated by Abu Hurayrah said:

"When a person dies, all his good deeds cease except for three: a continuous act of charity (Sadaqa jāriyah), beneficial knowledge and a righteous child who prays for him" [1].

Besides facing challenges in attracting people to contribute their money or property, the institutions of *waqf* face their biggest problem in managing property in the best possible way to maximize utilization and improve the institution's performance in providing good services and benefits to the beneficiaries. It is upheld that the conceptual understanding of *waqf* is to withhold the ownership of estates and to channel to charity its usufructs. Unfortunately, not all *waqf* institutions have the chance to develop their estates because of management and financially related problems[i]. Common problems may be a lack of expertise, poor management of *waqf* funds and the shrinking role of service as a result of bureaucratic management. This situation might burden all recipients and prevent them from enjoying great *waqf* benefits.

As an alternative in developing *waqf* estates, BOT can be a viable device for financing. This method, which has been practiced in the past, allows the private sector to operate, finance and manage public projects on *waqf* land. Through this type of project, the benefits from the *waqf* property are unlimited for everyone. In Malaysia, Majlis Agama Islam Wilayah Persekutuan (MAIWP), a *waqf* institution which is the authority in administering *waqf* for the federal territory, introduced and utilized the BOT contract as one mode of developing *waqf* property. Although BOT is a new application tool in *waqf* and its practice is very challenging, it has proven to be a useful mechanism for the institution's progress. All *waqf* related issues and problems faced by the *waqf* will be identified and a probable solution will emerge to maintain *Shari'ah* legitimacy and validity.

This new alternative (BOT Contract) can perhaps be an effective way to increase an institution's performance, and hence boost the great role of *waqf* institutions, which has been ignored by many people. The general concept of the BOT scheme should be analyzed before heading to the main focus of this study, which is the application of BOT in MAIWP, transactional flow, practical implementation, as well as *Shari'ah* compliance issues and challenges.

Definition of BOT

BOT is an English abbreviation for the term 'Build,' 'Operate,' and 'Transfer.' As for the exact meaning of BOT, many scholars differ upon the definition depending on their opinion concerning certain elements and conditions of BOT. However, none of the definitions deviates from the fact that BOT is a financing scheme entrusted by the government to a private company to set up a specific project at its own expense, where beneficial ownership remains in the possession of the project company for a fixed duration before the project is returned to the government [2].

Concept of BOT

The BOT contract involves government assignment of a construction project to a company for a fixed duration. The assigned company is responsible, at its own expense, for the construction and operation of the public infrastructure project for the benefit of the public. The responsibility of operating and managing the project is borne by the company, which is given the right to conduct services for public users within the specified time (compensating its costs with expected profits) before the project is transferred to the government [3].

The concept of the BOT contract can be summed up by the three words of its acronym: 'Build,' which carries the meaning of constructing or assessing a project; 'Operate,' which reveals the sense of managing and running the project; and 'Transfer,' which conveys the sense of returning the project to the authorized party at the end of the tenure. During the operation of the project, the company has the right to acquire all benefits and profits to cover all managing costs. Since its emergence, the BOT contract has been intended specifically to reduce the government budget needed to execute infrastructure projects. Nevertheless, currently, application has been extensively expanded not only to the government but to legal authorities, companies, and various organizations such as *waqf* institutions [2].

Maintaining *Waqf* Properties

Waqf is a righteous deed encouraged by *Shari'ah*. It is submitted that the general aims of *waqf* include helping the poor and needy, raising the standard of living for many people, facilitating the distribution of knowledge and technology, and administering and constructing schools, mosques, libraries, hospitals, clinics and shops for the benefit of people in general and Muslims in particular. By recognizing these significant aims of *waqf* in the development of human life, there is a need to enlarge and develop *waqf* properties, to maintain and expand them without neglecting their ultimate purposes as outlined by the *Shari'ah* (Islamic law). This attempt is to ensure that the assets of *waqf* are preserved and utilized at the maximum level so that they will not be left idle or inactive, which in turn would contradict the founder's main purposes and pious aims in assigning his or her properties as *waqf*.

There are many Quranic verses that generally encourage Muslims to maintain and increase properties. One of those verses is in Surah Al-Mulk:

"It is He who has made the earth tame for you, so walk along its slopes and eat of His provision and to Him is the Resurrection" (Quran, 67:15).

Another verse which indicates this encouragement is as follows:

"And when the prayer has been concluded, disperse within the land and seek from the Bounty of Allah and remember Allah often that you may succeed" (Quran, 62:10).

These verses of Quran indicate the encouragement of wealth creation and preservation. Upon this basis, it is important to conserve *waqf* property and to avoid all misdeeds that will lead to eradication of its usufruct. The *waqf* should be suspended for the right of beneficiaries as well as for ongoing charitable and religious purposes, whether required by the *wāqif* (founder) or not. Ongoing charity will be increased only through developing *waqf* property, thereby proving that the expansion of utilization of the *waqf* assets is certainly an obligation from the *Shari'ah* perspective. Enhancing and developing *waqf* assets are crucial to guaranteeing that the beneficiaries have continuous benefits and the founder have continuous rewards from Allah [4]. As for economic contributions, *waqf* will provide job opportunities and stimulate the circulation of wealth among members of society.

Validity of BOT from an Islamic Perspective

Since there is no specific Quranic verse or Prophetic tradition that clearly depicts and establishes the legitimacy of this contract, a brief study of the doctrine of freedom of contracts in Islamic law is necessary to establish its legality. Therefore, before proceeding to cite evidence by scholars for the legitimacy of the BOT contract, the theory of freedom in Islamic contracts needs to be discussed first.

It is submitted that everything prohibited or allowed has been clearly prescribed in the Quran and the Sunnah of the Prophet Muhammad (Peace be upon Him). There is no explicit text in the Quran or Sunnah (Hadith) establishing the legality of a BOT contract. Similar to *istisna'*, its legality is not stated in Al Quran and Sunnah but is based on the concept of *istihsan*, which focuses on needs and necessities as advanced by many Muslim jurists. Yet this basis is not very convincing [5].

In relation to the concept of freedom of contract, Islamic rulings illustrate that the contracting parties have full freedom to engage in whatever contract they desire as long as it is in line with the principles of *Shari'ah*. According to Ahmad Bukhait, adding a new contract is permissible under the freedom of contracts, but it must be compliant with the principles of *Shari'ah* [2]. Therefore it can be clearly stated that any contract that is not consistent with *Shari'ah* is considered invalid even though the contracting parties have freedom to form whatever contract they wish. Thus, Muhammad al Bashir says:

"...the will of contracting parties has played no role in forming a contract or in stipulating a condition which the Lawgiver has not allowed and explicitly authorized in the Quran and Sunnah" [5].

Nevertheless, it is worth noting that in the natural state of things, there is a presumption of legality (*Al-asl fi al-ashya' al-ibāhah*). Therefore all acts and deeds, including the making of contracts, are permissible (*mubāh*) unless they have been particularly prohibited. To support this point, the Prophet Muhammad is reported to have said:

"Every stipulation is lawful among the Muslims except one, which declares forbidden what is allowed or allows what is forbidden" [ii].

Thus, there should be clear principles of freedom of contract to ease transactions among the public, to be applied in any new emerging contract to ensure the contract's validity and legality. The principle might be on behalf of contracting parties where mutual consent may be expressed among them, and the contract must be free from any element of *gharār* (uncertainty) that may result in injustice among the parties and consequently lead to dispute. Therefore, for the BOT contract to be valid, it should have certain elements. Firstly, there should be mutual agreement between the contracting parties. Secondly, the nature and consequences of the contract should not be contradictory to *Shari'ah* principles.

Another core principle supporting the validity of the BOT is *Mashalih al mursalah*[iii] (public interest), which is based on the interest of the public. In general, *al-Mashalih al mursalah* is the derivation of a *hukm* (Islamic legal ruling), which is not stated in the Quran or Sunnah but the implementation of which aims to provide benefits to many people. This concept is applied primarily to promote a harmonious environment for people. In this case, the BOT contract is applied in the financing of an infrastructure project to develop current public facilities and necessities that would serve the interest of the people. The long-term result would be to ensure sustainable convenience benefits and services for the life of today's *'ummah*[iv].

The BOT contract is similarly applied in the development of *waqf* property with the purpose of gaining perpetual benefits and returns for the beneficiaries and other people. It also prevents the *waqf* property from becoming idle, which might bring damages to the property itself. The BOT contract is compatible with the saying of Imam al Ghazali regarding this concept. According to him, *maslahah* consists of considerations that secure benefit or prevent harm. Protection of life, religion, intellect, lineage, and property is *maslahah* [6]. Realizing this consequence, the BOT contract is applied even though the *hukm* is not there in the primary sources of Islamic law. This signifies that it is lawful to apply the BOT contract for the sake of *maslahah* of *'ummah*, as it will secure benefits and at the same time it goes not harm *waqf* estates. The BOT contract can be a breakthrough alternative effective in increasing the utilization and benefits of *waqf* properties, and it can perhaps be an efficient means of increasing the revenues of *waqf* property.

It is also possible to prove the legality of BOT by the legal maxim "harm must be eliminated" (*al-dararu yuzāl*), because the application of BOT may eliminate the harm of *waqf* or other lands being left idle. Leaving a *waqf* property idle is harmful in the sense that benefits may not be utilized at a maximum level since there is no yield from the property, while at the same time operational and maintenance costs may increase.

BOT has a special nature of reliability as compared to other Islamic financing methods such as *ijārah*, *istisna'*, and *ḥukr* contracts[7], which are known (nominated) contracts in Islamic jurisprudence. Muslim jurists allow the application of all these contracts with certain conditions and terms [8]. In this regard, the application of BOT can be claimed as valid since it is a combination of these known contracts. For example, it only combines *ijārah* and *istisna'*, which are previously known as valid and recognized contracts. In essence, the contracts and instruments used are legal and valid.

In developing *waqf* property, the *hukr* mode is applied to avoid any ruin of the land, but this approach is not entirely convincing since it involves utilization of the land for a long duration, which would possibly damage the usufruct of the *waqf*. As an alternative, the BOT contract is applied with certain conditions. According to Khalid Sa'ūd, BOT can be applied in developing *waqf* with a condition on the duration spent. The contract also should preserve the rights of the beneficiaries and the developer during the operation of the project [7]. The duration therefore should not be as long as for *hukr*, or otherwise it should not be applied, as it may ruin the *waqf* property itself. This idea is to preserve the *manfa'ah* and maintain the ultimate purpose of assigning one's property to *waqf* as well as avoiding harm among the contracting parties. The legal maxim pronounces that major harms may be replaced by lesser ones (*Yukhtāru ahwan al sharrain*) to preserve the *manfa'ah* and *maslahah* of the public.

In view of the structure of the BOT contract, it is possible to make a clarification on its legality as a contract comprised of more than one contract at once. The permissibility or impermissibility of the combination would provide an idea on the validity of its application. *Shari'ah* permits the act of combining more than one contract. For example, the BOT contract is comprised of many contracts, namely *istisna'* and *ijārah*, yet does not impose one contract as a condition on the other. The combination is also acceptable unless it involves a prohibited element, as stated in the Al Quran and Sunnah [9].

It is suggested that a combination of contracts is permissible if all combined contracts are legal upon any basis and there are no elements of *gharār*, no harmful conduct or behavior among contracting parties, and no dealing or practice of *riba'*. A combined contract would be prohibited if its utilization or application were an excuse for dealing in *riba'* [9]. In other words, the combined contract should not be a trick for committing *riba'* or anything that is prohibited. With regard to *riba'*, it has been reported that Prophet Muhammad (P.b.u.h) instructed one of his employees to sell low quality dates before buying those of high quality to avoid *riba'* [9] and forbade the act of exchange between low and high quality dates, which would cause injustice among the contracting parties. The Quran prohibits the demand and receipt of interest, as was mentioned in the following terms:

"*O ye who believe!Devour not usury* [v], *doubling and quadrupling (the sum lent). Observe your duty to Allah that you may be successful.*" (Quran, 3:130).

The application of the BOT contract in the development of *waqf*, in our opinion, is prone to be valid and permissible based on the earlier discussion, since there is no restriction in the Quran or Sunnah that entails its prohibition. Furthermore, evidence of the lawfulness of BOT in practice is the fact that the ultimate purpose of the contract is to develop *waqf* property and to prevent the usufruct of *waqf* assets from being misused. The BOT contract could possibly be the financing mode most likely to protect the *waqf* from *dharar* (harms) even though its application should perhaps be frequently reviewed. Although there are a number of possible issues pertaining to BOT and *waqf* rules and regulations which would bring minor difficulties in practice, this is not a prudent reason totally to reject the application of BOT, as this approach has the tendency to secure *waqf* properties. All minor issues can be solved and readjusted as long as they remain within *Shari'ah* limitations, which will ensure that the BOT contract turns out to be an absolute *Shari'ah* compliant mechanism. Moreover, Prophet Muhammad (P.b.u.h) has said:

"*What is good for Muslims is good in the sight of Allah*" [vi].

Therefore, the BOT contract should be considered as valid in practice.

Background of MAIWP

The only *waqf* institution in Malaysia that is currently managing a BOT project to develop its property is *Majlis Agama Islam Wilayah Persekutuan* (MAIWP). MAIWP was incorporated on 1st February 1974 in commemoration of the enforcement of the Federal Constitution for the federal territory. MAIWP is an organization responsible for all Islamic administration affairs in the Federal Territory. In addition, it is the policy maker and the Islamic religious advisor of *Yang di Pertuan Agong*. For this reason, it is worth noting that MAIWP is a statutory or legal unit with the authority to create and amend the law, established under the enactment of Islamic legislation.

MAIWP is currently undertaking a significant endeavor to develop *waqf* land using a BOT contract on a project known as Menara MAIWP (MAIWP Tower), which is now under construction and is expected to be completed in May 2011. The construction is being conducted on the land of lots 168 and 169 at Jalan Perak, Kuala Lumpur. The total area of both land lots is equivalent to 52,796 square feet, or 1.2 acres. The land area of Lot 168 is 28,620 square feet while Lot 169 is 24,176 square feet. The land has been entrusted to MAIWP for 16 years, starting from 14th May 1993, as endowed by Allahyarham Ahmad Dawjee for philanthropic purposes. As of the end of year 2006, the value of lots 168 and 169 is approximately RM31.915 million, comprised of both lots' value of RM17.29 million and RM14.625 million, respectively.

In a special meeting of the Islamic affairs council, federal territory, held on 5th July 2005, a decision was made to develop this particular *waqf* land. Construction of 'A' class office buildings consisting of 34 levels is to be carried out by *Lembaga Tabung Haji,* who will operate the project for 25 years after construction has been successfully completed[vii].

It is interesting to note that the development of this *waqf* land is being conducted under BOT contract, the core subject of this study, although this type of contract has never before been used by the council.

The application of BOT contract in developing *Waqf* land

The development of Menara MAIWP on the land of both lots contributed by Allahyarham Ahmad Dawjee is being carried out by *Lembaga Tabung Haji* (LTH) through its subsidiary company TH Technologies *Sdn. Bhd.* TH Technologies has the right to assess the project, and LTH has assigned the company the construction work according to *istisna'* and *wakala* concepts. Under *istisna'*, TH Technologies is to construct the building and obtain returns after completion. Under the *wakala* contract, TH Technologies will act as a representative of LTH Construction to develop the Menara project. In this case LTH is considered the contractor as well as the financier of the total cost of the project.

The project is estimated to cost RM151 million, and the estimated construction period is 46 months, equivalent to 3 years and 10 months. In consideration of LTH's efforts to construct the office building, LTH is authorized to manage building operations for a period of 25 years upon completion of the building. Thus, the concession received by LTH is actually issued by the MAIWP itself as the concessionaire of the project.

It is worth noting that ownership of the building remains with MAIWP during the operation period since ownership was never transferred to the LTH. The building will be leased by LTH from MAIWP under the concept of *ijārah*. The *ijārah* fees are in the form of deferred payments to be remitted at the end of the concession period. Most of the time, payment is made by returning the project to MAIWP after the operation segment. Mr Hanif, Deputy Director of Development and Investment Unit, Federal Territory Islamic Religious Council (MAIWP), said that Bank Islam Malaysia Berhad will be one of the tenants other than TH Technologies *Sdn. Bhd.* itself. All construction costs are being financed by LTH, which is therefore entitled to all revenues and profits obtained during the initial operation phase.

The structure of BOT application

Diagram 5.1: Typical BOT organizational structure

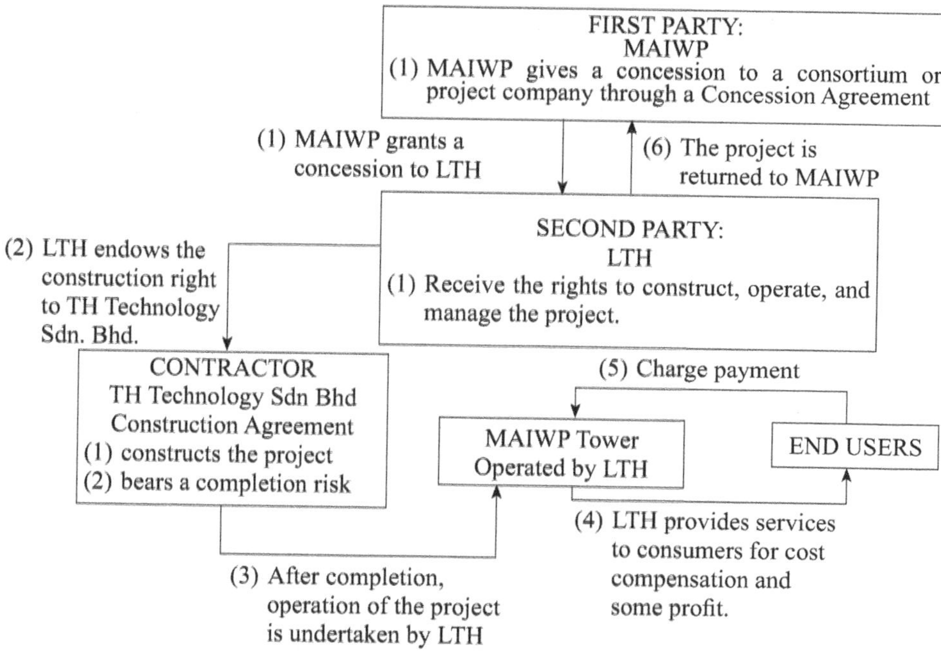

The development of *waqf* land through the BOT contract involves several parties, each with their own responsibilities, including the MAIWP itself, *Lembaga Tabung Haji* (LTH) and TH Technologies *Sdn. Bhd.* (LTH's subsidiary company). LTH is the only party that covers all development costs and operates the project for revenues and profits, while TH Technologies *Sdn. Bhd.* is responsible for project construction and thus bears construction risks, including the requirement to complete the project on time. MAIWP is the primary party and is also the issuer of the grant to allow LTH, through the concept of *istisna'*, to construct the project and operate it for a certain duration. Under this concept, MAIWP seeks financing from the financier, LTH, to construct a building on waqf land for expansion in accordance with agreed specifications at a determined price and for a fixed date of delivery [10].

After construction, LTH will subcontract the performance of the project work to a third party (TH Technologies *Sdn. Bhd.*) to construct the building through the concept of *istisna'* and *wakala* [viii]. The *istisna'* concept allows TH Technologies *Sdn. Bhd.* to be prepared for construction. The estimated cost of RM151 million will be compensated at the end of the concession period. Under the *wakala* agreement, the financier (LTH) appoints the constructor (TH Technologies *Sdn. Bhd.*) as its agent to construct the project on its behalf. All responsibilities relating to project completion are borne by the agent [10].

LTH leases the building from MAIWP and operates it for a 25-year concession period under an *ijārah* agreement. This duration is for LTH to operate the project and gain revenues as compensation for construction costs and also to derive some profits. Under the *ijārah* agreement, MAIWP as the owner of the project leases the building to LTH, which then subleases the building to lessees, for example Bank Islam Malaysia *Bhd.* and TH Technologies *Sdn. Bhd.*, to run and operate their own businesses.

LTH would earn rental payments, which would in turn be used to pay MAIWP. As a result of the entire project, MAIWP will obtain approximately RM56.6 million in lease payments at the end of the concession period, together with the land and the project, which will have a value then estimated at about RM600 million[vii].

Strengths of the BOT Contract

The BOT contract reduces the burden on MAIWP's resources and budget, since funding for the construction is not derived from MAIWP's budget, and the utilization of assets is not restrained for construction. MAIWP also does not have to bear the burden of debt because LTH will fund all construction costs, a situation which would prevent the imposition of new taxes as well as maintain existing tax rates. Through BOT, LTH assumes responsibility for funding and operating the project, releasing MAIWP from the burdens and risks it would otherwise have to assume. LTH will bring investment into the development of MAIWP through its involvement in the construction and operation of the project [3].

Therefore, the application of BOT offers a solution to MAIWP for its financing problems. Through the BOT contract, MAIWP would eventually acquire the project without increasing its expenditure or borrowing from other financial institutions or banks to accomplish the project.

Through this scheme, MAIWP will improve its performance through the new concepts instilled by LTH in terms of skills in operating and managing the project and its related assets. The United Nations Industrial Development Organization (UNIDO) aptly states that the involvement of private sponsors and experienced commercial lenders is an additional sign of project practicability that ensures technology transfer, enhanced training of local personnel, and the advancement of the national capital market for the economic development of a nation. The project company assigned to construct an infrastructure project always seeks to achieve the greatest possible economic gains. This could be attained by the use of modern means and advanced techniques. Modern tools and mechanisms assist the company in gaining the greatest possible profits, reducing construction and operating costs to a minimum level [11].

The Menara MAIWP constructed under the BOT contract will become a milestone in improving MAIWP's image as an advanced Islamic agency that is able to move forward in tandem with other globally renowned agencies. Indirectly it shows the integrity and advancement of Islam to the world as a whole. Throughout construction, the community will be acquainted with MAIWP's proactive effort in expanding *waqf* land, which is in line with Malaysia's Ninth Plan encouraging the development of land for the benefit of the public.

MAIWP will acquire a return of RM56.6 million after 25 years of leasing the land. This sum will be used by MAIWP to double its efforts to protect Muslims, especially those located in the federal territory. When the lease concession expires, the land and the building will be returned to MAIWP with an estimated value of RM600 million [vii].

Various *Shari'ah* Issues Related to the BOT Contract

There are several possible ambiguous *Shari'ah* issues related to the BOT contract. This, however, does not mean that the issues have not actually been addressed according to *Shari'ah* regulations. Among *Shari'ah* issues are the following:

First issue:

It is mentioned that the BOT contract is formed on the basis of *istisna'* and that the payment for it is the *manfa'ah* (usufruct) of the project, which does not exist during the contract due to the subject matter's (a thing, i.e., a building) not being constructed yet. The act of giving usufruct as compensation has been approved by Islamic jurists in situations of selling and leasing [12]. The issue is that in *istisna'* the usufruct of the item is non-existent and this is different from cases of selling and leasing, where the usufruct is in existence during the contract. So, is it permissible to consider a non-existent usufruct as compensation for the cost of *istisna'*?

According to Sheikh Muhammad Taqi 'Usmān, it is permissible even though the *manfaah* (usufruct) is not present in the contract, because there is a true need for the development of *waqf* land. Furthermore, there is no element of ambiguity, as the utilization of the usufruct is based on an amount of time agreed upon. This is also valid in order to ensure construction work under the concept of *istisna'* [5] is accomplished on time because the usufruct comes together with the construction. The contractor will not obtain the benefit until the construction is fully complete. In other words, the benefit depends on the contractor bringing it into existence.

Second issue:

As mentioned earlier with regard to the first issue, the *thaman* (price) in *istisna'* is a *manfa'ah* (usufruct) of the subject matter (i.e., a building) that will be authorized to the developer for usage after the completion of construction. Some argue that the price should be in *'amwal* (a specific price and must be a valuable thing) [13]. In other words, is *manfa'ah* (usufruct) *'amwal*?

The *Hanafī madhab* does not consider the *manfa'ah* (usufruct) as *'amwal* because according to them, in order for something to be considered as *'amwal*, it must be material, tangible and desired by people for purchase even though it may be deemed unlawful by Islam. However, according to Ahmad Farāj, contemporary jurists from *Hanafī madhab* consider *'amwal* to be things that can be possibly exchanged for currency be they objects, usufructs or rights [14].

According to *jumhūr* (the majority of jurists), *manfa'ah* (usufruct) is deemed to be *'amwal* as long as its source is an object that can be kept and possessed even though people do not spend money or effort to attain it. This is because the *jumhūr* (the majority of jurists) view that an object is meant for the benefit it comprises, not for the object itself [14]. A thing is valuable if there is benefit to be used [ix]. For example, a car or house is valuable due to its usefulness, not for the object itself. Since *manfa'ah* (usufruct) can be considered as *'amwal*, Islamic jurists allow it to be the *'ujrah* (rental fee) in *ijārah* transactions [15].

From the researcher's point of view, opinions of the majority of Islamic jurists and contemporary jurists from the *Hanafī madhab* widen the scope of *'amwal* to encompass all the things unknown before, especially before this era of globalization. For example, intellectual rights and the rights to invent something, not discussed by former jurists, are deemed *'amwal* as long as they can be estimated in monetary terms [14]. Therefore, *manfa'ah* is considered *'amwal* and is applicable as the price of *istisna'* in a BOT.

Third issue:

If it is permissible for the *manfa'ah* (usufruct) to be the price in a BOT transaction, how can one ensure the *manfa'ah* (usufruct) will be the same amount as the construction cost? Sometimes the return of usufruct utilization made by a project company might exceed the cost of construction, or it could be less. This will cause uncertainty and injustice, causing a contract to be voided.

In the BOT contract, the project company is entitled to build and operate the project. If the revenues obtained during the use of the building exceed the cost of construction, these revenues are deemed profit for the project company. Moreover, such revenue resembles an *'ujrah* (wage), in the form of compensation by the landlord, paid to the company for operating the project. Likewise, the period of operation and utilization is known and agreed upon by the contracting parties. As such, elements of uncertainty and injustice can be tolerated.

Fourth issue:

In the practice of BOT, the contractor or manufacturer responsible for operating the project intends to receive financial resources from the fees imposed on those tenants who use or consume benefits from the project. However, the total amount of fees derived from consumers is unknown during the contract [12]. This situation will lead to ambiguity in the contract.

To counter this ambiguity, it is worth mentioning that the *thaman* (price) of *istisna'* is not the estimated financial resources, but the benefit from the project itself for a certain duration. This benefit is independent and not based on the payment imposed on the consumer. This situation can be clearly understood through the following example. A man buys a residential complex of rental apartments but during his purchase he does not know the fare or rental payment that he is going to receive from his future tenants. This unknown fare is not a major ambiguity because he bought only the usufruct of the complex. The usufruct of the second leg leasing/sublease is not the issue through which the financial resources are expected to be obtained.

Fifth issue:

Normally, the construction of a project under the BOT scheme requires a huge amount of funding to cover all costs, and the contractor is unable to afford such huge expenses on his own. This situation will force the contractor to seek financial support from any financial institution on the basis of *riba'*-based loans. This is obviously forbidden in Islamic law.

In this case, the contractor could benefit from any Islamic financing contracts such as *Mushārakah, Murābahah, ijārah*, etc., all of which are *Shari'ah* compliant. The application of BOT in MAIWP perhaps does not engage in (cause) this kind of problem since the contractor (LTH) is the financier of the project who can afford all the expenses independently and does not require support from other financial institutions.

Another alternative means is financing through the *Mushārakah bi tārīq al-tawrīq* (securitization) mode. This method permits the contractor to issue *ṣukūk* (Islamic bonds) to raise funds from the contributions of other companies [12]. It is practically acceptable to issue *ṣukūk* before or after the construction of the project. This enables all contributing companies to put in efforts to develop and operate the project, which finally would entitle them to a share of the financial resources gained from the utility of the project development.

Sixth issue:

It is assumed that the concept of Build, Operate and Transfer (BOT) contract resembles *Bai' al 'inah,* a type of trading where the seller sells his assets to the buyer at an agreed price to be paid by the buyer at a later date. After that, the buyer immediately sells back the assets to the seller at a cash price, lower than the agreed selling price. According to the majority of Islamic jurists (Malikī, Hambalī and Hanafī), *Bai' al 'inah* is not permissible because it is a *zarī'ah* (way) or *hiyāl* (legal excuse) to legitimize the practice of *riba'* (usury) [16].

First, it needs to be clarified that the BOT contract is not a form of selling where the seller sells his asset to a buyer. This is because the asset is non-existent during the agreement. BOT is nearer to the concept of *istisna'*, where the seller must bring the buyer's required item into existence under a specific description made by the buyer.

Second, the BOT contract does not have two separate contracts properly executed. In contrast, *Bai' al 'inah* consists of both a contract of sale and a repurchase, for example the contract of sale by person A to person B on deferred payment terms and the contract of repurchase by person A from person B on cash payment.

Third, the aim of BOT is to seek financial support for a costly construction project in order to reduce the financial burden of an organization. In return the financier will get profits. This is unlike *Bai' al 'inah,* where the purpose is *riba'*, with an exchange of money in unequal quanities taking place, with the asset treated only as a formality. The contracting parties are not interested in the transfer and acquisition of ownership but in the exchange of cash. In other words, the parties do not wish to own the asset but to own an amount of money in the form of cash.

Moreover, the BOT contract is not similar to *bai' al 'inah*, even though the owner of each contract will be getting back the asset at the end of the transaction. This is because BOT does not involve selling the asset but leasing it, which maintains its ownership, whereas *Bai' al 'inah* does involve selling, and a transfer of the asset's ownership occurs [16]. It is important to know that the objective of sale is for the seller to transfer ownership (*tamlīk*) and for the buyer to acquire ownership (*tamalluk*), which does not occur in BOT. In *bai' al 'inah*, both parties are uninterested in transferring and acquiring ownership but rather cash, as discussed earlier [x]. From the researcher's point of view it is invalid to have an asset that is not in one's ownership any more by purchasing it back from the buyer at a lower cost, hence opening a backdoor to *riba'*.

Seventh issue:

Other than *bai' al 'inah*, the BOT contract is considered to have a similarity with *bai' al wafa'* in terms of the buyer not being allowed to sell the asset to other parties. In *bai' al wafa'* a seller in need of cash sells his asset to a buyer, with the promise that the seller can redeem the asset from the buyer at the same price at which he previously purchased it, while the buyer does not have the right to sell it to other parties. Dr. Engku Rabiah Adawiyah Engku Ali said that this resembles a *qard* (loan) with *rahn* (pledging) [xi], meaning that when there is *bai' al wafa'*, the seller has the authority to claim the asset he has sold by paying the buyer the full price of the asset. It is known as *wafa'* because it is a promise of returning the asset sold when the seller claims it by paying the full amount.

BOT is not like *bai' al wafa'*, which is prohibited by the Islamic jurists of Malikī and Ḥanbali *Madhab*, as well as the earlier generation (*mutaqaddimūn*) [17]. However, the jurists of Ḥanafī and Shafi'ī *Madhab* consider *bai' al wafa'* as permissible on the basis that it is an effort to prevent one from committing *riba'*. Permission for the sale would give relief to people needing to borrow money without violating the prohibition against *riba'* [17]. For this reason, the Shafi'ī scholars considered this contract a type of pledge, which they called *rahn al-mu'ād* (a pledge whose object will be given back) [17]. Currently, *Majlis Majma' al Fiqh al Islami* in its 7th Session has resolved that *bai' al wafa'* is not permissible because it is considered "a loan that generates benefit" and a "legal trick for committing *riba'*" [xii].

First, the BOT contract is not a form of *bai' al wafa'* in which the seller is in need of cash but a form of *istiṣna'* where the buyer really needs the asset, not the cash.

Second, the asset in the BOT contract (i.e., the building) is not considered as *rahn* (pledging), in which the seller pays back the asset sold and the buyer returns the asset to the seller, but it is more like *ijārah* (lease) where the asset will be returned back to the owner after the use of its usufruct upon completion of a fixed period upon, along with a certain payment. As it is under the *ijārah* concept, the party that uses the building is not allowed to transfer ownership of the asset from the original owner to other parties.

Furthermore, the main features of *bai' al wafa'* do not suit the BOT contract. Those main features are the following:

(1) In *bai' al wafa'*, the seller and buyer can terminate the contract at any time. In contrast, in the BOT contract, full consideration must be taken before the contract can be terminated, as termination might burden both parties. Since the BOT contract is an agreement mainly between two parties, termination is valid only if there is consent by both parties. Usually, if termination occurs, it does not happen because of the project company. This is because the company usually spends a huge amount of money for project construction. A long-term operation period granted by the government to the project company is the essential element to cover all the costs. If termination happens on behalf of the government, for example, it will bear the burden of compensating all costs incurred by the project company [3]. In order to avoid this burden, termination is always strongly avoided.

(2) The buyer can utilize and benefit from the property bought through *wafa'*. This is similar to the BOT contract, where the building constructed by the company can be used for profit but is under the other party's supervision. One party (e.g. the government) grants property utilization rights to another (i.e. the project company) to pay off all construction costs before the property is returned to the government. Under *bai' al wafa'*, the buyer is allowed to take benefits from the property sold until it is redeemed by the seller. The property is considered as collateral for paying debt [18].

Eighth issue:

As in the application of the BOT contract, the first company (i.e. the government) gives the ownership of *manfa'ah* (usufruct) to the project company after the completion of construction. However, the question about its legality is raised. Is the ownership of usufruct (beneficial ownership) recognized by *Shari'ah*?

It is submitted that in Islam there are two kinds of ownership. Firstly, *milkiyah tām* (full ownership) is when the owner has the absolute right to transfer or dispose the ownership to another person by selling it, giving it as a gift, or contributing it as charity, as well as the right to give the *manfa'ah* without transferring ownership by leasing it, etc. The other form is *milkiyah nāqisah* (partial ownership), in which individuals have the right only to benefits and are not allowed to transfer ownership, as the object is not theirs. They are permitted to use the usufruct for profits but they are not allowed to sell it or contribute it to *waqf*, etc, as those practices cause ownership to be transferred [14].

From the above discussion, it is clear that *Shari'ah* recognizes beneficial ownership. Among the transactions highlighted by the *Shari'ah* to be reasons for having that kind of ownership are *ijārah*: (leasing), *i'ārah* (borrowing), *wasiyyah bi al manfa'ah* and *waqf* (on behalf of the *Nazīr*). Merely a trustee for *waqf* properties, the *Nazīr* has the responsibility of looking after their benefits for the good of Muslims [8].

Ninth issue:

What is the position of construction in the case of default?

A property whose construction is not completed belongs to the landlord on whose land the construction is being conducted. *Hukum Kanun Tanah* (The Lands Laws) states that anything built or constructed on a landlord's land is deemed to be in his ownership, even in the case of default [xiii]. It is necessary to identify the party causing a default and to clarify the possible effects of default on each party. Usually, default is due to natural disasters or termination on behalf of the landlord (i.e. the government) and very rarely happens on behalf of the project company, which must cover all construction costs and avoid loss [3]. If a disaster causes the default, the contracting parties will sign a new contract terminating the project or renewing it, a process which usually requires the mutual consent of two parties in reaching a conclusion. However, if the government causes the default, it must compensate for all the losses incurred by the project company. The government shall also obtain compensation from the project company if the project company fails to accomplish construction within the stipulated time or the project company breaches the *'aqd* (contract) without the permission of the other party [3].

Tenth issue:
The BOT contract is like two contracts in one, i.e. *istisna'* and *ijārah*. With regard to the BOT contract, the project company will construct a building according to the forms, designs, and plans desired by the first party (i.e. the government). The BOT contract is a kind of *istiṣna'* contract where a subject matter or object is requested by the so called buyer that does not exist at the time the contract is concluded. It can also be a contract with a manufacturer to produce a specific commodity in a specific way [5]. After construction is complete, the government grants the project company the right to utilize the project for a specified time through a second contract, which is *ijārah*. At the end of the period, the leased building will be given back to the government along with the lease payment. The question is: Is combining multiple contracts permissible from a *Shari'ah* perspective?

The concept of combining contracts involves two or more parties agreeing to more than one contract [9]. The combination is permissible by *Shari'ah* unless it breaches Islamic restrictions, which would prohibit its practice [9]. However, there are certain rules and regulations imposed by *Shari'ah* to regulate its permissibility:

(1) No case prohibited by *Shari'ah* can be included in the combination of contracts, for example combining sale and lending in a contract. This prohibition is based on the directives of Prophet Muhammad P.b.u.h which forbade combining lending with selling. Prophet P.b.u.h said: *"it is not permissible to give a loan and sell at the same time..."* [1].
(2) It should not be utilized as a trick for committing *riba'* such as *bai' al 'inah* and *bai' al wafa'*, as discussed above.
(3) It should not be used as an excuse for a means of practicing *riba'*. For example when both parties conclude a lending contract, but at the same time impose excess repayment in terms of the quality and the quantity on the borrower, or stipulate other conditions on the borrowing such as giving something to the lender, be it money, accommodations, a car, etc. The *fuqahā'* agree that when there is a stipulation in the loan contract that there should be a reward to the lenders, the contract becomes null and void, and is considered as *riba'*.
(4) The combined contracts must be free from the element of contradiction with regard to the ultimate goals and rulings, for example, granting a property to a person as a gift but selling it back to him simultaneously, or combining *Mudārabah* with lending the *Mudārabah* capital to the *Mudārib*. Contracts that contradict each other in their rulings and effect cannot be combined in the same transaction [9].

From the above discussion, it can be seen that there is no element of dealing in *riba'* as far as the BOT contract is concerned. It is a financing contract based on the concept of *istisna'* and *ijārah,* which are both permissible in practice [13]. It is observed that the above mentioned conditions are not violated in the implementation of BOT. Furthermore, the purpose of the BOT contract in general is not in contradiction with the objectives of *Shari'ah*, such as preservation of *mal* (property), transparency, justice, and mutual concerns among the contracting parties where all are in a win-win situation.

Eleventh issue:

The costs of delays during construction due to *force majeure* should be covered by insurance; otherwise it is borne by the project company. However, a large number of contemporary Muslim jurists regard modern commercial insurance invalid and incompatible with the injunctions of Islamic Law [13]. In insurance, the buyer of the insurance policy does not know what he has bought with his premium. The time of occurrence of events is also uncertain for the parties. As such, insurance is primarily a *gharār* contract.

Instead of insurance, the *takāful* scheme can be applied. *Takāful,* or "mutual guarantee," is generally considered as an Islamic alternative to the modern insurance business. *Takaful* is also known as solidarity of *mudārabah,* since contributions in *takāful* are invested on the basis of *mudārabah* [19]. A participant, while entering *takāful,* concludes two contracts:

(1) The contract of donation, or *tabarru',* whereby a person undertakes to donate a portion of his contribution in a *tabarru'* fund established to provide *takāful* benefits to any participant who suffers some loss or damage.
(2) The contract of *mudārabah,* whereby a person undertakes to pay a portion of his contribution to the company for the purpose of business on the basis of *mudārabah.*

Both the Islamic concepts of *mudārabah* and *tabarru'* are satisfactory from the *Shari'ah* point of view. Furthermore, *takāful* is free from the element of *riba'* because the participant or the policyholder does not get a fixed return on the Participant's Account (P.A.), instead he gets a varied income. His capital is subjected to the principle of profit and loss sharing as enunciated in the Prophetic tradition "profit goes side by side with the risk." The amount accumulated in the P.A. is invested in various forms of business strictly in accordance with Islamic investment law [13].

It is submitted that *takāful* is not a contract of *gharār* [13] The element of uncertainty and lack of knowledge is not inherent in *takāful* yet is inherent in insurance because the uncertainty relates to the occurrence of events (i.e. the subject matter of the contract, the acquisition of amount of insurance, its quantity and time of payment).

Twelfth issue:

BOT contracts must be supervised by *Shari'ah* Committee executive officers for *Shari'ah* compliance.

The application of BOT in developing *waqf* land by MAIWP has been discussed and approved in the meeting of the Federal Territory *Shari'ah* Laws Consultative Committee (*Jawatankuasa Perundingan Hukum Syarak Wilayah Persekutuan*) at Menara Pusat Islam, Kuala Lumpur [xiv]. This committee was established on 19th February 1974 and among its jurisdictions is to make decisions on rules governing problems or issues requiring an Islamic opinion. The committee consists of twelve members, among whom are Dato' Haji Wan Zahidi b Wan Teh (*Mufti* of Federal Territory) the chairman, Tan Sri Datuk Sheikh Ghazali bin Hj Abdul Rahman, Dato' Haji Md. Salleh bin Haji Ahmad, Prof. Madya Dr. Muhammad bin Arifin, Tuan Haji Mohd Alwi bin Haji Yusoff and Tuan Haji Ghazali bin Mohd Taib [20].

Conclusion

The above discussion indicates that the method of Build, Operate and Transfer (BOT) in developing *waqf* assets is a viable *Shari'ah* compliant means. The BOT contract could be the solution to the problem of inadequate resources in financing the development of *waqf* properties. It may also be a new and enhanced alternative for *waqf* institutions in expanding property for beneficial gains in revenues and profits. As a result, other development projects can be accomplished through the revenues obtained for the continuous benefits of the people. Through the BOT scheme, many companies will collectively combine their efforts for development, consequently easing the burden of *waqf* institutions and providing them with the skills to run projects efficiently until the end of their concession periods. It is worth noting that the combination of other *Shari'ah* compliant contracts such as *istisna'* and *ijārah* would contribute to the success and validity of the BOT scheme, although certain *Shari'ah* issues might arise during practice. There are a number of identified *Shari'ah* concerns for BOT transactions, as discussed in this chapter. Those issues and their solutions should be taken into consideration in order to ensure there is no violation of *Shari'ah* principles.

References

[1] Al-Tarmidhi, Muhammad bin Isa (2008) *al Jami' al-sahih.* Dar Ihya al Turath al 'Arabi, Beirut.

[2] Bukhait, Ahmad Muhammad (2009) *Tatbiq 'aqd al-bina wa al-tasyghil wa al i'adah* (BOT*) fi ta'mir al awqaf wa al-marafiq al-'ammah.* Fiqh Academy Conference, UAE, 19th April.

[3] Al-Bahji, 'Isom Ahmad (2008) *'Uqud al-BOT al-toriq li bina'marafiq al-daulah al-hadithah.* Al Azaritoh, Dar al jami'ah al jadidah.

[4] Dibani, Abd Majid (1792) *Manshurat Jami'ah Qaryunis, Banghazi. Muhadarat fi Fiqh al-mu'amalat. 1st edt.* Manshurat Jami'ah Qaryunis, Banghazi.

[5] Al Amine, Muhammad al Bashir, (2006) *Istisna' (Manufacturing contract) In Islamic Banking and Finance Law and Practice.* A.S NOORDEEN, Kuala Lumpur.

[6] Omar, Ismail (1999) *Manhaj aqidah ahl al Sunnah wa al-jama'ah.* Percetakan Hiza, Selangor.

[7] So'ud, Khalid (2009) *Tatbiq 'aqd al-bina' wa al-tasyghil wa al- ia'dah "BOT" fi ta'mir al-marafiq al-'ammah wa al-awqaf.* Fiqh Academy Conference, UAE, 19th April 2009.

[8] Umar, Muhammad Abduh (2009) *Tatbiq nizam al-bina wa al-tamlik BOT fi ta'mir al-waqf.* Fiqh Academy Conference, UAE, 19th April 2009.

[9] Shari'a Standards (2003) Accounting and Auditing Organization for Islamic Financial Institution.

[10] Ali, Rahail. Islamic Finance (2008) *A Practical Guide,* edited by Rahail Ali. Globe Business Publishing Ltd.

[11] Solah, Hani (2011) *al-Tanzim al-qanuni wa al-ta'aqudi li mashru'at al-banniah al- asasiah al-mamulah 'an toriq al-khas,* 1st edn. Dar al Nahdhah al 'Arabiah.

[12] Al-Usmani, Mohd Taqi (2009) '*Uqud al-bina' wa al-tasyghil wa naql al-milkiah min an nahiah al-syar'iah.* Fiqh Academy Conference, UAE, 19th April 2009.

[13] Mansuri, M. Tahir (2007) *Islamic Law of Contracts and Business Transactions.* Adam Publishers and Distributers, New Delhi.

[14] Husayn, Ahmad Farāj (1986) *Al-milkiyyah wa nazariyyah al-'aqd fi al-Shari'ah al-Islamiah.* Dar al Jāma'iyyah, Iskandariyyah.

[15] Ahmad, Ahmad Muhyiddin (2009) *Tatbiq nizom al-bina' al-tamlik (BOT) fi ta'mir al-awqaf wa al-marafiq al-'ammah.* Fiqh Academy Conference, UAE, 19th April 2009.

[16] Abd Razak Abd Mutalib (2007) *'Uqud al- muamalat al-maliah wa tatbiqatiha al-mu'asarah fi al-Fiqh al-Islami*. Dar al Fikr al Jami'ie, Iskandariah.
[17] Buang, Ahmad Hidayat (2000) *Studies in the Islamic Law of Contracts: The Prohibition of Gharar.* Kuala Lumpur, Malaysia 2000.
[18] Niazi, Liaquat (n.d) *Islamic Law of Contract,* Research Cell, Dyal Sing Trust Library, Lahore.
[19] Chaudry, Muhammad Sadiq (1995) *Islamic Insurance (takaful) concept and practice, Encyclopedia of Islamic Banking and Insurance.* Institute of Islamic Finance and Banking, London.
[20] 2008 Annual Report, Majlis Agama Islam Wilayah Persekutuan.

Endnotes

(i) Based on information given by Mr Hanif bin Haji Salim during the interview session held on 2nd April 2010. He is an Assisstat Manager of Property Development Unit, Federal Territory Islamic Council (MAIWP), Kuala Lumpur.
(ii) Sahih Muslim Muslim, Kitab al-Shurut, Hadith No 2876.
(iii) The most important element of Imam al-Syatibi's legacy in his Muwaqaf, a tretise that contributes "immensely towards the making modernist" conception of Islamic law, particularly the concept of maslaha, which is its central ingredient.
(iv) ZuhayIE, Wahbah, Usul al_fiqh al-Islami, (Beirut: Dar al Fikr, 2004), p.756.
(v) The Quran uses the words *riba* in all places. This word is sometimes translated as usury and sometimes as interest. There is now a legal difference in the English language, usury being exorbitant interest. But this difference is unknown in the Quran.
(vi) Musnad al Imam Ahmad, Hadith No 3660.
(vii) Mohd Hanif b Salim, interviewed by Saidatolakma bt Mohd Yunus, (7th Floor, PERKIM Building, No. 150, Jalan Ipoh, Kuala Lumpur, 2nd April 2010).
(viii) *Wakala* is the Islamic juridical concept of agency. It has a variety of different applications in Shariah-compliant financing arrangements, but it is fair to say that it has not been the product of choice in recent project financings.
(ix) Zuhayli, Wahbah, *Al-Fiqh al-Islami wa adilatuhu*, 5th edn, (Damsyik: Dar al Fikr, 1989), p. 493.
(x) *Bay' al inah dan Tawaruq, Isu-isu dan penyelesaiannya dalam Konteks Kewangan Islam*, (Muzakarah Cendiakawan Syariah Nusantara, 2006), p. 144.
(xi) *Bay' al inah dan Tawaruq, Isu-isu dan penyelesaiannya dalam Konteks Kewangan Islam*, p.149.
(xii) Engku Ali, Engku Rabiah, Bay' al-'inah and Tawarruq: Mechanisms and Solutions, in *Bay' al-'inah dan Tawarruq: Isu-isu dan Penyelesaiannya dalam Konteks Kewangan Islam*, (Muzakarah Cendiakawan Syariah Nusantara, 2006), p.146.
(xiii) Mr Sanusi, Interview by Saidatolakma bt Mohd Yunus, (11th Floor, PERKIM Building, No. 150, Jalan Ipoh, Kuala Lumpur, 2nd April 2010).
(xiv) Eidey El Nemeiry, interviewed by Saidatolakma bt Mohd Yunus, (11th Floor, PERKIM Building, No. 150, Jalan Ipoh, Kuala Lumpur, 14nd April 2010).

SECTION 2:
GOVERNANCE

Joint Ventures in Construction 2
ISBN 978-0-7277-5783-8

doi: 10.1680/jvc2.57838.067

ice
Institution of Civil Engineers
publishing

Chapter 6

Foreign Equity Investment in India: Trends, Policy Priorities and the 'Legitimacy Capital'[i]

Gautam Ray, Kyoto University

Introduction

India presents an intriguing story of foreign equity investment unraveling in recent years in an interesting manner. The aggregate FDI inflow in fifteen years between 1991-92 and 2004-05 was a meager $33.3 billion-an amount Russia and Brazil attracted in about two years and China attracted in less than six months during the middle of last decade. Such a low level of FDI inflow defied easy explanations, particularly because during this period India had been growing at an astoundingly high average annual rate of 6%. In 2006-07 India saw a sudden steep increase in foreign equity inflow. India's annual FDI inflow increased two and a half times, from $8.9 billion in 2005-06 to $22.1 million in 2006-07. The annual inflow quadrupled in five years since 2005-06 to reach $37.76 billion in 2009-10. Although FDI inflow dipped in 2010-11 in India as it did globally, it got back the buoyant momentum with 50% increase in India, $20.76 billion in the first eight months of 2011. One intriguing aspect of this sharp increase in FDI level is that neither India's economic fundamentals nor its FDI policy underwent any major improvement over the last five years to warrant such a steep increase in FDI inflow.

Notwithstanding the recent spurt, India's overall performance in attracting FDI remains much below that of Brazil, Russia, China and Mexico. Between 2000 and 2010 India's aggregate FDI inflow was $170 billion compared to Brazil's, China's and Russia's aggregate inflows of $350 billion, $390 billion, and $390 billion respectively. India's FDI level during this period was also lower than Mexico's $230 billion. It is not easy to see how India could have been such an underperformer in attracting FDI inflows despite its impressive growth dynamics driven by domestic consumption, vast reservoir of human capital resources and gigantic market size. Much of this underperformance is normally attributed to its infrastructure and governance deficits. But if that is the only explanation India could not have attracted such high levels of FDI as it did in the last five years. In fact there are other important reasons why aggregate FDI stock in India has been lower than that of other emerging economies. First, foreign investors are often unable to comprehend the major policy objectives of the local and central governments which make it difficult for them to align their competitive strategy with the public policy objectives.

The second reason is that many foreign investors often struggle to find their feet in a uniquely Indian marketplace and end up in not investing substantively even after setting up their subsidiaries in India. The lessons they have learned in other developing countries such as China, Brazil, Mexico and Thailand are often not very useful in India's context. This is so because foreign investors mainly target the local markets in India while in other developing countries they usually invest to integrate low cost manufacturing facilities in those countries with their global supply chain networks. The third important reason for India's underperformance is foreign investors' inadequate understanding of the longer-term benefits of investing in a country like India where domestic market has been growing steadily because of the growth of consumption among the middle class as well as bottom of pyramid (BOP) population [1]. Finally, foreign investors continue to harbor doubts about their capabilities to navigate through Indian ways of doing business and survive in competition with local players.

A significant part of India's 'missing' foreign equity investments has much to do with foreign investors' unfamiliarity with Indian ways of doing business. A major objective of this chapter is to attempt to inform potential foreign investors about the dynamics of India's business environment, current thrust areas in FDI policy and the key legal perspectives relating to foreign direct investment in India. With that objective in view, the next section of this chapter traces the evolution of FDI policy and at the end summarizes the shifting policy focus over the years. Section 3 examines and explains the trend in FDI in India over the last two decades. The case study presented in Section 4 provides certain key legal perspectives arising out of the conflict over management control of the joint venture, Haldia Petrochemical Company Ltd. (HPL). Section 5 studies the case of Suzuki Motor Corporation (SMC) Japan's successful investment in Maruti Udyog Ltd. (MUL). Section 6 concludes the discussions with lessons learned from the two case studies and suggests a set of strategic recommendations for potential foreign investors.

FDI Policy: Evolution of India's Regulatory Environment

India's FDI policy in the first four decades since its independence in 1947 was highly restrictive as it wanted to protect infant domestic enterprises and direct the flow of foreign equity investments to high technology and export intensive industries. There was a 40% ceiling on foreign equity holding and all foreign investment proposals were subject to government scrutiny and approval. FDI policy was liberalized for the first time in the middle of 1980s to allow foreign investors to own 100% equity holding in '100% export oriented units' (EOU). At that time the most important policy objective in FDI was export promotion as India's balance of payments position had been deteriorating after the liberalization of imports of capital goods and electronics.

FDI policy underwent a major structural overhaul in government of India's New Industrial Policy on July 24, 1991 that did away with the industrial licensing system. It gave foreign investors opportunities to set up joint ventures in 35 priority sectors with majority shareholding up to 51%. For the first time foreign investors were allowed to invest in specified sectors and activities without government approval under the automatic route in which only the Reserve Bank of India (RBI) is required to be notified of the details of such investment within thirty days. The second round of FDI policy reforms came in 1997 when foreign companies were allowed for the first time to operate in a few top-priority sectors through its wholly owned subsidiaries with 100% equity holding (100% FDI). In the same year FDI was opened up to about 100 new sectors, which was divided into three ceiling categories with 74%, 51%, and 50% ceilings on the share of foreign equity ownership in the total paid up capital of a domestic company.

The third major push in FDI policy reforms came in 2000 when government of India allowed 100% FDI in a large number of sectors. A list of other sectors where ceilings in foreign equity participation were prescribed and also a list of eight prohibited sectors (mentioned later in this section) were brought out. Since 2000 FDI policy has been clearly mentioning the sectors/

activities where foreign equity investments are permitted without prior approval of government (automatic route) and the sectors/activities which require foreign investors to seek government of India's prior approval. Since then government of India has been liberalizing FDI regime from time to time by increasing the ceiling in the share of foreign equity ownership or by doing away with the need of government approval and putting more sectors under the automatic route.

The government of India carried forward FDI liberalization in 2001 to open up for foreign investment new sectors such as defense industries, mass rapid transportation system, banking in the private sector, printing of scientific and technical journals, petroleum product pipelines, telecom and e-commerce. Further initiatives included enhancement FDI limit from 40% to 49% in domestic airlines under automatic route with no participation from foreign airlines. In 2005, FDI up to 100% foreign equity ownership (100% FDI) was allowed under the automatic route for Hosing & Real Estate Sector. FDI limit in basic and cellular telecom services was also increased to 74%.

The transfer of shares from resident to non-resident including acquisition of shares in an existing company was placed in the automatic route in 2006. Other reforms included 51% for single brand product retailing under government approval route, 100% FDI under automatic route for green-field airport projects, export trading and cash & carry wholesale trading. Also allowed under automatic route was foreign technology collaboration involving payment of lump sum amount up to $2 million and/or royalty at the rate of 5% on domestic sales and 8% on exports. The policy on use of trademarks and brand names of foreign collaborator without technology transfer was liberalized to allow without government approval royalty payment at the rate of 1% on domestic sales and 2% on exports.

FDI policy was further liberalized in 2008 in civil aviation, petroleum and natural gas and industrial parks. FDI up to 74% was allowed in domestic airlines and ground handling services. 100% FDI was allowed in sectors such as industrial parks and maintenance and repair organizations. In a major reform measure in 2011, government of India (GOI) withdrew the press Note 1 of 2005 that prohibited foreign investors to enter in any activity if they had an existing joint venture or technology transfer/trademark agreement in the same or similar field in India. Now they require only a no-objection certificate from their domestic partner before investing independently in the same activity in India. In the last week of November 2011, GOI had decided to allow FDI up to 51% in multi-brand retail and 100% in single brand retail, but in the face of strong opposition in India's Parliament, the decision had to be put on hold till the implications of the decision were discussed in an all party meeting.

India's FDI policy lists eight activities/sectors in which foreign direct investment is not allowed. They are: (i) Multi-brand Retail Trading, ii) Atomic Energy, iii) Lottery Business, iv) Gambling and Betting, (v) Business of Chit Fund, (vi) Nidhi Company, (vii) Trading in Transferable Development Rights, and (viii) Activity not opened to private sector investment. Barring these activities, FDI is allowed in all other sectors and activities. FDI policy classifies all these sectors and activities open to FDI in three categories: (i) sectors/activities in which there is no ceiling in foreign equity investment and no government approval is required, that is, 100% FDI is allowed under automatic route; (ii)sectors and activities where ceilings on foreign equity shareholding have been prescribed, but no government approval is required if the investment is below the prescribed ceiling limit; and (iii) sectors where ceilings have been prescribed and government approval is required. The consolidated FDI policy of Gover 1, 2011 is available in http://dipp.nic.in/English/Policies/FDI_Circular_02_2011.pdf [2].

Figure 6.1 shows how the share of FDI inflows through automatic route increased during the last fifteen years. The share of investment under automatic route increased as FDI policy progressively opened up new sectors and activities under this route. It jumped from 30% in the second half of 1990s to 53% in the first half of last decade with the share further increasing to 87% in the last five years. The progressive liberalization of FDI over the last two decades proved to be effective in attracting foreign investment. This is evident from the progressive increase in FDI through the automatic route.

Shift in policy priorities

The history of FDI policy evolution narrated above suggests that the policy priorities have been shifting over the years. Before 2000, FDI policy focused on manufacturing industries and deriving potential benefits of export competitiveness and technology spill over. The policy of promotion of exports through FDI, which was the focus in early 1990s when India was recovering from the acute balance of payment crisis, ceased to be the policy priority after 2000. Instead, the FDI policy gave more emphasis on promotion of competition in the domestic economy and generation of employment opportunities.

The policy shift is evident from the decision of the government of India to withdraw Press Note 1 of 2005 and allow foreign investors having a pre-existing joint venture or technology transfer/trademark agreement to invest on similar activities independently or with other partners even if such activities are not for promotion of exports. The dividend balancing and related export obligation conditions that were applicable for foreign investment in 22 consumer goods industries were also withdrawn in 2000. In addition, 100% export oriented units were allowed to sell a significant part of their output in domestic market. To sum-up, the thrust areas in FDI policy evolved from protection of domestic enterprises before 1990 to promotion of export and technology intensive industries in 1990s. In recent years, the policy thrust has shifted from technological spill-over and export competitiveness to generation of employment opportunities and increasing efficiency of domestic enterprises by promoting competition.

Figure 6.1: Growth in the share of FDI inflows through Automatic Route since 1991 (Source: Author's calculation from the data in RBI Monthly Bulletin [3])

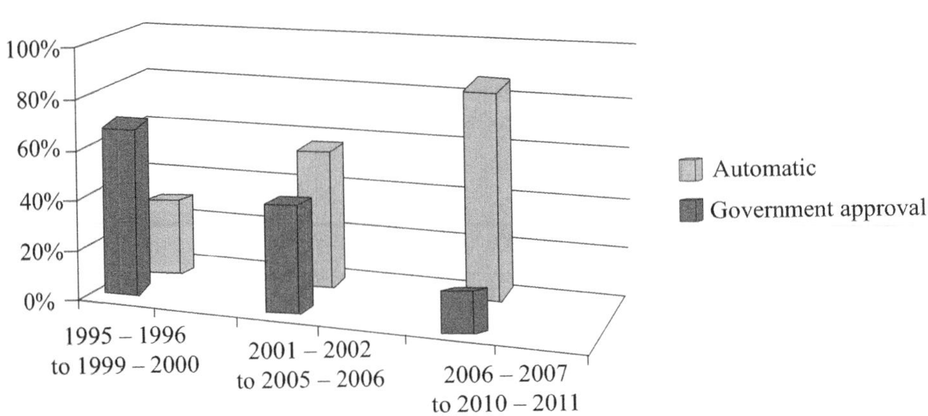

Trends in Foreign Direct Investment Inflows in India

Figure 6.2 shows the trend of foreign equity capital inflow from 1991-92 onwards while Table 6.1 gives the annual FDI flow and its break-up among three components: equity capital inflow, reinvested earnings from an existing venture, and other capital such as intra company loans [ii]. It is clear from the Table 6.1 and Figure 6.3 that historically FDI inflow in India had been low until 2006-07. The number of joint ventures of MNCs in India also plummeted from 120 in 1992 to a low level of 10 in 2005 according to Kale and Anand [4]. Even after sharp increase in FDI inflow and joint ventures over the last five years, India's share of FDI stock in 2010 as a percentage of GDP stood at 12.5% of its GDP, nearly half of Russia's or Brazil's FDI: GDP ratio in 2010. Table 6.2 also shows that India's FDI stock as a percentage of GDP in 2010 was about one third of the countries in south-east Asia such as Thailand and Malaysia. Interestingly, in terms of FDI stock in 1990 and 2000, India figured last in rank among all major emerging economies in Asia, Africa and Latin America as shown in Table 6.2. What explains such a history of low level of FDI inflow in India?

From the sector-wise cumulative FDI inflow till April 2006 shown in Table 6.3, it is clear that during the period up to April 2006, 90% of FDI inflow came in the manufacturing sector while about 10% came in the services sector. In sharp contrast, nearly 50% of the cumulative FDI inflow in April 2011 came in non-manufacturing sectors, namely, services, construction, housing & real estate, and computer software. This fact pretty much explains the trend. India's FDI inflow had been low historically because global multinational corporations (MNCs) did not find India an attractive destination for setting up their manufacturing facilities. This was so because MNC's strategy was to integrate such facilities with their global supply chain networks and India did not figure in that strategy mainly because of its poor transport infrastructure, exchange rate risks, rigidity in labor market, and governance deficits. Besides, targeting domestic markets in foreign countries had not been the focus of global MNC's business strategy until the recent economic downturn in the West.

What explains the rapid decline in number of joint ventures between 1991 and 2005 shown in dark shaded line in Figure 6.2 even as annual FDI increased, albeit slowly, during these years as shown in the light shaded line? Kale and Anand [4] surveyed 69 JVs of MNCs and identified two factors:

(1) Progressive liberalization of FDI policy since 1991 allowed 100% FDI in most manufacturing sectors which in turn allowed MNCs to break-up existing joint ventures and establish 100% subsidiaries in India; and
(2) Foreign partners were ahead in the learning race than domestic partners which significantly reduced resource complementarities between them. Their increasing familiarity with local market condition and government regulation enabled them to break-up joint ventures and set up independent ventures in India.

Another important reason that prevented potential foreign investors to take the joint venture route was the regulatory restriction about forming independent ventures for foreign investors who had earlier entered into a JV in the same or similar field with domestic entrepreneurs in India.

Table 6.3 and Figure 6.4 clearly show that since 2006-07 much of the FDI inflow came in the services sectors, and construction intensive as well as knowledge intensive sectors. Before 2006-07 services sector was 4th in rank and contributed 9.45% in the total stock of FDI since August 1991. In April 2011, services sector ranked 1st with 21% share in FDI stock since April 2000 suggesting huge FDI inflow in financial and non financial services sectors over the last five years. Knowledge intensive industries such as computer software & hardware (second in rank) and telecommunications (third in rank) were the second most attractive area of foreign equity investment in April 2011. Together they attracted 16% of the total FDI inflow.

The third major area was construction intensive sectors comprising real estate and housing and construction including roads and highways. Construction intensive sectors did not figure among the 40 sectors that attracted cumulative foreign investment worth $50,000 and above between August 1991 and April 2006 [5]. But they came right on top in 2010-11. Real Estate and Housing climbed to the 4th rank in aggregate inflow between April 2000 and April 2011 with an inflow of $9.66 billion, while construction including Roads and Highways shot up to the fifth rank with an inflow of $9.49 billion during the same period. Together the two construction intensive sectors-Construction including Roads and Highways, and Housing & Real Estate- attracted third highest FDI inflow with an aggregate share of 14% in the last decade. The FDI inflow in these sectors has been consistently high throughout the latter half of the decade which is evident from the fact shown in Table 6.3 that they did not attract significant level of FDI as on April 2006.

With the huge surge in equity investment since 2006-07 in financial and non-financial services, knowledge intensive sectors and construction intensive sectors, the number of foreign joint ventures (JVs) increased sharply from 10 in 2005 to 72 in 2010 even as FDI was progressively liberalized during this period to allow 100% FDI. This can be explained from the following two facts: i) high financial risks associated with huge investments in mega construction projects necessitate incorporation of joint ventures with government agencies and large corporations in India; and ii) large resource complementarities among foreign and domestic partners exist in all these sectors. It is also true that ceilings on foreign equity investments still exist in many services, which is why foreign investments take the JV route.

What explains the sudden trend breaking steep jump in foreign equity inflow since 2006-07? A look at the broad trend of FDI inflows in different countries available in World Investment Reports published annually by UNCTAD will show that global FDI capital flow has been shifting from the developed economies towards emerging economies over the last ten years, and the shift was pronounced by the economic downturn in developed economies particularly after the 2008 financial crisis. As the global capital market shifted its radar towards emerging economies, India as a FDI destination loomed larger than others. This is captured in Table 2 as it shows that India's rank in FDI stock among major emerging economies jumped to the fifth rank in 2010 from the lowest rank in both 2000 and 1990.

India's sustained growth over the last two decades at an annual average rate of about 6% and the recent quarterly growth trend of about 8% since 2007 have drastically changed global capital market's perception about India's economic fundamentals. In the World Economic Forum's Global Competitiveness Report, India's ranks in the last five year since 2006 stood at 27,27,29,30 and 31, higher than those of Brazil and Russia. Interestingly India's ranks were higher than even Korea's till 2008 and they were closer to Japan's ranks 24, 22, and 27 in 2007, 2008, and 2010 respectively. It was not therefore surprising that India figured among the three global destinations for foreign investment along with China and USA in UNCTAD Survey Reports and A.T.Kerney Confidence indices in recent years. The steep jump in FDI equity inflows since 2006-07 reflected global capital market's growing confidence in India's sustainable growth story and the important contributions its skilled workers made in the growth of knowledge intensive service industries across the globe. Although FDI inflow declined last year as shown in Figure 6.3, it rose again by 50% to $20.76 billion during January to August 2011. Services (financial and non financial), telecom, housing and real estate, construction and power continue to be the sectors that attracted maximum FDI during the first eight months of 2011. According to an estimate of KPMG, FDI inflow is expected to cross US$35 billion mark while a Morgan Stanley research suggests that India could attract $80 billion in the next 1-2 years (http://ibef.org/india/economy/fdi.aspx estimate).

While round tripping seems to have increased in India mainly through transfer pricing and money laundering mechanism, it has not yet reached the level in China and Russia according to the international watchdog Global Financial Integrity's (GFI) December 2011 Report [6] that featured in The Times of India story, "$128 billion siphoned out of India in a decade" dated December 26, 2011. From the steep rise in FDI inflows in recent years in services and knowledge intensive sectors, it is apparent that the changed perception about India's economic fundamentals and future growth outlook is arguably the most important driver of the growth of FDI inflow in recent years. This argument is buttressed by the survey report published in December 2010 by the Federation of Indian Chambers of Commerce and Industries. The survey showed that 62% of the respondent foreign investors reported profit from their India operations while 17% reported break-even return and the remaining 22% reported loss. 91% of respondents reported that they were considering expansion of Indian operations [7].

Table 6.1: FDI inflow in India since April, 2000(In US Millions)
(Source: Government of India (GOI) FDI Statistics [5] various years; and Reserve Bank of India (RBI) Monthly Bulletin, November 2011 [3])

Fancial Year (1)	Equity Inflow (2)	Reinvested Earning (3)	Other Capital (4)	Total FDI (2)+(3)+(4)
1991 – 1992 to 1999 – 2000	16,703	NA	NA	16,703
2000-01	2,400	1,350	279	4,029
2001-02	4,095	1,645	390	6,130
2002-03	2,664	1,833	438	5,035
2003-04	2,229	1,460	633	4,322
2004-05	3,778	1,904	369	6,051
2005-06	5,975	2,760	226	8,961
2006-07	16,481	5,828	517	22,826
2007-08	26,864	7,679	292	34,835
2008-09	28,431	9,030	777	37,838
2009-10	27,149	8,669	1,945	37,763
2010-11	20,304	9,424	652	30,380
Cumulative Total (from April 2000 to April 2011)	143,191	51,582	6,518	201,291

Figure 6.2: Trend in number of joint ventures and annual FDI inflow (Source: Kale and Anand [4]; and Singh and Vaidya [8])

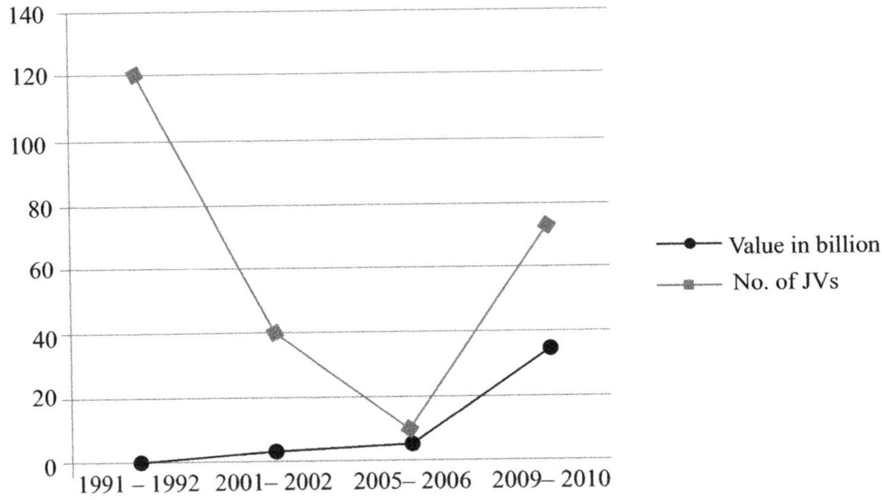

Figure 6.3: Foreign Direct Equity Investments in Indian ventures since 1991 (Source: FDI Statistics, DIPP, Government of India [5])

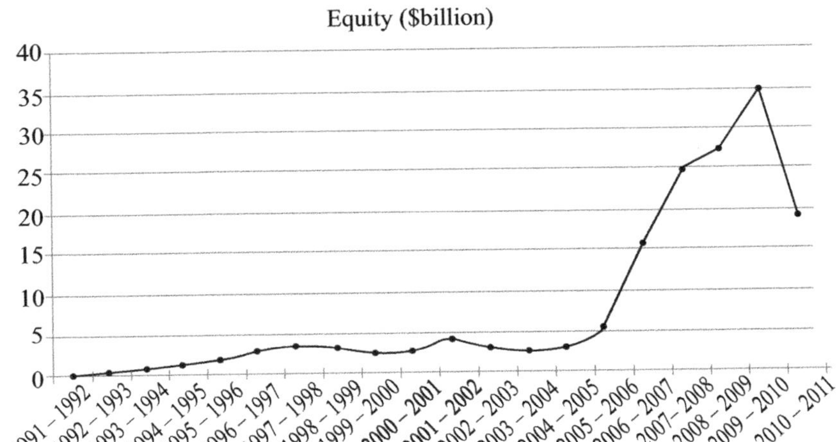

Table 6.2: FDI stock accumulation in emerging economies (In US Millions): FDI stock as percentage of GDP in 2010
(Source: UNCTAD, World Investment Report for FDI stock [9]. Annex table 1.2. Nominal GDP figures of 2010 are sourced from IMF, World Economic Outlook, September 2011)

Country	1990		2000		2010		(FDI Stock/ GDP) in 2010
	Amount	Rank	Amount	Rank	Amount	Rank	
Brazil	37,143	1	122.150	2	472,579	2	22.6%
Russia	NA	NA	32,204	7	423,150	3	28.6%
China	20,691	3	193,349	1	578,818	1	9.8%
India	1,657	11	16,339	12	197,939	5	12.1%
Mexico	22,424	2	97,170	3	327,249	4	31.6%
Turkey	11,150	4	19,209	11	181,901	6	24.7%
Indonesia	8732	9	25.060	9	121,527	9	18%%
Malaysia	10,318	6	52,747	5	101,339	10	42.4%
Thailand	8,242	10	29,915	8	127,257	8	39.8%
South Africa	9,207	7	43,451	6	132,396	7	36.4%
Egypt	11,043	5	19,955	10	73,095	12	33.55%
Argentina	9,085	8	67,601	4	86,685	11	23.4%

Table 6.3: Top ten FDI sectors from August 1991 through March 2006
(Source: Government of India, DIPP, FDI Statistics April 2006 [5])

Ranks	Sector	Cumulative FDI Inflow (in US $ million) from August 1991 to March 2006	% age of Cumulative total
1	Electrical Equipment	5496	17.49
2	Telecommunications	3372	10.58
3	Transportation Industry	3178	9.82
4	Services Sector (financial and non financial)	3091	9.45
5	Fuels (power & oil refinery)	2581	8.10
6	Chemicals	2143	6.33
7	Food Processing Industries	1179	3.47
8	Drugs and Pharmaceuticals	1007	3.18
9	Cement and Gypsum	747	2.38
10	Metallurgical industries	655	2.08

Figure 6.4: Sectoral share in cumulative foreign investment in India up to 2010 – 2011 (Source: GOI [5], DIPP, FDI statistics, April 2011)

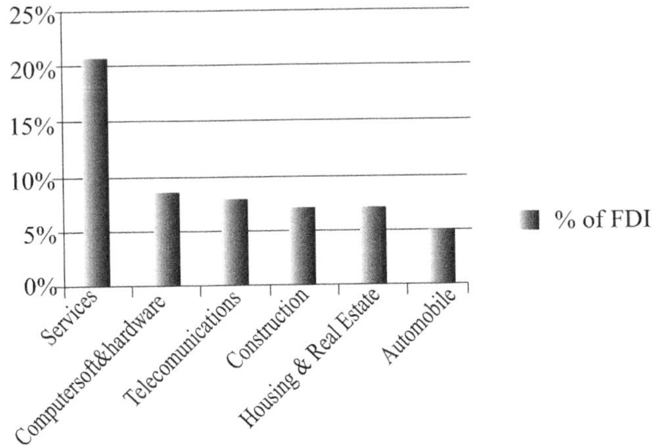

Haldia Petrochemicals Ltd: A Joint Venture with Government of West Bengal[iii]

Haldia Petrochemicals Ltd. (HPL) is a public-private joint venture green field petrochemical complex based on naphtha as feedstock. It started in 1997 as a flagship project for industrial resurgence in the state of West Bengal that would generate more than 150,000 employment opportunities. The three promoters of this company were: (i)West Bengal Industrial Development Corporation(WBIDC), a wholly owned undertaking of government of West Bengal (GoWB), (ii) Chatterjee Petrochem (Mauritius) Company Co.(CPMC) and (iii) the Tatas. Their equity holdings were 25%, 25% and 8.33% (3:3:1) respectively. The remaining 41.67% equity capital came from financial institutions, foreign institutional investors, NRS and public.

In August 1994, the promoters signed a joint venture agreement (JVA). It provided that in case of disinvestment by WBIDC, the same would first be offered to CPMC. Right from the beginning of the project, the Indian American promoter (PC) owning CPMC had been working towards acquiring controlling interest of 51% share so that they could exercise complete management control of HPL. Towards that end CPMC exchanged four letters between January 30, 1994 and January 5, 1995 with GoWB and WBIDC for sale of at least 60% of WBIDC's shareholding in HPL to CPMC at Rs 14 per share within two years of commencement of commercial production and for restricting government's role to promotion and guidance through the initial phases of the project. The operative parts of these agreements were not however incorporated either in the JVA, or in the article of association.

Before the commercial production could start in August 2001, HPL's debt service problem came to a head and it approached the principal lender, Industrial Development bank of India (IDBI) for debt restructuring. On September 6, 2000, HPL issued rights share for Rs.2.50 billion. While WBIDC and the Tatas subscribed their portions, CPMC did not invest its share of Rs.1.07 billion. The lenders were not satisfied with the equity infusion of Rs.1.43 billion by WBIDC and the Tatas. They were demanding induction of Indian Oil Corporation (IOC), ranked 98th in Fortune Global 500 list, as a strategic investor who could supply the key input, naphtha. But IOC refused to enter the joint venture merely as a portfolio investor without management control. They wanted 26% share and management control.

It was in this background GoWB, WBIDC, CPMC and HPL signed an agreement dated January 12, 2002 . In terms of this agreement, WBIDC undertook to transfer requisite number of its HPL shares to CPMC that is sufficient to increase the latter's holding to 51% of paid up equity of HPL. The agreement put an obligation on CPMC to invest Rs.1.07 billion for the rights issue share that it did not subscribe earlier and infuse Rs.5billion equity or equity like instruments or advance by March 31, 2002.

CPMC, however, could not invest Rs.1.07 billion to acquire its share of the rights issue, nor could it infuse Rs.5 billion worth equity or equity like instruments. TCG could however arrange for a loan of Rs.1.07 billion from HSBC. A loan agreement dated March 8, 2002 was signed between WBIDC as lender and TCG's Chatterjee Petrochem (India) Pvt.Ltd., CPIPL. According to this agreement, TCG would infuse equity capital of 1.27 billion through its company, Winstar. On July 30, 2004 Winstar took over the HSBC loan of Rs.1.07 billion and it acquired 107 million equity shares at par.

Since the lenders' deadline of financial restructuring had long passed and IOC refused to invest merely as a portfolio investor, a meeting was held on September 13, 2004 between GoWB, HPL management and IOC in which the latter finally agreed to invest Rs. 1.50 billion as a portfolio investor. With IOC's investment in the equity capital of HPL, the total shareholding of two government companies crossed 50% while TCG's shareholding was 47.33%. For acquiring the controlling share holding in HPL, TCG had to acquire WBIDC's shares in HPL as per their agreements with WBIDC and GoWB, but WBIDC was not ready to sell its HPL shares at Rs.14 per share. Alleging oppression and mismanagement, CPMC, CPIPL, Winstar and ITML under the Chatterjee group of companies (TCG) filed a petition before the Company Law Board (CLB) under Sections 397 and 398 of the Companies Act, 1956. The petition sought for cancellation of the allotment of shares to IOC and for a direction to WBIDC/GoWB to transfer to the petitioners 36% shares held by WBIDC in HPL. The petitioners alleged that WBIDC/GoWB did not complete the transfer of 155 million shares even after receiving payments, and that WBIDC/GoWB did not honor the agreements for sale of WBIDC shares in HPL to enable TCG to secure management control of HPL.

The CLB decision dated January 31, 2007 [10] went in favor of the petitioners even though it held that allotment of shares to IOC was legal and proper. Invoking the doctrines of legitimate expectation and quasi partnership, it directed GoWB and WBIDC to transfer the ownership of 155 million shares to the petitioners at par value of Rs.10 per share. CLB also directed WBIDC to transfer to the petitioners 36% shares held by WBIDC so that TCG could acquire 58% shareholding in HPL and acquire management control of the company. Being aggrieved by the decision of the CLB, Government of West Bengal and WBIDC appealed to the Calcutta High Court [11]. The High Court upheld the appeal and observed that the doctrine of legitimate expectation and quasi partnership could not override the three statutory conditions mentioned in Section 397(2) (b).

In paragraph 59 of the judgment, the High court observed:

"If the Chatterjee group was really after management control, it was free to invest suitably in terms of the agreement dated January12, 2002, for ensuring its absolute majority status...In terms of agreement dated January 12, 2002, the CPMC was free to call upon WBIDC to sell its entire shareholding. It never did that, and instead gave up that right on July 30, 2004. It therefore appears that what CPMC actually wanted was to acquire the majority shareholder status by

making the minimum investment and not to permit any strategic investor to come in with funds for HPL. I wonder how on the facts of the case a person of reasonable prudence can say that the Chatterjee group was entitled to get the benefit of the doctrine of legitimate expectation...".

TCG appealed to the Supreme Court. The Apex Court rejected the appeal and upheld the Calcutta High Court's order. The relevant portion of the order of the Supreme Court dated September 30, 2011 [12] is reproduced below:

"Although the Chatterjee Group has complained of the manner in which it had been reduced to minority in the company, it is also obvious that when the Company was in dire need of funds and the Chatterjee Group also promised to provide a part of the same, it did not do so and instead of bringing in equity, it obtained a loan from HSBC through the Merlin Group, which only increased the debt: equity ratio of the company. Furthermore, while promising to infuse sufficient equity in addition to the amount that would have been brought in by way of subscription to the Rights Issue, the Chatterjee group imposed various pre-conditions to do so, which ultimately led GoWB and WBIDC to terminate the agreement to transfer sufficient number of shares to the Chatterjee Group to enable it to have complete control over the management of the company and also to retain its private character. It is at a stage when there was a threat to the supply of Naphtha, which was the main ingredient to used by HPL for its manufacturing process, that it finally agreed to induct IOC as member of the Company...".

As regards the failure of WBIDC and GoWB to register 155 million shares transferred to CPIPL, the Apex Court held that it was not a failure of the company, but of one of its members, and the remedy of such failure did not lie under section 397 of Companies Act. The Court therefore ruled that CLB had no jurisdiction to make any order under 397 or 402 and give directions to WBIDC and GoWB to transfer 520 million shares to the Chatterjee Group (TCG) of companies. The net result of this five year long litigation was a loss-loss outcome for both sides as they lost about Rs 1 billion in an unproductive litigation. In addition, TCG also lost the valuable 'legitimacy capital', a notion that is developed in the concluding section of this chapter.

Suzuki Motor Corporation's Joint Venture with Government of India[iv]

SMC entered into a joint venture agreement with government of India (GOI) on October 2, 1982 for investing 26% equity capital in Maruti Udyog Limited (MUL). MUL's Project Report dated 27 May 1982 describes major objectives of the venture to be the following: modernization of automobile industry; production of fuel efficient vehicles; production of 'people's car'; and generating employment through establishment of ancillary industries. In early 1980s, India had only three other domestic car manufacturers. These cars were 'gas guzzlers' consuming gasoline at twice the rate of Suzuki's small car model.

MUL completed the project in 1983 and began commercial production in 1984. It had hit the ground running: the number of cars booked in the first year 1984-85 was more than six times the production target that year. The rest is history. MUL became an icon of fuel efficient vehicles in India, and it spawned development of a world class ancillary industry.

The pace of indigenization however lagged behind the pace committed by SMC in the joint venture agreement. Realizing that the target may be difficult to achieve, SMC had incorporated in the agreement the conditionality that components supplied by indigenous manufacturers must be of the same quality as those supplied in Japan. Seen in this background, the achievement of cumulative indigenization of 27.15% in 1984-85 was perceived by GOI to be commendable.

By 1988-89 achievement in indigenization was 86.23% for cars (target: 95.30%), 82.60% for Van (target: 84.50), and 58.32% for Gypsy (target: 57.30) [13].

MUL also increased the price much too frequently, ten times during the first five year period. But every time it increased the price, it came out with detailed analyses that showed that rise in input costs was largely responsible for the price hike. Eventually the company was allowed to drift away from the objective of manufacturing 'people's car' because such an objective could not be enforced in Indian company law jurisprudence. This is so because according to Companies Act, 1956, any business strategy of a company is governed by the decisions of its Board of Directors, not by the objectives in the project report.

GOI allowed SMC to acquire 40% equity in 1987 following which a Japanese manager was appointed Joint Managing Director. SMC eventually took control of the management of MUL in 1992 acquiring 50% shareholding in the company. A new agreement was signed between GOI and SMC which stipulated that both partners were entitled to nominate the Managing Director (MD) for five years in turns. In 1993, SMC faced for the first time a conflict situation when government of India did not approve Suzuki's proposal of an expansion and modernization plan of Rs.22 billion on the ground that the project cost was inflated by SMC. The transfer of gearbox technology was another bone of contention between the two partners. In August 1997, another issue cropped up. This time SMC was on the offensive. GOI's appointment of MD was challenged by SMC whose representatives in MUL Board voted against the appointment.

SMC-MUL challenged the government appointment in the court on two grounds: i) Article of association of MUL was not amended to incorporate the terms of the SMC-GOI agreement in 1992; and ii) MUL Board is not bound by an agreement between two members of the company in their individual capacities. Following the change in government of India after parliamentary elections in 1998, an out of court settlement was made in June 1998. It was agreed that GOI nominee will step down in December 1999 and SMC will appoint its nominee from January 2000. In 1999, GOI announced its decision to divest its stakes in MUL. In 2002, GOI divested all its stakes with SMC acquiring a part through rights issue and the remaining equity was divested through public offering.

Suzuki-GOI partnership in MUL was an epoch making event because it transformed the automobile industry and brought Indian auto components industry to global forefront. It also brought about radical transformation in total quality management among Indian industrial ventures in general. As a result, by 2004 Indian companies bagged more Deming Prizes than firms in any country other than Japan [14].

Conclusion

What went wrong with the foreign investor's strategy in the HPL case? The most important mistake on the part of TCG was to refuse to contribute their share in the rights equity issue at a time when such equity infusion was necessary to mitigate HPL's debt problem. The second mistake was to act contrary to the terms of their agreement dated January 12, 2002 with HPL and GoWB and instead arrange for HSBC loan, which compounded the debt problem. And the third mistake was to keep insisting on acquiring controlling share and management control even as the company was reeling under the pressing debt crisis. It was also not clear to any stakeholder of HPL how TCG's acquiring controlling interest and management control would have eased HPL's financial problems. TCG could not point out a single act of mismanagement that could have resulted in HPL's debt problem. On the contrary, their sequence of acts-at a time when HPL's lenders were pressing for equity infusion by strategic investors- suggested to all stakeholders that TCG might not have a 'lasting interest' in the company. Any claim of legitimacy in FDI has to be founded upon this fundamental attribute of 'lasting interest' that distinguishes FDI from portfolio investment. This is not to say that a foreign equity investor cannot exit; but the investor's exit strategy must not have the effect of crippling the company and making it bankrupt after the exit.

The agreement dated January 12, 2002 between HPL, GoWB and TCG was fair and legitimate as it sought to align TCG's objective of acquiring management control with the GoWB's objective of industrial rejuvenation and employment generation in the state. By failing to act in terms of this agreement, TCG lost immense goodwill and 'legitimacy capital' that they had earlier developed by joining the HPL project for industrial resurgence of West Bengal. It is not therefore surprising that Calcutta High Court and the Supreme Court of India could not find much substance and validity in TCG's claim for acquiring the controlling shareholding and management control on the ground of legitimate expectations, fairness, equity and justice.

In contrast, SMC succeeded in India as the sequence of their strategic acts was consistent with their legitimate business objective and at the same those acts demonstrated SMC's 'lasting interest' in MUL. The frequent increase in the price of cars, non achievement of indigenization target and non transfer of gearbox technology could have resulted in counterproductive long drawn conflicts as happened in HPL case. But the conflicts could be effectively managed and were nipped in the bud because SMC's acts were perceived to be fair, just, legitimate and consistent with GOI's basic policy objective of developing fuel efficient cars and competitive automobile components industry in India. SMC could thus develop its legitimacy capital in MUL that made it easier for them to acquire controlling interest in MUL shareholding and, consequently, the management control. In contrast, TCG actually lost its legitimacy capital as it attempted to push vigorously its agenda of acquiring management control at an inopportune time when the company was in dire financial condition.

The two case studies bring out a number of key legal perspectives on joint ventures in India. First, no agreement between the partners can bind an incorporated company if the operative part of the agreement is not expressly incorporated in the articles of association. In other words, no petition under the Companies Act, 1956 is likely to succeed in a court of law in India if the claims in the petition come entirely from a separate agreement between partners in which the joint venture company was not a party. Second, if there is an inconsistency between provisions of Companies Act, 1956 and those in Articles of Association of an incorporated joint venture company, the former shall prevail. Third, if a dispute is on acquiring management control of a company incorporated in India, enforcement of the terms of a separate agreement between partners cannot be sought through arbitration under the Indian Arbitration and Conciliation Act, 1996. A company incorporated under the Indian Companies Act cannot be treated as a quasi partnership and a petition for relief under the said Act cannot be allowed in contravention of a statutory provision in the Act. Fourth, under Indian jurisprudence, courts in India can go beyond the legal technicalities and literal construction of statutory provision to deliver equitable justice on the ground of legitimacy. However such a decision on legitimacy ground is subjected to close scrutiny by higher courts.

An important perspective that emerges from the two case studies is that a foreign investor must be able to align its legitimate business strategy with the avowed public policy objectives behind the foreign investment. In the context of FDI, a foreign investor's business strategy is considered legitimate as long as it is consistent with the definitional attribute of 'lasting interest' that distinguishes FDI from portfolio investment. 'Legitimacy capital' is developed through a series of acts that demonstrate an investor's 'lasting interest' and the efforts to align investor's legitimate business objectives with those of all other stakeholders. In the present context of foreign equity joint ventures with government agencies, 'legitimacy capital' can be defined as the asset a foreign investor develops from a series of acts that promote such an alignment of legitimate private business strategy with the socially legitimate objectives of such agencies.

Foreign equity investors would do well to develop their 'legitimacy capital' by demonstrating lasting interest in the investment and aligning their strategic business objectives with the socially legitimate policy objectives of governments in India. Any strategic move that seeks to undermine a legitimate public policy objective is likely to adversely affect foreign investor's 'legitimacy capital' that is so crucial for doing business effectively in India.

References

[1] Ray, Gautam (2011) Doing Business in India: opportunities and Challenges. *Journal of marketing Development and Competitiveness* **5(4)**: 77-95.

[2] Government of India, Department of Industrial Policy and Promotion, (2011) "*Consolidated FDI Policy* (Effective from October1,2011) Available in http://dipp.nic.in/English/Policies/FDI_Circular_02_2011.pdf.

[3] Reserve Bank of India (RBI) (2011) *RBI Monthly Bulletin November 2011* Available in http://www.rbi.org.in/scripts/BS_viewbulletin.aspx.

[4] Kale, P.,& Anand, J. (2006) The Decline of Emerging Economy Joint Ventures: The Case of India. *California Management Review* **48(3)**: 61-76.

[5] Government of India, Department of Industrial Policy and Promotion, *(2011, 2005, 2000) FDI statistics: Fact sheet on foreign direct investment from August 1991 to April 2011.* Available in http://dipp.nic.in/fdi_statistics/india_fdi_index.htm.

[6] The Times of India, December 26, 2011, *$128 billion siphoned out of India in a decade.* Subodh Verma's coverage of GFI Report, December (2011). Available in http://timesofindia.indiantimes.com/india

[7] Federation of Indian Chambers of commerce and Industries (FICCI) (2010) FDI Survey. Available in http://ficci.com/SEDocument/20097/FICCI_FDI_Survey_Dec_2010.pdf.

[8] Singh A.& and Vaidya, Y. (2011) Joint Venture or Misadventure?. *The Economic Times*. July 16. Available in http://economictimes.indiatimes.com/articleshow/9243194.

[9] UNCTAD (2011) *The World Investment Report, 2011*. Available in www.unctad.org.

[10] Company Law Board (2007) *Chatterjee Petrochem (Mauritius)* Co. *and others vs. Haldia Petrochemicals Ltd. and others.* CLB Decision reported in [2008]143CompCas 726(CLB).

[11] Calcutta High Court. (2007) *Government of West Bengal vs. Chatterjee Petrochem (Mauritius) Co and others.* High Court Judgment reported in [2008] 143 Comp cas 837 (Cal).

[12] Supreme Court of India (2011) *Chatterjee Petrochem (I) Pvt.ltd. v. Haldia Petrochemicals Ltd.* Supreme Court Judgment delivered on September 30, 2011. Available in http;//indiankanoon.org/doc/1814805 downloaded by author on December 16, 2011.

[13] Venkataramani, R. (1990) *Japan Enters Indian Industry: The Maruti-Suzuki Joint Venture.* Radiant Publishers, New Delhi.

[14] Khanna, T., Palepu, K.G., and Sinha, J. (2005) Strategies that Fit the Emerging markets. *Harvard Business Review* **83(6)**: 63-76.

Endnotes

(i) An earlier version of this chapter was presented at the JV International Conference 2011 at ITB Bandung on September 29. I am grateful to Prof. Kiyoshi Kobayashi for his generous support in this research. I am also thankful to Prof. Khairuddin Bin Abdul Rashid for his helpful interventions during the presentation.

(ii) Before 2000, FDI statistics database of the Department of Industrial Policy and Promotion, Government of India, reported only equity inflow, and the annual FDI statistic did not include reinvested earnings and other capital such as intra company loans.

(iii) The facts of this case are culled out by the author from the court judgments reported in [2008] 143 Comp Cas 726(CLB) and {2008} 143 Comp Cas 837(Cal).

(iv) The facts of this case have been taken mostly from Venkataramani (1990) [13].

Joint Ventures in Construction 2
ISBN 978-0-7277-5783-8

doi: 10.1680/jvc2.57838.083

Chapter 7

Reflexive Capability Maturity Assessment for Firm Management of Multi-partner International Projects

Kerry London, RMIT University
Jessica Siva, RMIT University

Introduction

In recent times, the landscape for international construction business environment has been fast changing and a number of key developments have led to this. The most dramatic development has been the unprecedented technological advancements in information and communication systems. The introduction of the Internet to the public in the 1990s has had a major impact on global construction business. Since then, collaborative systems have been rapidly improved thus enabling project team members to collaborate in virtual project environments much more easily and efficiently. Such virtual project environments have provided opportunities for instant global interactivity on a 24-hour basis where the relationship between time and distance has been revolutionalised [1]. Construction firms are no longer constrained to working on domestic projects but are instead highly mobile and have considerable potential to seek out work wherever required and operate on a global scale. Instant access is now possible and firms can come together to respond to problems collaboratively, irrespective of geographical location.

Trade liberalisation has also enabled an increase in the exchange of goods and services across international boundaries [2]. The official establishment of the United Nations in 1945 was a useful first step in encouraging international organisations to cooperate on specific matters to solve world problems. Since then, a number of organisations and bodies have been set up around the world to help develop and regulate the world economy and international trade including the International Monetary Fund (IMF), World Trade Organisation (WTO), Organisation for Economic and Cultural Development (OECD), Asia-Pacific Economic Cooperation (APEC) and the World Bank. More recently, the OECD, WTO and United Nations Conference on Trade and Development (UNCTAD) have called on the G20 countries to make a stronger commitment to open trade and investment as the global economy begins its recovery from the financial crisis [3].

Recent projections estimate that emerging market countries will spend US$22 trillion on infrastructure and facilities in the next decade alone and many of these projects are considered to be mega projects [4]. These projections represent tremendous opportunities for global firms in the architectural, engineering and construction (AEC) sector; however, these firms are also faced with a range of uncertainties and challenges [5]. International mega projects are often large-scale and characterized by a high level of complexity and risks, which are not found on domestic projects [6]. The skills, knowledge and expertise required to design, construct and manage these highly complex undertakings necessitate effective collaborations and partnerships between multiple participants from various countries working in foreign contexts who are crossing not only geographical boundaries but also political, professional, social, cultural, linguistic and financial boundaries [7]. As the need for large-scale international mega projects grows, firms in the AEC sector are increasingly seeking to share resources and human capital through partnerships and joint ventures to achieve increased strategic flexibility. The assembly and management of partnerships within such an intense environment pose significant challenges for firms, requiring them to have extensive experience and exceptional foresight in understanding the critical success factors for international collaborative partnerships.

This chapter reports the findings of a research project which was funded by the Australian-Malaysian Institute (AMI), aimed at developing and promoting a greater understanding of the uniqueness of successful Australian-Malaysian partnerships and joint ventures undertaken on complex multi-partner mega construction projects. The aim of the investigation was to identify the critical success factors specific to such partnerships. The theoretical framework was grounded in a previous empirical study on design firm internationalisation which demonstrated that long-term success in international markets is enhanced by the management of integration of transformation of social, cultural and intellectual capital within the firm [8], [9]. Social capital is the creation of personal relationships and networks based on trust built over time and has relevance for the project team networks and firm and client networks [10]. Cultural capital entails physical 'dispositions' such as 'building visible buildings, winning design competitions, or obtaining important tenders' [11]. premised on the reputation of the firm in that the success of marketing architectural services 'depends upon the firm's ability to sell and deliver a credible promise' [12]. Intellectual capital is a firm's collective skills, experience, competences and knowledge and is critical to the sustainability of firms, particularly in international markets [13].

A generic adaptive performance framework mapping the key performance indicators for each of these dimensions was developed which can be modified to unique contexts related to specific geographical localities. In the present AMI study the framework is tested within the context of firm and shared partnership capital. The broad interpretive framework of adaptive performance was refined as a result of the findings of the five case studies of Malaysian firms to develop a partnerships capacity performance measurement framework for Malaysian firms working on international projects. The reflexive capability maturity assessment tool offers firms a way of evaluating the firm and shared partnership's capacity or level of capital in relation to international collaborative partnerships towards effective management of partnerships on complex multi-partner megaprojects.

Research Strategy

A case study methodology was employed through the study of six organisations, which included four architectural firms, one property developer and one industry association. Data collection was carried out in Malaysia in 2009. Six in-depth stakeholder interviews were conducted with ten participants who represented six Malaysian organisations with an involvement with the property and construction industry both locally and internationally (refer to Table 7.1). The duration of the interviews was between 1-2.5 hours. Participants were asked to describe their experiences either working in international markets or working with international partners and/or clients. Participants were asked to identify from their perspectives key challenges associated with working across international boundaries and successful strategies used to overcome the challenges. Participants also described the international markets and/or partners they were involved with and the various market entry strategies, factors affecting entry mode and reasons for internationalising. The interviews were recorded, transcribed and subjected to two stages of analysis:

- Open coding: involved the loose association of themes as revealed by individual transcripts.

The barriers to internationalisation and the strategies used for overcoming barriers were identified and categorised accordingly. The strategic management of non-economic capital was identified as a key strategy to overcoming barriers to internationalisation.

- Axial coding: involved the arrangement of data according to dominant themes that emerged. At this stage a comparative analysis was conducted in order to ascertain common themes and irregularities and to enhance the potential for generalisation of resultant theory. This stage identified the dominant themes which demonstrate the importance of understanding the dynamics of non-economic capital acquisition which lead to the development of the firm's increasing reflexive capability.

It should be noted that M-I1 is the same person as M-PD1. The interview with MI1, although not included in the axial coding, served to obtain general contextual information relating to the exporting activities of Malaysian firms in the property and construction industry. Due to his roles as both the Executive Director of the property development firm, M-PDF1 as well as the past president of the Real Estate and Housing Developers' Association (REHDA), he was able to discuss his experiences in relation to both roles. Furthermore M-I1 was also the Vice Chairman of the Malaysian-Australian Business Council (MABC) and was able to offer his observations on the general working relationships between Australian and Malaysian firms from the perspective of the MABC. M-I1 highlighted that some of the difficulties in developing relationships between Australian and Malaysia businesses was a result of the mismatch of scale between project and firm size, which often limited the opportunity for Australian and Malaysian firms to collaborate on projects despite being in similar industries or sectors. In relation to his experiences being the past president of the REHDA, he raised the importance of internationalising particularly in relation to how it contributes to improving existing design and construction practices and/or methods in Malaysia. Table 7.1 presents an overview of the case study participants interviewed, the type of firm and the international markets they are involved with.

Table 7.1: Overview of case study participants

Case study	Firm type	Participants	Markets
M-AF1	Architectural design; office in Malaysia	2 participants: Project architect, M-A1 Principal architect, M-A2	Australia, Japan
M-AF2	Architectural design; offices in Malaysia, Vietnam, Singapore	3 participants: Project architect, M-A3 3D CAD visualisation consultant, M-A4 Project architect, M-A5	Vietnam, India, China
M-AF3	Architectural design, offices in Malaysia, China, UK	3 participants: Development manager, M-A6 Senior Design Manager, M-A7 Project architect – intern, M-A8	Australia, Singapore, Hong Kong, India, Vietnam, Canada, UK, China. Middle East
M-AF4	Architectural design, office in Malaysia	Managing director, M-A9	Australia
M-PDF1	Property Development, office in Malaysia	Executive Director, M-PD1	Australia, India, China
M-IA-1	Industry Association – Real Estate and Housing Developers' Association	Immediate Past President, M-I1	N/A

Internationalisation Process

The first stage of analysis identified the firms' internationalisation and project processes including market entry strategies and factors influencing choice of entry mode as well as motivations for internationalizing and these are briefly discussed in this section. The primary reasons for the case study firms working in international markets were associated with the firms' desire to grow and enhance their financial viability as well as an ability to shield away from fluctuations and downturns in the local market. All the firms acknowledged that the opportunities presented by working on international projects were higher levels of diversity and complexity, and that the significantly larger scale was more challenging. In addition, M-AF2's decision to internationalise was also guided by the firm's motivations to increase its competitiveness in preparation for the future possibility of foreign firms' ability to practice freely and obtain relevant certifications without involving local architects in Malaysia:

"I think the reason we're going international is that our boss sees that ... with globalisation and all that and I think in 2 years – in 2012 – that Malaysia will open up and foreign architects are able to come in and practice here so he wants to be prepared for that – and be quite robust" [M-A5].

The firms were highly dependent on social networks and joint ventures to gain entry into markets. The firms' primary mode of entry into international markets has been through joint ventures, that is, by structuring a relationship with a local partner in the international market. This choice of entry was largely determined by the firms' small-to-medium--sized scale and therefore joint ventures as the primary mode of entry provided the firms not only an increased ability to win projects and minimise financial risks, it also provided solutions to issues of geographic and cultural distance. A large percentage of the firms' projects were also the result of repeat clients and referrals whereby the firms have a number of ongoing relationships with architectural firms in various international markets including Singapore, Vietnam, Hong Kong, India and the Middle East. As explained by M-A7 and M-A6, such ongoing relationships with both clients and other partner firms are beneficial as they may lead to new projects in international markets:

"We do have some ongoing relationships with local firms where they'll want to establish a relationship when they have another project they come back to us to bring us on board...One project we were collaborating with Firm X and that led directly to another project which is ongoing so sometimes once we establish relationships with local offices they then turn back to us" [M-A7].

"So there are situations where a client or developer who's Malaysian who'll come to us and they want to do something overseas. We've done a project with a Malaysian client and they were developing a resort in South Australia. So that happens...Its both the client and the architect who may ask for us to come on board" [M-A6].

Firms M-AF2 and M-AF3 have also set up offices in a number of international markets. In particular, for M-AF2, "having branches" in a number of international markets is seen as presenting various positive practical outcomes in developing deep local engagement with an international market in terms of site locality, local culture and values as well as bureaucratic systems and processes. Therefore, the final selection of entry mode is largely dependent upon market specifics as well as chance and opportunity whereby there tends not to be a strict adherence to a specific entry mode strategy for these firms, but rather a more reactive approach is undertaken when entering markets.

Barriers, Success Factors and Reflexive Capability

Further to the identification of the firms' international processes, the barriers to internationalisation and the strategies used for overcoming barriers were identified and categorised accordingly. The analysis revealed an array of challenges or barriers to internationalisation which can broadly be classified into two key themes including dislocation from local context and differences in cultural understanding. A series of coding matrices were developed through the analysis but is not the intention of this chapter to discuss this part of the results. Table 7.3 (refer to **Appendix**) provides an overview of the key themes and sub-themes in relation to the barriers to internationalisation. A number of themes arose from the analysis in relation to the strategies used to overcome barriers to internationalisation. Success factors tended to be discussed in the interviews as the various strategies used to create, develop and maintain the different forms of non-economic capital which contributed to firm success.

As we proceeded to the next stage of analysis and 'revisited' the data the dominant themes were identified in relation to how the firms utilised each type of non-economic i.e; human capital to develop financially sustainable business models in the different international markets, which in turn led to the firms' increased flexibility and adaptivity in those markets. In relation to the firms' management of social, cultural and intellectual capital, there were four, two and two key themes identified respectively. This section is focused on examining the ways that the firms' strategic management of human capital led to increasing reflexive capability to generate financial sustainability in international contexts.

Reflexive capability model

Findings from the past study [8]. identified that the societal, cultural and economic differences across various geographic regions brought about extreme differences in institutional arrangements and responses. A reflexive capability model was developed based on the varying levels of firms' ability to respond to extreme differences and local conditions. Reflexivity has its derivation in sociological research [14]. It is based on a positive interpretation of change and a continual responsiveness to change by participants in the system. There are three key dimensions to reflexivity:

- Awareness: Staff members need self-awareness about what international process practices they are embedded within and to be perceptive of constantly changing international markets;
- Responsiveness: Staff, individually and collectively need to have openness to changes in the firm's international practices and procedures. The firm business culture also needs to be supportive to change; and
- Adaptability: Firms need to consciously identify the core principles which are non changeable versus changeable and adapt accordingly their practices and procedures based on the staff adaptive skills and the firm value, culture and procedural mechanisms that facilitate institutional change. To facilitate this adaptability, firms need to actualise a pro-reactive attitude towards change.

The reflexive capability approach is appropriate to all design processes but what past findings [15]. have demonstrated is that reflexive capability is particularly intrinsic to AEC firms who work globally and who are successful in that process. Findings of the case studies identifies that barriers are primarily the result of a low degree of reflexive capability, and that success factors demonstrated by the firms are the product of increasing reflexive capability. The firms' strategic management of social, intellectual and cultural capital to develop the three key dimensions to reflexivity was the primary means through which barriers of dislocation from local context and differences in cultural understanding were overcome. Three key themes were identified in relation to the case study firms' levels of reflexive capability including market entry, knowledge management and design management.

Reflexive capability for market entry

The case study firms' generally demonstrated a high level of reflexive capability in relation to developing market entry strategies for working in international markets. For example, M-AF3 was clearly aware of the common perception of clients that there is a certain amount of prestige associated with international firms and developed appropriate strategies in response to this as M-A6 highlighted:

"...sometimes Asian clients think of the UK firm as a higher standard...it's just a perception that ... if we bring in a foreign architect there's more prestige to the project. So we work that way as well but we will still share the work between the offices... it also happens the other way around, firms from say the Middle East think that if we do it from London its more expensive and if we do it from KL its cheaper so different strategies for different clients..." [M-A6].

Therefore allocation of projects and more specifically the decision on which staff members to work on which project is largely dictated by the specific needs of the client and project and can vary from one to the next. This is of course only enabled through the firm's organisational structure which allows a degree of flexibility between the various firm offices located in different markets. However, M-AF3 clearly recognised the advantages offered by the international nature of the firm and strategically utilised the ability to be flexible when allocating projects to different offices when attracting new clients.

Another example which demonstrates the firms' high level of reflexive capability maturity in relation to market entry is M-AF1's strategic management of social and intellectual capital in the development of relationship with a partner firm, A-AFI on a large-scale university campus project. A-AF1 is a prominent Australian architectural firm with experiences of having worked on Australian university projects and also with the Australian client seeking to build the university campus in Malaysia. The firm's ability to win the project and gain access to the Australian client was enabled by their reputation disseminated through social networks. The capacity to then respond to the project requirements was again afforded by the firm's ability to gain access to valuable skills and knowledge by structuring a relationship with the Australian partner firm through a joint venture. By entering into a partnership with A-AF1, the firm was not only able to increase the level of social capital but also was offered instant access to intellectual capital in the form of A-AF1's past experiences working with the client and proximity to the client as well as knowledge and skills in relation to Australian university requirements, which helped them overcome difficulties associated with dissociation from the client and users.

"So we had an arrangement with A-AF1 and that is that they will deal with the Australian user and we deal with the Malaysian developer... So that's where we cant be traveling back and forth so that's where A-AF1 played an active role in the initial setting up of the brief, the design needs and all that. So that's how we managed to get the local support over there. Basically more on the engineer, the landscape, its more the local side here. So the only other foreign support are the specialists. So we also consulted their specialists on lab design for mechanical services because of the exhaust and treatment and classification of rooms. But structural was local and because of our statutory submission, we need to have our local professionals as what we call the submitting person has to be local so foreigners cannot practice or submit. They can be our associate but cannot be the submitting person for the project. So that's how we managed to get the consortium group for the project" [M-A2].

Therefore, by structuring a relationship with each other through a joint venture, both firms were provided the opportunity to share resources and develop and build capacity in order to respond to the project requirements. M-AFI was able to overcome difficulties associated with briefing and liaising with the Australian client to develop the design whereas A-AF1 was able to rely on M-AF1 to gain the relevant project certification and approvals. The firm clearly demonstrated an awareness of the various methods used to create, use and maintain social and intellectual capital during market entry. The firm was aware of the changeability of strategies for different markets and responded accordingly based on this understanding. This was demonstrated through their strategy to develop a new relationship with A-AF1 on the university project to access the firm's intellectual capital instead of working with A-AF2 despite having worked with them before.

Reflexive capability for knowledge management

Market knowledge in the form of both informational and experiential knowledge was considered crucial by the firms in achieving success on international projects. Cultural understanding is a valuable resource in providing the capacity for 'deep' rather than simply 'surface' level communication through awareness of culturally specific nuance and subtlety and therefore not every person can manage the difficulties of cross cultural communication related to direct contact with international clients or collaborators. The individual who has the contact with overseas clients or collaborators occupies a centrally important position as the linking interface between internal design team and international networks (clients or partners). This highlights the value in the firm having an in-depth understanding of the interrelating forms of social, cultural and intellectual capital required in any given position within a firm's process. Significant findings were that those individuals with direct contact with the international client or partner were responsible for capturing and developing essential informational knowledge as well as utilizing experiential knowledge to facilitate relationships for more effective work on projects. Consider the following quote by M-AF1:

"Like I'm working with these Japanese – the way they write their letter its very diplomatic, very personal...So you try to work out what is it ... so I say you have to be very direct and they're not and you have to try and understand and read between the lines and sometimes contractual is different from personal... So I think I understand how the Japanese work differently. So not to say they don't know how to run projects but we learn from them and they learn from us. Oh you're Japanese clients, we need to understand them. When they want to tell you they're not happy they won't tell you directly. We're trying to guess what they're trying to say...So a lot of times you cannot be too direct or not too direct cos some of them are very personal" [M-A1].

M-AF1's accumulation of intellectual capital in the form of cultural understanding was seen as a valuable commodity which was useful for minimising difficulties arising from working with international collaborators and clients on projects. The firm's awareness of the differences between markets structures or systems and in the case of the Japanese clients, the differences in the way Japanese and Malaysians communicate, afforded the ability to adjust its own communication practices to integrate with those of the Japanese client they worked with. This aspect of cultural understanding demonstrates the highly fluid nature of these forms of capital and how the accumulation of such capital is highly beneficial because it can so easily be exchanged or transformed into useable capital. For this firm, an understanding of the subtleties of the client's national culture was instrumental in the ability to respond appropriately to the client's requests. Intellectual capital in this way also provides an understanding as to why certain problems might arise and how best to manage them as they occurred. It can thus be related to the formalisation of experiential market knowledge into informational resource for the firm, which can be used again on the same project or any other project as required to improve project performance.

Although this discussion centred on the Japanese culture, it is important to note that the underlying principle of utilising deep cultural understanding to develop appropriate responses or to manipulate situations for strategic advantage is one that cuts across all cultures. In this sense, cultural understanding is considered as a form of intellectual capital that exists as a general understanding of "how culture works" rather than a culturally or market specific understanding.

"... the challenge is global you need to know everything so probably that I think is important, how people expect you to present or run or implement the project. Its all expectations – and to know what to expect and if you know what you expect then you need to educate them to see in your perspective. Its not that you're better or that you're more competent – it's just opening up. So basically that's how I feel we managed to work with a lot of consultants is that I understand their constraints, their shortcomings..." [M-A1].

As M-A1 highlighted, this form of intellectual capital as a general understanding of how culture works is strongly linked to expectations and in particular the expectations that firms have of each other and the quality of materials and workmanship when working in international markets. A1 attributed the firm's successful relationships in the past with various consultants to their ability to "open up" to understand the other party's constraints and shortcomings in order to respond appropriately to achieve successful outcomes. This ability to understand and develop shared expectations is crucial to the development of effective working relationships on international projects.

Reflexive capability for design management

The efficiency of each firm's design management practices is largely dependent on the firm's understanding of the array of social, cultural and intellectual capital possessed by individuals occupying the various positions within internationalisation processes. The firms demonstrated a general understanding in relation to the value of reflexive capability and the measures taken to develop this capability in individual staff and within the firm as a whole in terms of the arrangement of relationships between roles that have specific requirements related to the unique characteristics of the particular project. Furthermore, it implies a need for an awareness of the skills, and capabilities of individual staff in order to place them in appropriately suitable positions to facilitate the capacity for individuals to respond accordingly and instantaneously in any unique situation that may arise. The following quote by M-A7 clearly shows a high level of understanding demonstrated by the firm in relation to the value of reflexive capability in design management practices:

"... there are different working cultures in different countries and cities. So on one hand things are relatively defined, on the other hand there is a certain amount of flexibility required because ... we have to adjust to local conditions. A project that is managed in Hong Kong is going to have to be managed very differently in many cases to a project managed in India or Australia... there's different relationships with local authorities ... local cultures and also ... within the field itself. So a project that is built in China may tend to have many more decisions happening on site whereas a project that is happening in Germany would have much more detailed design. So we lay out a certain strategy when we start the project but there's a certain amount of flexibility that's demanded as well...you cant necessarily take one experience internationally and then apply it to a different context... when you're dealing with the authorities there are certain principles that are more universal and so its not that you start from scratch every time but

you do need a certain amount of flexibility... there's a lot of variety but I think there's different learning curves for different clients, architects. It's about applying your scope of knowledge within different contexts, with adjusting your approach to complement the approach of your client, your local architect. And often times a client can teach you as much about how to adopt certain things as you can them" [M-A7].

A reflexive model is most effective when the project is considered a unique entity, and when the process is understood as structured yet malleable. As explained by M-A7, each project undertaken in any international market will differ from one to the next and therefore will need to be treated as unique, "so you can't necessarily take one experience internationally and then apply it to a different context". That is not to say that each project requires that firms "start from scratch" in the development of market knowledge for the project. It does however imply that there is a need for the development, enhancement and maintenance of experiential knowledge as a resource or "scope of knowledge" and the development of reflexive processes and systems to "complement the approach of the client, site and project requirements. The significance of practicing a degree of flexibility is emphasized in order to transcend individual projects and promote the development of the firm's reflexive capability. Such an ability to be reflexive thus enables firms to continuously improve processes and practices within the lifecycle of a project and not simply between projects.

Another example of the firms' reflexive capability maturity is M-AF1's accumulation of social capital through an understanding of the "people, the language, the culture", which was critical to its ability to work with international clients and partners. The firm's awareness of cultural understanding as a form of intellectual capital afforded the ability to strengthen business relationships and respond instantaneously to different situations experienced and in turn the capacity to manipulate situations to promote outcomes desirable for the firm. The opportunity for the firm to use various social strategies to build up rapport with project team members to strategically leverage power within a situation to achieve favourable outcomes could have been lost if the firm lacked the necessary intellectual and cultural capital required to respond accordingly.

The rigid nature of linear development processes in response to the lessons learned from previous projects is a useful tool of reflection, however, strict reliance on this process alone can itself prevent the maximisation of reflexive capability. The reflexive capability model implies that lessons learned from each project are applied within that project to immediately adjust the functional parameters, just as how M-AF1 was able to respond instantaneously to manipulate situations to achieve desirable project outcomes based on an understanding of "the people, the language, the culture". The implication here is that firm operating procedures and organisational systems are 'open' to this micro-managed change. Furthermore, lessons learned – as effective responses to a specific problem encountered – are collected as resources or referred to as "scope of knowledge" by M-A7, to be used in future projects according to the situational need, "its about applying your scope of knowledge within different contexts, with adjusting your approach to complement the approach of your client, your local architect". Therefore a process is constructed for the specific project in question that fulfils its requirements, and as new requirements come to light the potential exists for the process to be altered to accommodate new needs.

Reflexive Capability Maturity Assessment Tool

The reflexivity capability maturity assessment tool is the product of research into the barriers and success factors encountered by design firm's experiences of internationalisation [8]. It is a conceptual device for the measurement and management of a firm's levels of reflexive capability. It is capable of measuring the levels of reflexive capability present within any specific firm practice or activity. For example, these may include; market entry strategies, design management practices, knowledge management practices, communication practices, client management, briefing and feedback gathering practices, etc. The tool has been developed for three themes; market entry, knowledge management and design management and Table 7.2 is an example for design management (refer to Table 7.2). The tool has been developed to allow firms the means to gain a more detailed understanding and appreciation of an activity, practice or process both within a firm as well as in a partnership. By providing a tool for self-management rather than a series of recommendations, it is anticipated that the firms can achieve greater and more useful levels of reflexive practice.

It has been a central theme of this research study, and therefore the reflexive capability maturity assessment tool, that change is dynamic and takes place along a continuum rather than a series of linear categories. This is aimed at a clear representation of firm practices in terms of its level of maturity compared to the ideal scenarios for reflexive capability. Through this representation firms can identify the different areas in a firm or partnership that require improvement if the firm or partnership's success in international contexts is to be maximised. The advantage of understanding reflexive capability as a dynamic continuum lies in the realisation that influencing any particular component of a practice can also impact upon some other firm practice indicating the interdependency between the various forms of capital and how it may be utilised to a firm's advantage.

Table 7.2: Reflexive capability maturity assessment tool for design management

	AWARENESS			RESPONSIVENESS		ADAPTABILITY	
	Awareness of the need to create, use and maintain SCI capital for design management	**Awareness of key strategies needed to create, use and maintain SCI capital for design management**	**Awareness of changeability of key strategies for design management in different markets**	**Responsiveness to changing approach to creating, using & maintaining SCI capital design management in different markets**	**Firm support to be responsive in the creation, use and maintenance of SCI capital for design management**	**Firm policies and systems in place which define what is adaptable in relation to design management practices**	**Communication on how to re-strategise and adapt in design management practices**
Explicit	Comprehensive understanding of unique SCI capital needs to develop appropriate design management practices for international environments	Staff members within firm clearly aware of the need to adapt design management practices based on inter-changeability of SCI capital	Clear understanding of differences between markets enabling an ability to adjust design management practices and procedures to international markets	Consistent design management practices developed and well aligned with firm SCI capital competencies	Staff members fully trained to have adequate decision-making skills and levels of SCI capital in order to resolve design management issues as they occur on projects	Fully coordinated design management policies expressed including key principles and rationale behind principles adopted	Fully effective communication of design management practices based upon clearly defined project roles and responsibilities
Developing	Informal understanding of SCI capital needs for design management, i.e. importance of SCI mostly understood	Changeability of design management practices based on situational needs is mostly understood & aligned to staff levels of SCI capital	Attempting optimisation in developing design management practices based on inter-changeability of SCI capital	Seeking alignment of design management practices to firm SCI competencies	Formal arrangements between staff members in firm created to facilitate integrated project delivery through the use of SCI capital	Design management policies shared in common format with task allocation based on known competencies	Project roles are defined and sometimes communicated
Implicit	The importance of SCI capital for developing effective design management practices sometimes understood	Fragmented approach to design management in selected markets based on individual staff capability	Recognition that different markets present various situational needs & associated design management practices	Recognise that existing design management practices require development and alignment to SCI competencies	Some recognition that the use of SCI capital contributes to integrated project teams and processes thus project success	Recognise that design management policies need to be clearly defined & aligned to firm competencies and strategic objectives	Individual staff members define own design management practices based on personal observations & experiences
Lacking	No awareness of the significance of effective design management on project performance	No recognition of the need for SCI capital for design management	No awareness of differences between design management practices in different markets	No understanding of the relationship between firm SCI competencies and design management practices	No understanding of contributions of SCI capital to the development of integrated project teams and processes	No attempts to define firm policies in relation to design management practices with no recognised link between project & firm objectives	Individual staff members operate in isolation with silo mentalities dominating firm culture

Conclusion

The findings highlight that through reflexive capability it is possible to become consciously and explicitly aware of changes in market conditions and client requirements and absorb those changes in a manner that facilitates the firm or partnership's strategic objectives. This study serves to confirm the significance of the reflexive capability model which is highly relevant for mapping the capabilities of Malaysian construction firms in relation to developing and maintaining international collaborative partnerships. Contrary to the common view that the internationalisation process necessitates extensive country-specific knowledge and significant financial resources, the discussion in this report has demonstrated a longer-term perspective afforded by the firm's reflexive capability which enabled them to strategically manage the different forms of non-economic capital to ultimately overcome difficulties associated with internationalisation.

The firms' strategic management of social, intellectual and cultural capital to develop the three key dimensions of reflexivity was the primary means through which the firms were able to overcome barriers to cross cultural communication, dislocation from international client and users and lack of resources and skills. Understanding firm reflexive capability and how firms achieve this is significant since it can lead to successful international collaborative partnerships.

Within the matrix provided by the reflexive capability maturity assessment tool it is possible to identify components of project performance and to place them in a continuum of increasing reflexive capability. This application both encourages the objective rationalisation of firm practices and the evaluation of project performance as a group of inter-related components. Through this it is more likely that under-performing aspects of practices within a firm or partnership can be identified. Furthermore it is possible to then measure how far away from best practice a particular firm or partnership practice is and set in motion changes to practice, procedure or policy that will facilitate the elevation of that component towards achieving increased firm or partnership reflexive capability.

References

[1] Howes R. and Tah J. (2003) *Strategic management applied to international construction.* Thomas Telford Publishing, London.

[2] Ling, F. and Kwok, D. (2007) Enablers for Singapore contractors to internationalise construction services. *Construction Management and Economics* **25**: 267-275.

[3] Organisation for Economic Cooperation and Development (OECD) (2009). www.oecd.org (accessed September 2009).

[4] Economist (2008) *Economics Focus: Building BRICs of growth.* The Economist 88.

[5] Javernick-Will, A. and Levitt, R. (2009) Acquiring local knowledge for international projects. *Construction Research Congress: Building a sustainable future.* April 5-7 2009, Seattle.

[6] Fiori, C. and Kovaka, M. (2005) Defining megaprojects: learning from construction at the edge of experience. *ASCE Construction Research Congress: Broadening perspectives*, April 5-7, California.

[7] Han, S., Asce, M., Park, S. Kim, D., Kim, H. and Kang, Y. (2007) Causes of bad profit in overseas construction projects. *Journal of construction engineering and management* **133:** 932.

[8] London, K, Chen, J & Bavinton, N. (2005) *Internationalisation of Australian Construction Firms, Final report.* CRC_CI, Icon.Net.Pty.Ltd, Brisbane.

[9] London, K., Chen, J. and Bavinton, N. (2005b) Adopting reflexive capability in International Briefing. *Facilities* **23 (7/8)**: 295-318.
[10] LCohen, D. and Prusak, L. (2001) *In Good Company*. Harvard Bus Sch Press, Boston, MA.
[11] Skaates, M, Tikkanen, H. and Alajoutsijarvi, K. (2002) Social & cultural capital in project marketing service firms. *Scandinavian Journal of Management* **18**: 589-60.
[12] Lowendahl, B. (2000) *Strategic management of professional service firms* (2nd Ed). Copenhagen Business School Press, Copenhagen.
[13] Stewart, T. (1997) *Intellectual Capital: The New Wealth of Organizations*. Nicholas Brealy Publishing Limited, London.
[14] Giddens, A. (1991) *Modernity and Self Identity: Self and Society in the Late Modern Age*. Stanford University Press, Stanford.
[15] London, K. and Siva, J. (2007) Role of cultural capital towards the development of a sustainable business model for design firm internationalisation, *CIB World Building Congress, CIBW96 Architectural Management*, 14-18 May 2007, Cape Town, South Africa.

Appendix

Table 7.3: Barries to internationalisation

Case study	Dislocation from local context		Differences in cultural understanding		
	Codes, specificatios, standards	Geographic distance	National cultures (language, social customs, etc)	Project culture (fields of expertise, niche specialisation)	Business culture (management style, methods, processes)
M-AF1	Steep learning curve for Malaysian firm to meet Australian certification standards	Malaysian firm's lack of involvement in briefing makes refining design at later stage difficult	Different preferences between Australian designers and Malaysian users	Different fields of expertise and one conduit channelling information	Mismatched expectations between Malaysian developers and Australian firm's in terms of work scope and payment
	Different tax systems between Australia and Malaysia – increased monitoring	Malaysian firm's lack of involvement at initial stage causes confusion/lack of clarity at later stage	Malaysian users' response to unfamiliar air conditioning system	Different social customs in the use of libraries and different views towards security issues between Australian and Malaysian users	Different work methods/approach in terms of level of detail when designing
	Australian firm specifying to Australian suppliers' codes – different from Malaysian codes	Malaysian firm's lack of involvement at initial stage causes confusion		New area of learning – specialist knowledge – ie tubing system for air conditioning	Different expectations between Japanese and Malaysian contractors in terms of work quality
	Different by laws and guidelines between Australia and Malaysia – time consuming				
M-AF2	Different regulations between China and Malaysia	Inability to discuss project issues face-to-face a difficulty	Different terms used in design specifications/ descriptions resulting in longer discussions, ie time consuming & frustrating	Different genius loci across cultures	
	Difficult to stay up to date with regulations – locally and internationally		Linguistic capacity limits the selection of staff who can be involved on projects in certain markets		
M-AF3		Not having a presence in international markets to steer projects a challenge	Not speaking the local language a barrier	Pre-conceptions about green buildings and idea of sustainability varies across cultures	
		Difficulty ensuring design intent carried through post handover to local firm in international market			
M-AF4					Dislike for corruption as standard of practice
M-PDF1					Different work approaches between Malaysian and Australian firms – Australians more headstrong about their design ideas
					Australian firm's tendency to adhere closely to contractual agreements – Malaysians more on verbal – 'trust'

Joint Ventures in Construction 2
ISBN 978-0-7277-5783-8

doi: 10.1680/jvc2.57838.097

Chapter 8

Strategic Implementation of International Joint Venture in Taiwan: A Case Study on THSRC

Tsung-Chieh Tsai, National Yunlin University of Science and Technology
Hung-Chi Chen, Chiayi County Government, Taiwan
Min-Lan Yang, JU-MING Construction Co. Ltd.

Introduction

Since 1969, and most especially in 1975, Taiwan's government has opened its market to foreign firms. With the Government Procurement Act released in 1998 that ratifies the legal status of JV, many large foreign firms with JV contracts are able to participate and invest in Taiwan's public projects. During the years between 1988 and 1994, JV between foreign and domestic firms only amounted to 3 billion US dollars. Starting 1999, Taiwanese government has invited private firms to tender their bid for major public projects by applying B.O.T. (Build-Operate-Transfer), B.O.O. (Build-Operate-Owned) and B.O.O.T. (Build-Operate-Owned-Transfer). Foreign corporations from such developed countries as France, Japan, USA, and Germany have formed short term strategic alliances with Taiwanese firms to meet the demands of the proprietors, while investing more than 40 billion US dollars in Taiwan's public projects, among which is the highest amount and the biggest BOT project worldwide for THSRC (Taiwan High Speed Rail Corp). It is evident that in the last few years, public projects by JVs in Taiwan have become a trend.

JV projects, usually large in scale, and complex, technologically, involving construction time frames from a short period of three years, to long periods of five, ten or even more than ten years. Therefore, before applying JV, firms must search for an appropriate alliance partner from which they can acquire complementary techniques. Numerous studies have suggested that these firms actually benefit from the selection of a partner who can provide complementary techniques or competence that are expected to help the firm attain its strategic objectives [1], [2]. Partner selection might also influence alliance performance, as indicated by the findings of prior research [1], [3], [4] that the choice of a partner is an important variable influencing JV performance; thereby, highlighting the importance of selecting the right partner. A firm might make trade-offs between alternative complementary techniques or resources, and this concept is critical to success may be useful in overcoming this problem. These factors are determined by the underlying characteristics of the firm's industry [5]. Due to the geographical characteristic of the construction industry, several world class firms, such as Shimizu Corporation from Japan and Bilfinger and Berger from Germany, must form joint ventures with domestic firms to acquire project contracts. On one hand, this is due to government regulations; on the other hand, even if these world class firms have abundant capital and advanced construction technique, they must consolidate themselves with the management strengths of Taiwanese firms. Taiwanese firms are familiar with the environments for domestic construction and geography; the joint procurement of public projects will bring economic benefits for the firm. Comparatively, the limited economic scale of domestic firms contributes uncertainties to operations. Therefore, they must assess their strengths and weaknesses and form coexisting relationships with large firms. However, few prior studies have discussed the strategies of JV taken by foreign firms in selecting alliance partners in Taiwan, nor the strategies taken by domestic firms in forming effective alliances.

Even though there are no specific criteria for the selection of alliance partners [6], the decision on the seeking of complementarity could be examined by SWOT analysis (the strengths, weaknesses, opportunities and threats) for each partner. Through pilot case studies discussion, citing the current biggest BOT project in the world as an example, and using SWOT analysis to investigate the factors of the strengths, weaknesses, opportunities and threats of participating foreign and domestic firms in the Taiwan High Speed Rail project, as well as applying Analytic Hierarchy Process (AHP) in the quantitative analysis of the importance of each groups and factors, this study aims to achieve the following purposes: (1) Analyzing the strengths, weaknesses, opportunities and threats and SWOT factors of both foreign and domestic firms to explain their strategies in searching for appropriate JV partners; (2) Analyzing the relationship of SWOT, SWOT factors and project management elements (cost, schedule, quality and safety) of both foreign and domestic firms; (3) Analyzing the relationship of SWOT groups and factors of project cyclical stages (bid, design, construction and maintenance) as a reference on how both partners achieve complementarity of resource contribution over different project stages. The research flowchart is shown in Figure 8.1.

Figure 8.1: Research flow chart

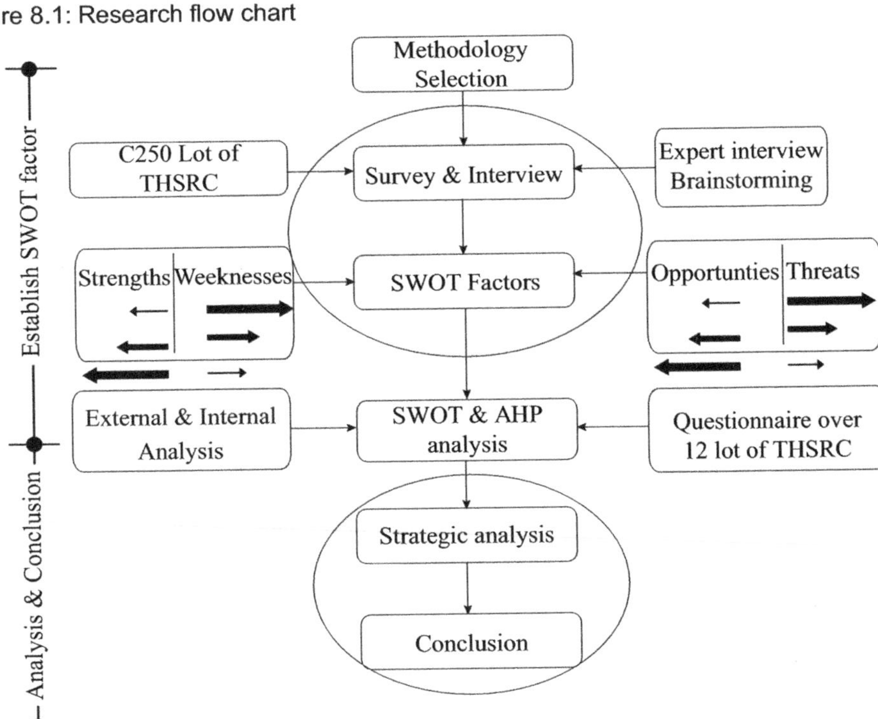

Strategic Alliance and Joint Venture

James [7] posits that strategic alliance is a mutually beneficial process where both parties depend on and keep their commitment to each other; and any change in a partner's benefits might weaken the alliance. Harrigan [2] considers strategic alliance as a key strategic choice in considering a firm's decision-making strategy, to ensure, maintain or even improve the firm's competitive advantage. Killing [1] points out that strategic alliance refers to two or more organizations that form alliances to achieve a common goal. Lewis [8] states that strategic alliance results from the mutual needs and risk sharing of firms; thus, they cooperate together to meet a shared goal.

The advantages of JV cooperation in large scale public projects refer to lowering risk, achieving economic scale, saving on development cost, accumulating experience, improving technique and so on [1], [2], [7]-[10], while its disadvantages are incurred also from the participation of its contractors, which include not only the obvious organizations of entrepreneurs, but also the invisible organizations of governments and nations (as shown in Table 8.1).

Table 8.1: Joint venture as viewed by advantages and disadvantages

Evaluation	Advantages	Disadvantages
Proprietor	1. Less managerial work for integration interface 2. Reduced project risk 3. More successful contracting	1. Concern over paperwork boycott 2. Concern over liability from job inefficiency 3. Vendor evaluation requires extra work
Joint venture contractor	1. More procurement opportunities 2. Higher cost estimation accuracy 3. More successful fund acquisition 4. Lower risk 5. Acquiring techniques and managerial skills 6. Learning the strengths of other contractors 7. Better project performance 8. Full utilization of idle resources 9. Eliminating unethical behavior of license imposture 10. Allowing partner selection by demand 11. Acquiring advanced techniques 12. Training for integration talents	1. Contractor cooperation hard to promote 2. Operation efficiency hard to improve due to provisional nature of organizations 3. Unequalled competence becomes obstacle for contractors 4. Contractors with financial problems 5. Contractors not duly fulfilling obligations 6. Increased operating cost 7. Complicated accounting practice
Government	1. Construction industry allowed more expansion 2. Contractors allowed international competitiveness 3. Reduced financial expenditure 4. Reduced project risk	1. Regulation or law changes 2. Interest rate increase 3. Volatile political situations

Before adopting JV, contractors must first assess their own characteristics and capabilities to seek for appropriate alliance partners and combine each other's limited resources and share the cost and risk. Yet, studies show that one of the most significant success factors is selecting the "appropriate" or "right" partner, "particularly when JVs involve a firm's core markets or techniques" [1], [3], since good partner can prompt the smooth progression of JV, change the category group in the strategic environment and lower the extent of uncertainty [5]. Several empirical studies also have discussed partner selection criteria with research category group from various perspectives and dimensions [3], [11], [12]. However, it can be observed from literatures of prior studies that the there is great difference in the partner selection criteria as observed in different countries with different extent of development, commercial traits, and benefits and traits of partnership; given the same selection criteria, due to the difference in priority, for instance, there is difference even in the selection of criteria. Therefore, the strategies on how foreign firms entering Taiwan's construction market as JV contractors deal with the selection of the right partners, or how Taiwan's domestic firms use their strengths to become JV alliance partners will be brought forward in the study findings to be discussed below.

Research Methodology

SWOT analysis

This study utilized SWOT analysis in examining the internal strengths and weaknesses, and external opportunities and threats factors faced by both domestic and foreign firms in Taiwan. There are differences in the weight and priority in the factors of the four groups (strengths, weaknesses, opportunities and threats), which may be due to the firms' trait, and market environments. To quantify the priority extent of the groups and factors, certain methods are required for verification. Therefore, this study applied Analytic hierarchy process (AHP) to render in quantitative values the extents of priority of the four groups and factors, while allowing the results to be used as basic data in subsequent analysis.

Analytic hierarchy process

This study utilized SWOT analysis in examining the factors of strengths and weaknesses internally, and opportunities and threats externally that are confronted by both domestic and foreign firms in Taiwan (as shown in Figure. 8.2), while using AHP analysis to divide complicated data to be studied into relationships of hierarchical category group, and, after evaluation and analysis by opinions of experts, to show the extents of priority or contribution of the factors in each tier of the hierarchical group and factors so that there is a theoretically consistent criterion to check against the extent of consensus among experts, so as to allow the conclusion to be more consistent and objective, while the cumulative priority of SWOT groups and SWOT factors represented in quantitative values can be used in subsequent analysis.

Figure 8.2: An integrated group hierarchical analysis

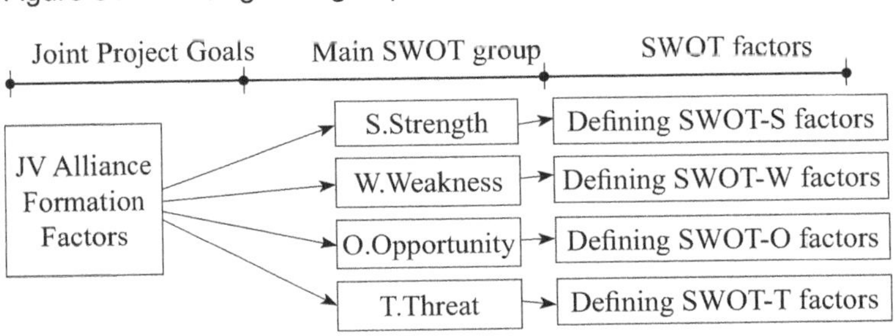

Industrial Background

External environment analysis

Despite of its great contribution to the domestic economy, the construction industry, more than any other industries, experiences limitations on its overall economy by the administrations of the government, which results in the instability of this market that fails to provide a sustainable environment for the firms operating business in it. For instance, the laws and regulations concerning the operation standards for construction industry are administrated by the Construction and Planning Agency (CPA) of the Ministry of Domestic Affairs, while business operation realm of firms is regulated by Company Act and Investment & Planning Act stipulated by the Ministry of Economic Affairs, and the need for funds is often influenced by the policies of the Ministry of Finance in their monetary restriction and relaxation adjustment in attempt to regulate domestic economy, which also shows the lack of comprehension and consistency in the policy-making and law-making of the government. Moreover, the shift in social value system also makes younger generation unwilling to join the construction industry, resulting in severe shortage of construction labor to such an extent that professional skills and experiences are hard to be carried on. The production value of the construction industry, for instance which has been declining since 1995 dipped down to a gross production value of 1.1 billion US dollars in 2005, accounting for only 2.15% of the gross domestic product (GDP), while due to the stagnant real estate construction and limited progression of major construction projects, the index for building-related economy still lingered around the level of all time low in 2005. From this viewpoint, the overall economy of Taiwan's construction industry, is developing in a bearish trend, and in order to maintain project performance and prompt gross profit, firms are gradually reducing the revenue ratio of conventional projects, while switching to those for high-tech intelligent factories and environmental-friendly projects.

The world economy has since 1999 gradually recovered after the Asian financial crisis, with the growing starting to slow down in 2005, though at a rate not quite significant, while the growth is expected to be moderate for 2006 and 2007. Despite of ever-increasing energy prices, the overall economies in the US and Japan are doing well, while strong growing is expected. In the US, the economy remains strong, apart from oil price hike, federal interest rate increase, and hurricane disturbance which impose insignificant impact on the economy; over in Europe, the economy maintained steady growth, while the unemployment rate are gradually being improved; the economy in Japan, with economy-boosting measures being continuously implemented and the recovering economy of semiconductor manufacturing industry, is slowly coming back up from the bottom. This strong development of world economy are even more visible among foreign firms, particularly in their ventures overseas in developing countries, such as China, India, Russia, and Brazil, etc. With WTO accession, the contracting of various projects, the sourcing of labor brokerage, and procurement of major materials in Taiwan are all required to be made accessible for foreign vendors, and the possible impacts on Taiwan's construction industry may include the following: (1) deregulating the construction market to foreign investment (2) opening the market of domestic public projects, and (3) adjusting tendering procedures and operation modes, and so on.

In terms of operation scale, techniques required for construction and capital, the firms in Taiwan still lag behind their counterparts in Europe, Japan and the US; the entry to Taiwan's construction by these foreign firms, with their abundant financial supports, superior managerial and technological skills, are likely to cause great impacts on local firms.

Internal environment analysis

The advantages of firms in Taiwan, in terms of the practical aspect of operation, include the acquaintance with local construction systems, the knowledge about domestic law and regulations, familiarity with the administration of local affairs, the acquaintance with the geology and weather environment, and so on, but the lack of clear, long or mid-term industrial policies has jeopardized the development of the construction industry, leaving such problems as the shortage of financial power, imperfect construction equipment, and even the lack of competence in research and development; the advantages of foreign firms venturing in Taiwan's market, in contrast, include customer proof construction performance, strong financial power and superb construction techniques, while facing great risk relating to such factors as different culture and language, inconsistent policies of the government and the fluctuation of currency exchange incurred by the world economy, etc.

Background of Case Study

The project for THSRC (Taiwan High Speed Railway Corporation) is by far the most significant policy for the Taiwanese government to prompt private investment in the construction of infrastructure and public projects. The Construction and Operation Agreement (C&OA) was signed with the THSRC on July 23, 1998, by which THSRC was authorized to invest, build and operate by the BOT (Build-Operate-Transfer) mode. The authorization also extends to projects related to railway construction and operation (35 years after signing), train station development (35 years after signing) and business development in the vicinity (50 years after reclamation). THSRC has a total distance of 345 kilometers. The constructions officially began on March 1, 2000, and commence service after its completion on October 31, 2006. Project constructions include land works, core electro-mechanical engineering, station construction, railway construction, site maintenance and station zone development. The THSRC construction is divided into 12 projects. Each project adopts the D/B (Design and Build) method to award contracts, and uses the JV model for construction (See Table 8.2). The main structures of the whole system include a total distance of 252 km of elevated railway bridges (73% of the total distance), 63 km of mountain tunnel or cut and cover tunnel (18% of the total distance) and road embankment/slope sections for the remaining 30 km (9% of the total distance).

Table 8.2: JV biddings of lot construction of THSRC (by domestic/foreign firms)

No.	Lot	Joint venture contractors	No.	Lot	Joint venture contractors
1	C210	Futsu (Taiwan)/ Obayashi (Japan)	7	C260	Continental Engineering Corporation (Taiwan)/ Bilfinger Berger (Germany)
2	C215	Futsu (Taiwan)/ Obayashi (Japan)	8	C270	Continental Engineering Corporation (Taiwan)/ Bilfinger Berger (Germany)
3	C220	Jiu Tai (Taiwan)/ Guo Kai (Taiwan)/ Daiho Corporation(Japan)	9	C280	IE&C (Taiwan) / Samsung (Korea)/Doosan (Korea)
4	C230	Chung Lin (Taiwan)/ Hyundai (Korea)/Zen Pacific(Hong Kong)	10	C291	Evergreen (Taiwan)/ Shimizu (Japan)
5	C240	Chung Lin (Taiwan)/ Hyundai (Korea)	11	C295	Evergreen (Taiwan)/ PEWC (Taiwan)/ Italian-Thai (Thai)
6	C250	Pan Asia (Taiwan)/ Hochtief (Germany)/Ballast Nedam (Holland)	12	C296	Evergreen (Taiwan) / Shimizu (Japan)

Establishing Factors for SWOT Analysis

After analyzing the construction industry by its internal and external environments and the background introduction by the case study, this study did a primary investigative interview with the C250 lot work project, to help establish questionnaire items. The Taiwan contractor is Pan Asia Corporation, and the foreign contractors are Germany-based Hochtief AG vorm. Gebr. Helfmann, and Ballast Nedam Corporation from Holland. Through collective brainstorming and pair-wise discussions, the domestic and foreign engineers from the C250 project have gathered and eliminated identical or similar factors, shown in Table 8.3.

In this survey, the expatriated engineers of the foreign firms also defined 28 factors in four SWOT groups, such as the strengths of "S1. The contractor is well-experienced", "S2 Sound financial support". The weaknesses of "W1. Not acquainted with local construction systems", "W2. Different culture and language", the opportunities of "O1. Increase market share of workload in local construction market" and "O2. Undertake both local and major construction work", and the threats of "T1. Lack of raw materials or fluctuation" and "T3. The local legal system is not perfect".

Survey Analysis

Survey samples

The design of the questionnaire for this study is based on the four SWOT groups and the corresponding factors established described above (as shown in Table 8.3), and the subjects selected for this survey were the twelve contractors for the lot construction of THSRC projects (as shown in Table 8.2). After the survey results were compiled, 92% of the respondents were found to have 10 or more years of experience in engineering, while some respondents were found to be in senior management position. A total of 72 questionnaires were sent out, with 38 copies being returned with responses. To increase the credibility of analysis results, the criterion for the consistency index was set at 0.2 and, after such an adjustment, all the returned survey questionnaires passed the consistency test and can be regarded as valid responses.

Table 8.3: SWOT groups and factors

Domestic firms		Foreign firms	
SWOT groups	**SWOT factors**	**SWOT groups**	**SWOT factors**
SD. (0.566)	**SD1. Be acquainted with the domestic construction system (0.209)** [1]	SF. (0.579)	**SF1. The contractor is well-experienced (0.233)** [1]
	SD2. Be acquainted with the domestic laws/regulations (0.120) [2]		**SF2. Sound financial support (0.109)** [2]
	SD3. Estimate the construction cost precisely (0.088) [3]		**SF3. Strong management skills (0.077)** [3]
	SD4. Be acquainted with politician (0.055) [6]		**SF4. Fully equipped for construction (0.049)** [5]
	SD5. Have good relations with politician (0.037) [9]		**SF5. Skilled/experienced construction team (0.046)** [7]

Domestic firms		Foreign firms	
SWOT groups	**SWOT factors**	**SWOT groups**	**SWOT factors**
	SD6. Be acquainted with the geology and weather environment (0.028)		**SF6. Strong design team (0.034)** [10]
	SD7. Easy integration interface of construction (0.028)		SF7 Arbitration experience (0.031)
	WD1. Financial power shortage (0.066) [4]		**WF1. Not acquainted with local construction systems (0.058)** [4]
	WD2. Domestic construction market is not extensive in general (0.047) [7]		**WF2. Different culture and language (0.047)** [6]
	WD3. The government policy is not clear (0.033) [10]		**WF3. Not acquainted with local legal system (0.034)** [9]
WD. (0.196)	W4. Inadequate construction equipment (0.020)	WF. (0.221)	WF4. Not acquainted with local customs (0.028)
	WD5. Lack support by domestic heavy industry (0.012)		WF5. Not acquainted with the norms of the domestic constructors (0.023)
	WD6. Lack ability for innovation, research and development (0.009)		WF6. Not acquainted with the domestic geology and weather environment (0.017)
	WD7. Shareholders with conservative mentality (0.008)		WF7. Not acquainted with government unit operation models (0.013)
	OD1. Be acquainted with construction market, easily to compete with other companies. (0.057) [5]		**OF1. Increase market share of workload in local construction market (0.034)** [8]
	OD2. Other professional realms of development, expand the market (0.037) [8]		OF2. Undertake both local and major construction work (0.025)
	OD3. Cooperation with foreign firms, introduction to foreign techniques and management experience (0.024)		OF3. Establish popularity (0.021)
OD. (0.157)	OD4. Experiencing extending market in alliance with different industry (0.016)	OF. (0.132)	OF4. Be acquainted with the domestic construction system(0.015)
	OD5. Enter global construction market through strategic alliance (0.011)		OF5. Flexible utilization of funds (0.014)

Domestic firms		Foreign firms	
SWOT groups	**SWOT factors**	**SWOT groups**	**SWOT factors**
	OD6. Promoting company's image (0.007)		OF6. Building connections with political and business circles (0.012)
	OD7. Acquiring BOT practical experience (0.005)		OF7. Generate construction markets through strategic alliance (0.009)
TD. (0.081)	TD1. Competition of foreign firms (0.022)	TF. (0.068)	TF1. Lack of raw materials or fluctuation (0.018)
	TD2. Civil rights are disregarded (0.019)		TF2. Not in accordance with local firm's principles (0.009)
	TD3. Government unit operation model (0.014)		TF3. The local legal system is not perfect (0.010)
	TD4. Lack of raw materials or fluctuation (0.009)		TF4. The local firm tenders too cheaply (0.009)
	TD5. Competitor offering sweep-off low price in bid (0.008)		TF5. The civil rights are disregarded (0.007)
	TD6. Failure to upgrade techniques (0.005)		TF6. Lack of connections with politician (0.007)
	TD7. Inconsistent government policy (0.004)		TF7. Inconsistent government policy (0.008)

Note: 1. Figures in brackets "( )" represent the overall weight of cumulative priority of the group and factors.
2. Items shown in bold print represent top ten factors regarded as first priority by domestic or foreign firms. Figures shown in superscripts represent priority ranks.

Factor Analysis

SWOT groups and factors for domestic firms

The four SWOT groups established by the domestic firms in the first level are shown in Table 8.3, with relative weight ratio of 0.566, 0.196, 0.157 and 0.081 for SWOT groups of strength, weakness, opportunity and treat, respectively, indicating that strength, with weight ratio over 50%, is highly considered by domestic firms, far more than any other SWOT group, while the weight ratio difference between weakness and opportunity is not significant, and the threat group is barely considered in the decision-making for domestic firms.

As shown in Figure 8.3 and Table 8.3, domestic firms regard the strengths they possess as their best qualifications and the core competence in the competition. Through joint ventures with foreign firms, their benefits can be expected to maximize, and their weaknesses to improve in the meantime. However, opportunity factors are not regarded important, mainly because most domestic firms are small and medium sized in scale, and hardly engaged in any negotiation directly with foreign firms to acquire techniques or enter international market. Given the sluggish economy in Taiwan's construction market, it is necessary for domestic firms to rely on alliances with foreign firms to get big projects for public construction or to develop in different markets. The finding that threat group is not highly regarded also tells why the implementation of risk management is still hard for domestic firms.

Figure 8.3: Graphic representations of pair-wise comparisons of SWOT groups and factors (domestic firm)

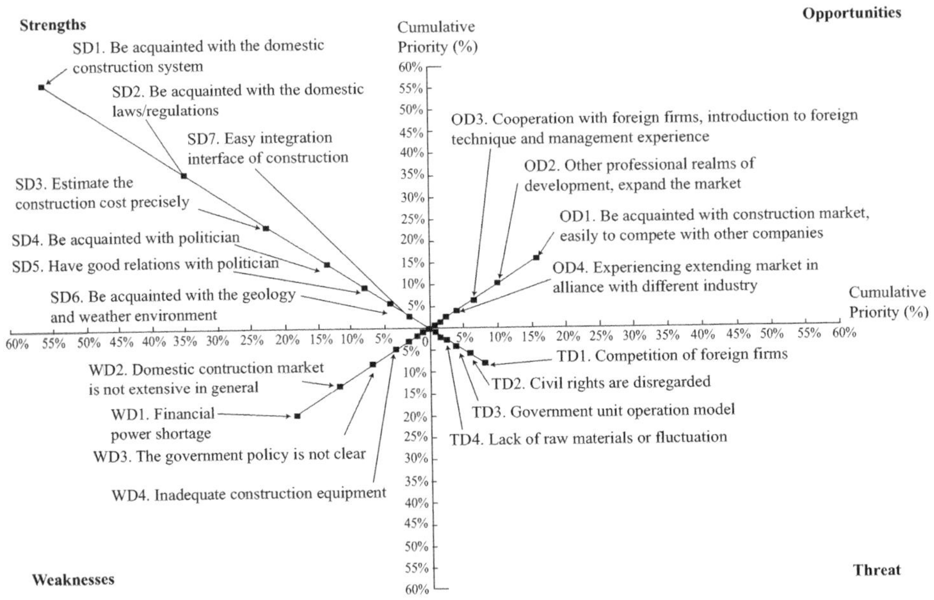

SWOT Groups and factors of foreign firms

The four SWOT groups established by foreign firms in the first level are shown in Table 8.3, with relative weight ratios being 0.579, 0.221, 0.132 and 0.068, for SWOT groups of strength, weakness, opportunity and threat, respectively, similar to the significance ranking by the domestic firms (of 0.566, 0.196, 0.157 and 0.566, respectively), with strength taking up over 50% of the weight ratio, followed by weakness and opportunity, while threat group is given the least regard by foreign firms.

As shown in Figure 8.4 and Table 8.3, foreign firms regard their strengths highly, relying on their strengths of experiences from multinational projects and sound financial support to venture into Taiwan's construction market while expanding their business territory, but have to rely on alliances with domestic firms to reduce their investment risk and other uncertainties, due to cultural difference and language barrier, and the lack of knowledge about Taiwan's construction market. The opportunity weight ratio did not turn out high, indicating the fact that project scales in Taiwan are not very extensive, and if foreign firms are to expand publicity to increase the opportunities for JV, the chance is not much. Furthermore, the finding of extremely low weight ratio on threat shows that foreign firms perceive no impact of threat or threat related factors on their decision-making on projects in Taiwan; moreover, given that almost all threat factors concern with government, it can be inferred that foreign firms are likely to rely to JV to transfer related risks to domestic firms. This will also be the strategy domestic firms are to adapt to acquire JV allies.

Figure 8.4: Graphical representations of pair-wise comparisons of SWOT groups and factors (foreign firm)

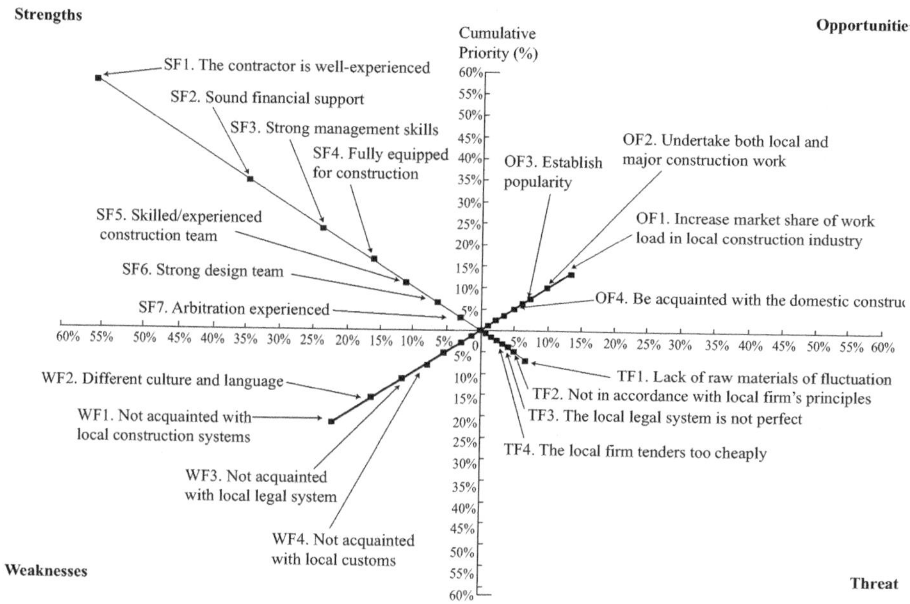

Analysis on project cyclical stage related factors

Domestic firms in over project cyclical stage

As shown in Figure 8.5, the analysis by comparing the number of times a factor is regarded significant over the four project cyclical stages shows that the domestic firms regard construction stage the most, with more concerns over factors related to weakness and threat factor groups, in contrast to only two strength factors referred for construction stage, "SD1. Be acquainted with domestic construction system" and "SD6. Be acquainted with the geology and weather environment", indicating that despite their high regard on construction stage, domestic firms lack the kind of ability of managing such uncertainties at construction stage as the factors of shortage of raw materials, and the disregard of civil rights, etc. The design and bidding stages are respectively influenced by the lack of ability for research and development in the internal environment, and the lack of clarity in government's policy in the external environment, while the maintenance stage is apparently not regarded by the domestic firms.

Figure 8.5: Project cyclical stage (domestic firm)

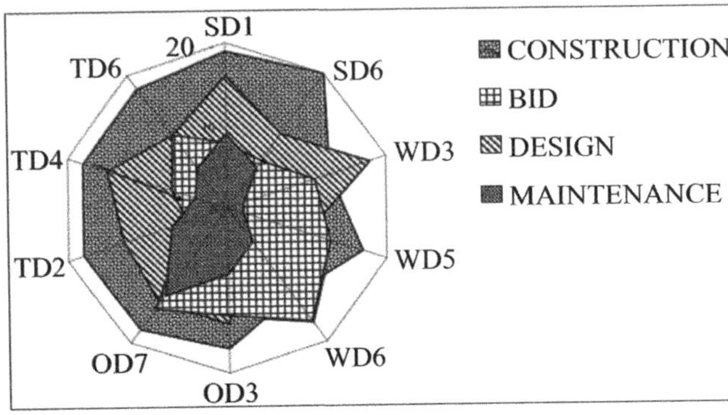

Foreign firms in over project cyclical stages

As shown in Figure 8.6, the analysis by comparing the number of times a factor is regarded significant over the four project cyclical stages shows that the foreign firms give more regards to construction stage than any other project stage, with more concerns over factors related to strength factor group, indicating that foreign firms regard highly on domestic firm's competence in construction and the construction stage, while the concern over the bid stage relates to the factor of not being acquainted with the domestic legal systems; they are certain of their research and development abilities in the design stage, but foreign firms share similar concepts with domestic firms regarding maintenance stage relatively lowly.

Figure 8.6: Project cyclical stage (foreign firm)

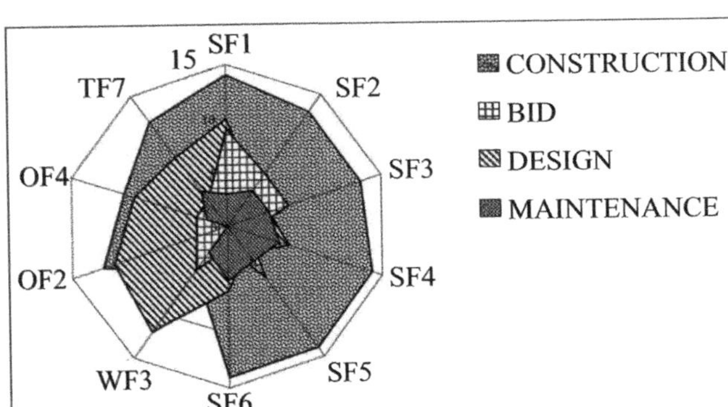

Analysis by project management related SWOT factors

Project management items for domestic firms

As shown in Figure 8.7, among the four project management items, cost is regarded as of the highest correlation, with five out of the top ten factors being cost-related, including "SD3. Estimate the construction cost precisely", "OD1. Be acquainted with construction market, easily to compete with other companies.", "OD5. Enter global construction market through strategic alliance", "OD7. Acquiring BOT practical experience" and "TD4. Lack of raw materials or fluctuation". Progress is correlated with three out of the top ten factors, "SD7. Easy integration interface of construction", "WD4. Inadequate construction equipment" and "TD7. Inconsistent government policy", while quality is correlated with two factors, "OD3. Cooperation with foreign firms, introduction to foreign techniques and management experience" and "TD6. Failure to upgrade techniques". Safety-related factors appeared to be regarded the least in terms of their correlation to project management. Among four SWOT groups, opportunity is most correlated with the cost item in project management, indicating that even though domestic firms regard cost highly, they do not possess the kind of strengths, and the way for them to acquire opportunities and achieve their goals is to undertake JV projects.

Figure 8.7: Main issues in Construction Project Management (domestic firm)

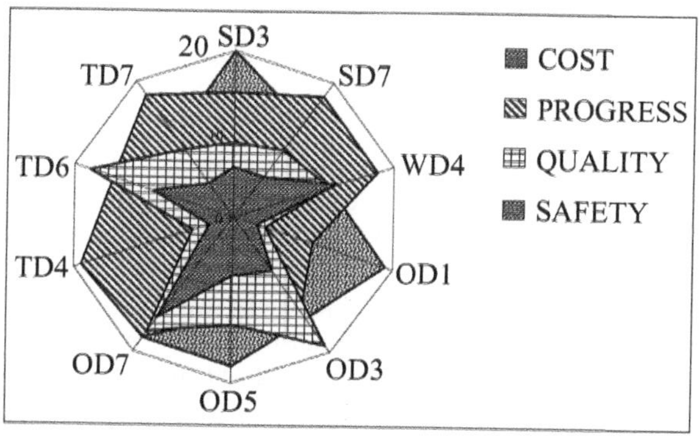

Project management item for foreign firms

As shown in Figure 8.8, among the four project management items, foreign firms regard cost as of the highest correlation, with seven out of the top ten factors items being cost-related, including "WF5. Not acquainted with the norms of the domestic constructors", "WF6. Not acquainted with the domestic geology and weather environment", "OF2. Undertake both local and major construction work", "OF5. Flexible utilization of funds", "TF1. Lack of raw materials or fluctuation", "TF3. The local legal system is not perfect" and "TF4. The local firm tenders too cheaply", while progress is correlated with two factors, "SF3. Strong management" and "SF4. Fully equipment for construction", and quality is correlated with one factor, "SF1. The contractor is experienced". It seems foreign firms tend to overlook safety among project management items. Among four SWOT groups, factors related to strength and opportunity groups are most correlated with items in project management.

Figure 8.8: Main issues in Construction Project Management (foreign firm)

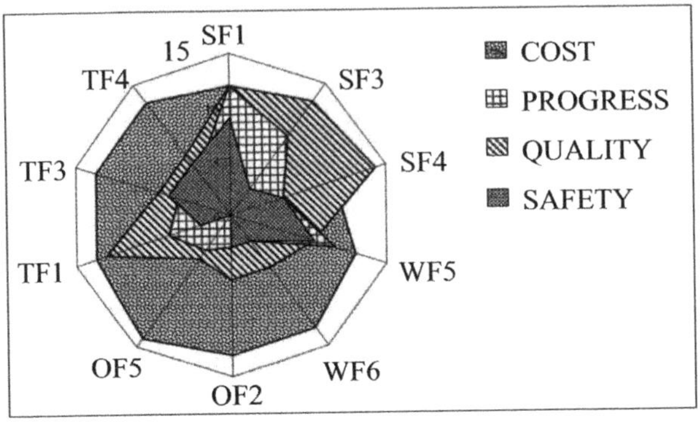

Conclusion

This study used the THSRC construction project as a case study example, in which firms from such countries as Japan, Korea, Germany, Italy, the Netherlands, and so on, formed joint ventures with Taiwanese firms to contract THSRC project. A SWOT analysis was done on the THSRC C250 bid JV of domestic and foreign firms. Experts' interviews and group brainstorming have yielded the four SWOT groups and factors of domestic and foreign firms - strengths, weaknesses, opportunities and threats, which then served as the basis for the survey questionnaire content; With the 12 lot bidding projects serving as the subject of this survey, AHP analysis was applied to quantify the weight ratios of each of the SWOT groups and factors, thus allowing the first objective of this study to be achieved. Besides, strength with related factors are highly regarded factors by both domestic and foreign firms in JV alliance partner evaluation, while strength factors of one party, more often than not, are found to be the weakness factors or threat factors of the other party in an alliance; therefore, domestic and foreign firms should first evaluate the strengths of the other party to see if these strengths can help improve the weakness and threats of their firm, or help increase opportunities for their firms to benefit, or if the strengths of their firms can provide commentary resources needed by the other party so as to enhance the feasibility of the alliance. Furthermore, the manifestation of the top ten factors show foreign firms have many distinct strengths to offer in their entering of Taiwan's construction market including such factors as possessing rich construction experience, adequate funds and strong management skills, and vice versa in the case of domestic firms, where strength factors of domestic firms are also the factors regarded highly for JV projects acquisition, thus allowing alliance with foreign firms to maximized benefits from both.

The pair-wise comparison of the four groups and 28 factors over project cyclical stage with project management items for both domestic and foreign firms can reveal how firms jointly achieve complementary resource contribution over different project stages or different project management items (See Figures 8.5 through 8.8), the second and third objectives of this study are thus achieved.

Foreign or domestic firms believe that in the course of project cycles, the strengths of a firm lie in construction and design stages, such as construction experience, adequate construction equipment or strong development and design abilities, but during the bid stage, firms are limited by not being acquainted with the legal system in Taiwan, and therefore in selecting a JV partner, domestic firms need to be evaluated by their strength in regard of this factor. Conversely, even though domestic firms evaluate more factors related to construction stage, it is apparent that they tend to use JV as a means to acquire project opportunities, experience or techniques, and for that reason, their being acquainted with the geology and weather environment or with domestic construction system are also items of complementary resources for foreign firms, and become their qualifications in their bid for JV partnership.

By means of above mentioned analysis and findings, this study is now able to provide alliance partner selection basis for foreign firms using JV to enter Taiwan's construction market, while it also serve as a basis for domestic firms in their bid for alliance partnership with foreign firms. After the section of the appropriate or right partner, both parties should consult with each other and discuss on how to effectively utilize each other's strengths in order to create more project opportunities, while reducing threats and risks. Such a process needs implementing early on to allow possible conflicts to be eliminated, so that such characteristics of JV as swiftness, complementarity and effect multiplication can be used for the purpose of integrating the limited resources of each other, cost sharing and risk reducing, economic scale construction, complementary techniques contribution, the breakthrough for investment limitations and trade barriers, and vertical integration.

References

[1] Killing, JP. (1983) Strategies for joint venture success. *Harvard Business Review.* **61**: 120-127.

[2] Harrigan, KR. (1988) *Strategic alliances and partner asymmetries.* Cooperative Strategies in International Business, Lexington Books.

[3] Tomlinson, JWC. (1970) *The joint venture process in international business*: The M.I.T Press, Cambridge, Mass.

[4] Medcof, JW. (1997) Why too many alliances end in divorce? *Long Range Planning.* **30 (5)**: 718-732.

[5] Luo, Y. (1997) Partner selection and venturing success: the case of joint venturing with firms in the People's Republic of China. *Organization Science* **8 (6)**: 648-662.

[6] Geringer, JM. (1998) Selection of partners for international joint venture. *Buisness Quarterly* **53**: 31-36.

[7] James, B. (1985) Alliances: the new strategic focus. *Long Range Planning* **19 (3)**: 76-81.

[8] Lewis, JD. (1990) *Partnerships for profit.* Free Pree, New York.

[9] Derek, HTW and Derick SJ. (2003) Construction industry joint venture behavior in Hong Kong- designed for collaborative results? *International Journal of Project Management* **21**: 39-49.

[10] Tam, CM. (1999) Build-operate-transfer model for infrastructure developments in Asia: reasons for successes and failures. *International Journal of Project Management* **17(6)**: 377-382.

[11] Tomlinson, JWC and Thompson M. (1977) *A study of Canadian joint ventures in Mexico.* Working Paper. University of British Columbia, Vancouver.

[12] Awadzi, WK. (1987) *Determinants of joint venture performance: a study of international joint ventures in the United States.* Unpublished Ph.D. dissertation. Louisiana State University.

Joint Ventures in Construction 2
ISBN 978-0-7277-5783-8

doi: 10.1680/jvc2.57838.113

ice
Institution of Civil Engineers
publishing

Chapter 9

Using Alternative Dispute Resolution for Effective Management of Disputes in Joint Venture Construction Projects in Malaysia

Ainul Jaria Maidin, International Islamic University Malaysia

"Persuade your neighbors to compromise, whenever you can. Point out to them how the nominal winner is often a real loser - in fees, expenses, and waste of time."

Abraham Lincoln, American Presiden [1]

Introduction

The construction industry is regarded as important not only because of its constantly evolving size but also because of its contribution to the national economy in terms of gross domestic product and large capital investment [2]. Joint venture projects are a common feature of the construction industry and are most useful in times of global economic downturn, when parties tend to embark on joint venture projects to lessen the risks involved in their undertakings in the construction industry. Most joint ventures arise from express written agreements, but could also be inferred from the nature of the parties' conduct of the venture. As the rights and duties of parties to a joint venture are typically governed by the same principles that govern 'partnerships', each party to the joint venture has the power and ability to bind the other party personally to unlimited liability to third parties, unless of course, the joint venture has been incorporated as a separate business entity.

Alternative dispute resolution has become a common recourse in the construction industry in the modern and global economy in the last decade. The focal point is not the type of dispute but the different methods for dispute resolution being used. Parties in the construction industry have always been worried about taking their disputes to a court of law because of the formalities to be complied with, expenses, time, complexity of issues involved, confrontation and delays. The characteristic of the adversarial system of justice applied in the courts, where parties introduce each piece of evidence (whether oral, documentary or real), argument or point against each other to prove their case on a balance of probabilities before a judge, who pronounces what is usually a win-lose decision. The adversarial process involves a method of adjudication in which active and unhindered parties, usually through their lawyers, contest with each other and present support in favour of their respective positions, usually through the examination and cross-examination of witnesses and the presentation of other evidence, to a neutral and independent decision-maker. The theory behind the underlying philosophy of the adversarial process is that the truth of a dispute will surface and thereby justice will be done to compensate the aggrieved person. The adversarial process is also used to buy time and more often than not results in a settlement unfavourable to the financially weaker party, as the process of determining winners and losers can be lengthy and costly [i].

In *Newacres Sdn Bhd v Sri Alam Sdn Bhd*[3] and *Hartela Contractors Ltd v Hartecon JV Sdn Bhd & Anor* [4], Gopal Sri Ram JCA found that in view of the mutual trust and confidence between parties to a joint venture essential for a proper working relationship, where there was reliance by one party upon the skill or expertise of the other, there was a duty upon the other to use his best endeavours to ensure the success of the venture. Equity would imply such an obligation in the absence of an express term in the joint venture agreement [ii]. Resorting to the use of equity by the courts is often very much within the discretionary powers of the judges and it is not a matter of the parties' rights in adversarial proceedings.

Alternative dispute resolution is not a new idea, having been around even before the adversarial system came about. Alternative dispute resolution advocates the principle that parties can resolve their disputes in a simple manner with the involvement of a skilled third party, usually an experienced professional in the particular industry. The benefits that ensue from the involvement of a skilled third party include improved communication, continued business relationships, effective management of disputes, better options for settlement, less formality, privacy in hearings, speedy resolution, reduced costs, and assurance of confidentiality. Skilled third party intervention ensures effective communication between parties, with a focus on solving problems and reaching amicable settlements.

The pressure within the millennium construction industry, where construction is focused on fast track building and other construction projects, is bound to create tension and disagreement over collateral and sub-contract terms and warranties. When construction contracts are not signed until after workmen have entered a site or, worse, when no contract is signed between parties to a joint venture, dispute resolution by the court is rather difficult. However, under various types of alternative dispute resolution processes, the third party, using his technical knowledge, skill, and experience in the construction industry, may be able to bring the parties together to achieve a settlement of the dispute, regardless of the written contract.

Joint Ventures in the Construction Industry

Joint ventures are established to take advantage of the economic, political and social conditions prevailing in a particular economy, especially at the global level. In Malaysia's case, international firms were attracted mainly by opportunities stemming from the country's planned economic growth, investment policies, political stability, relatively cheaper labour costs, the availability of raw materials, a good transportation system, and various incentives offered by the government. Construction joint ventures in Malaysia are popular and have involved multinational construction firms and local governments seeking to achieve their individual objectives. The Malaysian government has supported and encouraged local contractors to participate in regional and global markets based on the various parties' expertise and experience in the construction of buildings, infrastructure projects, highways, power generation facilities, ports and airports. There are already established joint ventures between two or more local indigenous contractors and also between indigenous local and foreign contractors [5].

In the Malaysian construction industry, construction organizations have extensively resorted to using international joint ventures as a mechanism to participate in construction markets around the world [6]. The rapid expansion of construction activities following industrialisation programmes has intensified the participation of foreign contractors. The Malaysian Government's encouragement of and incentives for such arrangements could also be considered as a significant factor in the development of joint venture projects. Foreign firms are often required to bid on large infrastructure projects with local partners, and a joint venture bid must have at least 30% bumiputera (indigenous

Malay) participation. Malaysia's open-door policy to foreign participation is evidenced by the large amount of payments made by the Malaysian Government for contracts and professional services. Major projects such as Kuala Lumpur International Airport, the Petronas Twin Towers, Bakun and Pergau Dams, the Multimedia Super Corridor, the Sepang Formula One Circuit, and various other mega projects were successfully realized with joint ventures.

The number of international construction joint ventures is growing worldwide at an unprecedented pace, especially in developing countries, and Malaysia is keen to benefit from these opportunities. The growth in joint venture dealing is due to the increasing magnitude of major construction projects, the associated complexities and risks of which make it necessary for organizations with diverse strengths and weaknesses to bid collectively and execute projects [7].

Thus, a joint venture is a procedure used to realize specific business phenomena such as to gain access to new markets, to promote specific government policy, to increase business capacity, to transfer technology, or to benefit from economies of scale. An international joint venture is a separate legal organizational entity representing the partial holdings of two or more parent firms, in which the headquarters of at least one is located outside the country of operation [5].

Parties seeking to enter into joint ventures must be careful to ensure that they seal contracts which will be beneficial to both. The key provisions in any joint venture contracts *inter alia* include the following [8]:

Clearly defined business objectives

The agreement must prescribe the purpose of the joint venture, the nature of the business, liabilities, and the interests of the parties to the joint venture.

Degree of participation and the management roles of each joint venture participant

The agreement must prescribe the roles, management responsibilities, duties and responsibilities of parties and the degree of participation of each party to the joint venture. The provision will be enforceable in contract, and the terms of the contract must be clearly drafted to define the roles accurately.

Contribution of capital and ownership rights/Division of the profits and losses

The agreement much set out the contributions and other resources each party has, the method and percentage of profit and risk, and loss sharing in the joint venture.

Dispute resolution mechanism

A dispute resolution mechanism for resolving any disputes that may arise between the joint venture participants should be incorporated. This mechanism is necessary for avoiding litigation, resolving any type of deadlock, and identifying the party responsible for resolving disputes.

Termination of the joint venture

Joint ventures are often entered into for one or more projects and may not last forever. The parties must specify when the joint venture will terminate, how it will be dissolved, how its assets will be handled, and other related matters.

Confidentiality/Intellectual property

The parties must set out clearly the need to protect confidential information and any intellectual property contributed during the formation of the joint venture or developed during the joint venture contract duration.

Indemnification

A joint venture agreement must contain an indemnification provision indemnifying joint venture participants against liability.

Despite all precautions taken by the parties, however, disputes are sometimes unavoidable and parties may have to refer the dispute for resolution. Construction is one of the most dynamic, challenging, and of course very rewarding industries [9], exposed to various predictable and unpredictable risks, among which are political, economic, technological and social risks. Risk is inherent to every construction project and is normally assumed by the parties in a joint venture unless it is transferred to or assumed by another party for fair compensation.

It is not within the purview of this article to delve in length into the nature of joint venture agreements. Instead, the focus will be on the problems faced in resolving disputes using adversarial methods and the possibility of using alternative dispute resolution methods for the amicable settlement of disputes arising from joint venture construction contracts.

Conventional Methods for Resolving Disputes in Construction Joint Ventures

The construction industry has become known as one of the most adversarial and problem-prone, with claims and disputes on construction projects frequently the rule rather than the exception [10].

In *McAlpine Humberoak Ltd v McDermott International Inc (No.1)*[11] Lloyd LJ has the following to say:

"It seems to be the practice in the construction industry to employ consultants to prepare a claim almost as soon as the ink on the contract is dry."

Lloyd LJ is correct in his observation, as disputes within the construction industry are unavoidable even when there is a well-drafted contract executed by both parties and sealed in accordance with the governing law. Courts are brought in to interpret construction contracts when the need arises. The court in determining the rights of the parties interprets the construction contract in accordance with the law which postulates what the words in the contract mean, what they say when read objectively and not in the way advocated by either or the parties, who are generally concerned only with their own subjective intentions arrived at in retrospect.

In *Modern Engineering (Bristol) Ltd v. Gilbert-Ash Northern* [12], Lord Diplock described a building contract as:

"... an entire contract for the sale of goods and work and labour for a lump sum price payable by installments as the goods are delivered and the work done. Decisions have to be made from time to time about such essential matters as the making of variation orders, the expenditure of provisional and prime cost sums and extension of time for the carrying out of the work under the contract."

The carrying out and completion of this contract, whether made using a standard form contract or others, differs from other manufacturing processes. HHJ Newey OR in *Emson Eastern* v. *EME Developments* [13] described the differences, within the context of practical completion of the work:

"I think the most important background fact which I should keep in mind is that building construction is not like the manufacture of goods in a factory. The size of the project, site conditions, the use of many materials and the employment of various kinds of operatives make it virtually impossible to achieve the same degree of perfection that a manufacturer can. It must be a rare new building in which every screw and every brush of paint is absolutely correct."

There is no special body of rules that applies to such contracts, whether they are described as building, engineering or construction contracts. Lord Reid said in Modern Engineering that where the parties enter into detailed building contracts there were 'no overriding rules or principles covering their contractual relationships beyond those which generally apply.' This principle was supported by Lord Lloyd of Berwick in *Beaufort Developments (NI) Ltd v. Gilbert-Ash (NI) Ltd* [14], where he stated that:

"Standard forms of building contracts have often been criticised by the courts for being unnecessarily obscure and verbose. But in fairness one should add that it is sometimes the courts themselves who have added to the difficulty by treating building contracts as if they were subject to special rules of their own."

Another reason for the complexity of the disputes within the construction industry is the heavy reliance on the use of standard form contracts, which are drafted and approved by professional bodies such as Lawyers, Architects, Engineers, Contractors or other professional bodies or institutions at a national or international level. In drafting agreements, the advantages of using standard form contracts are their convenience and familiarity, and the fact that they cut costs and save time.

A construction contract is best described as a complex web of competing interests. A particular problem in construction contracts is that there is little interest in building long-term relationships. The use of the standard form contracts has its advantages and disadvantages [15].

Some advantages of the standard form contracts are the following:
(1) The standard form is usually negotiated between different bodies that make up the industry. As a result, risks are spread equitably.
(2) Using a standard form avoids the costs and time associated with individually negotiated contracts.
(3) Tender comparisons are made easier since the risk allocation is same for each tenderer. Parties are assumed to understand that risk allocation and their prices can be accurately compared.

Some disadvantages of standard form contracts are the following:
(1) The forms are cumbersome, complex and often difficult to understand.
(2) Because the resulting contract is often a compromise, they are resistant to change. Much-needed changes take a long time to bring into effect.
(3) The terms of conditions in the contract are not easy to understand and interpretation between parties can give rise to disputes.

As problems arise, either or both parties will have recourse to lawyers or claims consultants for advice on how to protect themselves or advance their causes, sometimes even unmeritorious ones, before the dispute reaches full-blown proportions. Parties' energies and resources are thus diverted from what should be the mutual objective of the enterprise, that is to construct a project efficiently and at a fair and reasonable cost to the mutual benefit of both parties. Disputes are unavoidable and parties often tend to create an unpleasant situation for each other. Litigation and arbitration processes can be costly, time-consuming and at times unpredictable [16]. Cost overruns and schedule delays can be the subject of expensive and protracted claims and litigation, and pose serious risks for all parties to a construction project.

As a result, a clear majority of owners of construction projects rely heavily on alternative dispute resolution to prevent or resolve construction claims and disputes. Frequently, multiple alternative dispute resolution techniques are designated by contract, with binding arbitration the forum of last resort if other techniques are unsuccessful [10]. However, the adverse impact of old-fashioned litigation and arbitration processes adopted in the early days has led to the introduction of alternative dispute resolutions, as an alternate choice for resolving disputes [17].

Development of Alternative Dispute Resolution

Alternative dispute resolution or appropriate dispute resolution, as it is sometimes called, has achieved prominence in many legal systems because of the realisation that litigation does not always lead to a fair result, besides being costly and time consuming. Parties to litigation proceedings have to prove their case before a judge who will be burdened with the task of handing down a win or lose judgment [18]. Alternative dispute resolution is a term that refers to various procedures developed in the United States over the last 15 years or more in an attempt to overcome some of the weaknesses in the litigation and arbitration processes. The different alternative dispute resolution methods developed have one thing in common, that is the aim of blunting the adversarial attitude and encouraging more openness and better communication between the parties to a dispute [18], thus leading to early settlement in appropriate cases, saving time, including the court's time, as well as providing an atmosphere conducive to the parties' working together even after the settlement of a dispute [18].

Alternative dispute resolution enables disputing parties to choose flexible methods, such as mediation, to reach a settlement, rather than proceeding to formal adversarial litigation in a court of law. The concept has generated interest as a method of resolving conflicts and has been used in other areas of law, especially commercial and labour. Alternative dispute resolution in actual fact is a process that is not an alternative to court proceedings but only to the litigation process when circumstances render it necessary. Otherwise it may be complementary or sequential. In the event that the alternative dispute resolution technique is a failure, litigation may follow [19].

The five common types of alternative dispute resolution techniques are arbitration, mediation, early neutral evaluation, mini-trial, and judicial reference. Alternative dispute resolution also includes any other technique that will enable disputing parties to reach a compromise [20]. Parties in a particular situation are free to create and alter standard techniques to enable them to reach a fair settlement that best suits their individual needs [20].

Characteristics of Alternative Dispute Resolution

The typical characteristics of alternative dispute resolution methods are as follows:

Flexibility and informality

Alternative dispute resolution processes are less formal than judicial processes and lack the extensive written documentation of formal proceedings. They increase access and facilitate solutions for citizens who are intimidated by or unable to participate in more formal judicial systems.

Equity

Alternative dispute resolution is more concerned with equality than the rule of law. Resolutions are based on principles and terms that are determined to be equitable for each particular case rather than on uniformly applied legal standards. This becomes particularly important in societies in which citizens do not receive fair justice under the formal legal system.

Direct participation and communication between disputants

With alternative dispute resolution, disputants design their own settlements and therefore are required to engage in direct dialogue with one another in the interest of reconciliation. This participation is essential to lasting and solid resolutions.

Alternative dispute resolution is an important option in countries where the following conditions exist:

- Case backlogs or complex procedures impair court effectiveness;
- Illiterate and/or poor people cannot afford to navigate conventional legal channels; or
- Small informal systems can better reach a geographically dispersed population.

Alternative dispute resolution models can emphasize the importance of reconciliation and relationships over "winning" in dispute resolution. Mutually agreed-upon outcomes allow for creativity and increased satisfaction with judicial processes (thereby increasing court responsiveness to local conditions and compliance with court judgments). Alternative dispute resolution techniques, such as mediation, allow multiple parties to voice their concerns to reach an amicable, enduring decision efficiently and with compromise. In addition to fostering more direct participation by the parties, alternative dispute resolution encourages timely resolution of controversial issues and will likely yield more satisfying results to all parties involved [21]. The strongest motivation to using alternatives to litigation, though, is that such alternatives save financial resources.

Non-Suitability of Arbitration as Dispute Resolution Technique for Resolving Disputes in Joint Ventures

Arbitration is a popular alternative dispute resolution method that has been used in the construction industry in Malaysia for quite a long time. However, some studies have identified the disadvantages of arbitration, such as the inclination to use it as a delaying tactic, its cost, its adversarial (win-lose) nature, and its tendency to damage the relationships of the parties concerned [22]. Arbitration is a common and familiar mechanism among stakeholders in the construction industry, owing to the incorporation of an arbitration clause in the standard forms of Malaysian construction contracts. For instance, most of the standard forms of contracts currently include a series of formal steps to be taken to resolve any disputes through arbitration [22] - [25]. Use of arbitration in resolving disputes arising between parties in a construction joint venture project may not be the best choice.

Mediation is also being used in Malaysia in resolving disputes [23], [25]. The subsequent discussion will examine the use of mediation as a better choice of alternative dispute resolution method.

Using Mediation Techniques for Resolving Disputes in Joint Venture Construction Projects

Mediation, another method of alternative dispute resolution, is a process in which an impartial third party facilitates communication and negotiation and promotes voluntary decision making by the parties to the dispute [iii]. Mediation can be briefly defined as a process to resolve disputes where people get together with the assistance of a mediator (an impartial third party) to isolate issues, develop options, consider alternatives and reach an agreement everyone can live with, rather than having a settlement imposed on them by a formal body such as a court [iii]. The essence of mediation is the common-sense idea that the intervention, by invitation from the parties from amongst experienced, independent and trusted persons, can be expected to help the parties settle their quarrel by negotiating in a collaborative rather than adversarial way [26], [27].

In the past, civilizations such as Chinese [28], Hindu [29] or Islam always preferred compromise to confrontation. However, unlike other traditional legal systems that recognize amicable settlement of disputes in a generalized manner, Islamic law contains detailed rules for regulating disputes, with the Quran and *hadith* giving such rules a unique legitimacy and divine approval. The degree of detail found in Islamic rules dealing with the subject of dispute resolution is not found in other traditional legal systems. Islamic law presents alternative dispute resolution as the basic tenet of civil justice [30].

The principles of mediation are closely related to those of *Sulh*, which literally means "to end a dispute" or "to cut off a dispute," either directly or with the help of a neutral third party. The basis of the *sulh* can be found in the following Quranic injunctions:

"In most of their secret talks there is no good: But if one exhorts to a deed of charity or justice or conciliation between men, (Secrecy is permissible): To him who does this, seeking the good pleasure of Allah, We shall soon give a reward of the highest (value)" [31].

"If two parties among the Believers fall into a quarrel, make ye peace between them: but if one of them transgresses beyond bounds against the other, then fight ye (all) against the one that transgresses until it complies with the command of Allah; but if it complies, then make peace between them with justice, and be fair: for Allah loves those who are fair (and just)" [32].

Sulh recognizes the values that all parties bring to the table, and enjoins those parties to treat each other and others' ideas with respect.

There are many advantages to mediation over other forms of alternative dispute resolution (ADR) or civil litigation. Below are some of the major advantages that mediation and binding mediation offer to the construction industry as an alternative dispute resolution option. However, parties must be reminded and cautioned that a judge's responsibility is to interpret and make a decision according to points of law. With the alternative method, a mediators' responsibility is to assist the parties in settling their dispute and be fair and equitable to all parties in the case of a binding mediation. The following are amongst the advantages of mediation [iv]:

Voluntary involvement

The agreement to mediate can come from the parties, or be offered by the Judge. Any party may end the process at any time. The party should not feel pressured to be or stay involved. Preparation of the parties for the alternative dispute resolution process is critically important. Psychologist mediators have said that preparation not only assists with the logistics of the process but can also help the parties become psychologically prepared as well [33].

Active participation in good faith

Effective mediation is conducted face-to-face. Active participation with communication by all parties is essential if an effective agreement is to be reached. Mediation requires good intentions on the part of all parties, though sometimes the emphasis is better placed on good business and empowerment than goodwill, as parties may not believe the other side can be 'reasonable'.

Self-motivated

Mediation assumes that the parties are competent and informed and able to reach agreements that suit their needs. Each party must have at least one representative who is consistent in that role throughout the mediation process. Control of the dispute and the terms of settlement remain in their hands. The representative of each party must be fully authorized to participate. A resolution will occur only if the parties agree.

Inclusive

All parties must be present and all parties must agree to the process. The involvement of people with the authority to make settlements on behalf of others is essential.

Independent and impartial mediator

The mediator acts as a facilitator, communicator, motivator, and scene setter, creating the right environment for the process to be effective. He or she must be independent of both the parties and of the Judge who will review any agreement reached. The mediator must not give legal advice, offer opinions, or coerce parties into agreement. The mediator does check that all parties fully understand what they are agreeing to.

Mutual respect

The process allows the parties (including the mediator) to develop a degree of trust and confidence in themselves, in each other, and in the process.

Flexible outcomes

Parties may openly discuss matters unrelated to the appeal. The party may want to talk about business or personal relationships with the other parties. This openness can increase the chances of a satisfactory resolution. The affected party may agree on outcomes that the Court itself is unable to secure.

Confidential process

All discussions that take place in mediation must be completely confidential. Normally, in out of court mediations no formal record is kept, except an agreed decision (called a consent order) that, if approved by the Court, forms part of the public record. The mediator may meet separately with any party or parties and may be offered information which is to be kept confidential from other parties. The mediation process is conducted without prejudice to the dispute and shall not be referred to or relied upon in any other proceedings in the Court.

Protecting parties' rights

All verbal offers and discussions that take place during the course of mediation do not affect the rights of parties. The issues remain unresolved and must be brought to Court.

Finality of agreements

Where agreements are reached, they are final and treated as binding on the parties who have agreed. However, participation in mediation does not prejudice the existing legal rights of the parties.

Fairness and equity

All parties must be given a fair hearing and have equal access to information. The 'equity' suggested in this context carries a narrower meaning than amiable composition, which confers on the mediator and/or arbitrator broader powers, allowing him to apply his own sense of fair play, justice and good conscience [34].

Opportunities for Promoting the Use of Alternative Dispute Resolution in the Construction Industry in place of Adversarial System in Malaysia

The literature demonstrates that alternative dispute resolution's proper role can be as a supplemental and experimental tool to help reduce the time and costs involved in traditional litigation [35]. Alternative dispute resolution may also promote resolution whereby the parties to the dispute may continue to liaise with each other to resolve disputes amicably.

Alternative dispute resolution is not a new concept in Malaysia; arbitration of construction disputes and various other contracts have existed for some time [36]. The *Sultan* was the reference point and the pillar of justice, but did not play an active role in the administration of the legal process. He had ministers who controlled the government and dispensed justice in serious matters. For example, quarrels amongst relatives were settled with the help of village elders. The village headman called *Penghulu* handles disputes involving people from different families or villagers in the same area. The most common method of settlement is for the *Penghulu* to invite the parties to meet in his house. The *Penghulu* then serves drinks and some light refreshments to encourage the parties to eat together, which is a symbol of forgiveness and conciliation.[37] The respect for the Penghulu is an important factor influencing the parties who approach him to resolve their disputes.

As history shows, mediation has been a preferred option in dispute settlement in Malaysia [37]. Mediation has been used at the lowest possible level in attempts to resolve quarrels[v]. Therefore, it is submitted that it will not be difficult to promote alternative dispute resolution for settling environmental disputes since traditionally mediation has been the first option for Malaysians, who are not averse to litigation if matters remain unresolved [38]. Informal approaches to resolve conflicts are made through intermediaries who could be relatives, friends, business associates or professional arbitrators. The Arbitration Act 2005 (Act 646) governs the laws and procedures relating to arbitration and mediation in Malaysia. The Kuala Lumpur Regional Centre for Arbitration was established in 1978 to encourage international mediation and arbitration mostly in the area of commercial disputes. The Malaysian Institute of Arbitrators is a professional body established to disseminate knowledge and to educate the public on private dispute resolution processes and attitudes, with the aim of working towards a harmonious society [39].

Resolving construction disputes is a difficult task, especially when the available resources are limited and the dispute is complex. The use of alternative dispute resolution in resolving disputes between parties in joint venture construction projects is an attempt to overcome the shortcomings of litigation and arbitration. However, complex procedures that involve sequential use of a range of arbitration techniques and arbitration can destroy the originally designed positive effects, especially in terms of time and cost. Alternative dispute resolution methods can be useful in resolving construction disputes in Malaysia if they are implemented with the relevant procedural guidelines.

Mediation works for most disputes but not, of course, for all. Mediation fares well in cases of urgency, and in the maintenance of relationships, harmony, and confidentiality. It brings the parties together physically and requires them to negotiate with each other. If they are then able to agree on the terms of a settlement, they will be in a much more harmonious position than disputants who have gone through a litigation or arbitration process and been subject to a judgment or award. The mediation movement can be useful for early resolution of disputes. It is submitted that particularly in Malaysia there is indeed a grave need to espouse mediation in the court system and for the movement to provide rules for the court, goals which have been accomplished in many other developed nations.

Conclusion

Resolving construction disputes amicably without causing harm to joint venture parties is a difficult task, especially when the available resources are limited and the dispute is complex. The use of alternative dispute resolution processes in resolving disputes in joint venture construction projects is an attempt to overcome the shortcomings of litigation and arbitration. However, complex resolution procedures that involve sequential use of a range of arbitration techniques and arbitration can destroy the originally designed positive effects, especially in terms of time and cost. Ranking alternative dispute resolution features is a subjective exercise. Mediation appears to be one of the best alternatives to arbitration in resolving construction disputes. This can be confirmed further by undertaking detailed research in the construction industry to determine the relevance of mediation in resolving disputes.

References

[1] The Wisdom of Abraham Lincoln: 42. Internet edition accessed on 3rd May 2011. http://Ia600505.Us.Archive.Org/5/Items/Wisdomofabrahaml02linc/ Wisdomofabrahaml02linc.pdf.

[2] Hillebrandt, P. M. (1974) *Economic Theory and the Construction Industry*. The Macmillan Press Ltd.

[3] *Newacres Sdn Bhd v Sri Alam Sdn Bhd*[1991] 3 MLJ 474.

[4] *Hartela Contractors Ltd v Hartecon JV Sdn Bhd & Anor* [1999] 2 CLJ 788

[5] Hamimah Adnan (2008) An Assessment of Risk Management in Joint Venture Projects (JV) in Malaysia. *Asian Social Science* **4(6)**. www.ccsenet.org/journal.html

[6] Mohammed, S (2000) Risk Assessment in bidding for international projects – the Australian experience. *Asia Pacific Building and Construction Management Journal* **5**: 135-152.

[7] Kumaraswamy, M, Palaneeswaran, E and Humphreys, P (2000) Selection matters- in construction supply chain optimization. *International Journal of Physical Distribution & Logistic Management* **30(7-8)**: 661-680.

[8] Mark Warner, *Joint Venture Agreements - Key Drafting Issues*. http://EzineArticles.com/?expert=Mark_Warner.

[9] Mills A. (2001). Retrieved 15 August 2009 from http://www.emerald-library.com/ft.
[10] Richard H. Steen, Esq. *Alternative Dispute Resolution in the Construction Industry.* (Accessed from Internet on 20th August 2009) at http://www.njsba.com/activities/DRP-Steen-Constrresolutioninconstructionwithhinfomaster.pdf.
[11] *McAlpine Humberoak Ltd v McDermott International Inc* (No.1)58 BLR 1, p. 24.
[12] *Modern Engineering (Bristol) Ltd v. Gilbert-Ash Northern* [1974] AC 689, at p.717B.
[13] *Emson Eastern v. EME Developments* (1991) 55 BLR 114, p.125.
[14] *Beaufort Developments (NI) Ltd v. Gilbert-Ash (NI) Ltd* [1998] UKHL [1988] 1 AC 191.
[15] John Adriaanse (2007) *Construction Contract Law The Essentials*. 2nd Edition Palgrave Macmillan.
[16] *Intelek Timur Sdn. Bhd. v Future Heritage Sdn. Bhd.*[2004] 1 MLJ 401.
[17] Grossman, A. (2002). *Construction Disputes after Latham and Egan* [Electronic Version]. Centre for Effective Dispute Resolution. Retrieved 7 August 2007, from http://www.cedr.co.uk/ index.php?location=/library/articles/construction02.Htm& display.
[18] Bevan, A (1992). *Alternative Dispute Resolution.* Sweet & Maxwell, London.
[19] Rowan Robinson, J (1997) Alternative Dispute Resolution in Environmental Protection. Unpublished paper presented at *Conference on New Directions in Environmental Dispute Resolution*, on the 31st October 1997 in Aberystwyth Wales, p.1.
[20] Sinderbrand, C.A (1991) Alternative Dispute resolution in the Environmental Arena *WIS LAW*. Dec 1991, pp. 26-27.
[21] Walton Blackburn & Willa Marie Bruce (1995) *Mediating Environmental Conflicts: Theory and Practice 21.*
[22] Mohd Suhaimi Mohd Danuri, et.al (2008) Viability of Dispute Avoidance Procedure in the Malaysian Construction Industry. Paper presented at the *RISC Construction and Building Research Conference, COBRA 2008*. Dublin Institute of Technology, 4-5th September 2008.
[23] CIDB (2000) *CIDB Standard Form of Contract for Building Works*. Construction Industry Development Board Malaysia.
[24] IEM (1989) *Conditions of Contract for Civil Engineering Works.* The Institution of Engineers, Malaysia.
[25] PAM (1998) *Agreement and Conditions of Building Contract* (Private Edition Without Quantitites). Pertubuhan Akitek Malaysia.
[26] Michael Noone (1996) *Mediation – Essential Legal Skills Series* Cavendish Publishing Limited, London. p. 5.
[27] D.K. Sampath (1996). *Mediation Concept and Technique in Support of Resolution of Disputes*. National Law School of India University, India.
[28] Bobby K. Wong (2000) Traditional Chinese Philosophy and Dispute Resolution. *Hong Kong Law Journal* **3(1)**: 304-319.
[29] Justice RC Lahoti, Law of Arbitration 1996. *ICA Arbitration Quarterly*. **37(1)**: 3-4.
[30] Syed Khalid Rashid (2004) "*Alternative Dispute Resolution in the Context of Islamic Law.* **7 VJ**: 95-118.
[31] *Al Qur'an* Surah *An-Nisa*: Aya 114.
[32] *Al Qur'an Surah Al-Hujuraat:* Aya 9.
[33] Maureen Garwood (1999) Managing Quality of ADR for Commercial Disputes. *Australasian Dispute Resolution Journal* **3**:182.
[34] Rashid, S. K. (2004). Alternative Dispute Resolution in the Context of Islamic Law. *Vindobona Journal of International Commercial Law and Arbitration* **7**: 95-118.
[35] McDonald *J. (1994) The Application of Alternative Dispute Resolution Techniques to Environmental and Planning Disputes.* ENV. LIABILITY: 134-146.

[36] James A. Wall Jr. and Ronda Roberts Callister (1999) Malaysian Community Mediation. *The Journal of Conflict Resolution* **43(3)**: 343-365.

[37] Syed Hussin Ali (1975) *Malay Peasant Society and Leadership.* Oxford University Press, Singapore. p.129.

[38] Poh, L.T., Malaysia. In Poh, L.T. (ed.) 1997, *Asian Legal Systems - Law, Society and Pluralism in East Asia.* Sydney, Butterworths.

[39] Malaysian Institute of Arbitrators accessed from internet on 20th August 2009 at http://www.miarb.com.

Endnotes

(i) Refer, The Australian Law Reform Commission, Review Of The Adversarial System Of Litigation, Adr - Its Role In Federal Dispute Resolution, Issues Paper 25 - June 1998; Australian Law Reform Commission Issues Paper 20 Review of the adversarial system of litigation: rethinking the federal civil litigation system ALRC Sydney 1997 (ALRC IP 20), ch 1 – 2.Adversarial system, definition from the Free dictionary http://legal-dictionary.thefreedictionary.com/Adversary+System.

(ii) Equity is a system of law designed to furnish remedies for wrongs which were not legally recognized under the common law of England or for which no adequate remedy was provided by the common law.

(iii) The preamble of the 2005 version to the Model Standard of Conduct for Mediators that was approved by the American Bar Association's House of Delegates on August 9, 2005, the Board of the Association of Conflict Resolution on August 22, 2005, and the Executive Committee of the American Arbitration Association on September 8, 2005.

(iv) Derived from various sources, generally refer, American Bar Association, accessed from Internet on 25th August 2009 at http://www.abanet.org/publiced/courts/mediation_advantages.html and Construction Dispute Resolution Services, accessed from internet on 25th August 2009 at http://www.constructiondisputes-cdrs.com/advantages_of_mediation.htm.

(v) At the Penghulu's level, the *Penghulu's* efficiency in resolving conflicts has been formally recognised and a Penghulu court was established in 1948 when the organisation of the courts was formalised.

Joint Ventures in Construction 2
ISBN 978-0-7277-5783-8

doi: 10.1680/jvc2.57838.127

Chapter 10

The Mechanism of Dispute Arising in Construction Contracts

Masamitsu Onishi, Kyoto University
Toshihiko Omoto, Kyoto University
Kiyoshi Kobayashi, Kyoto University

Introduction

A construction contract can be characterized by a lot of unforeseen and indescribable contingencies; rendering it incomplete [1]. When an unforeseen contingency arises, the Contractor notifies the Engineers of necessity to modify the conditions of initial contract. Contractual disputes between the Contractor and the Employer often arise eventually resulting in bring extensional costs and time. As far as a construction contract is incomplete, the Contractor's serving notices to the Engineer is justified. However, Contractors are generally more accessible to the information about a contingency than the Engineer (Employer) and the outside observers. Due to this information asymmetry, Contractors may be stimulated to dispute frivolous claims of less merit, expecting erroneous judgments by the Engineers or adjudicators.

In 1999, the Conditions of Contact for Construction (1st edition) was published by Fédération Internationale des Ingénieurs-Conseils ("FIDIC") as the revision of the 1987 edition of FIDIC Conditions of Contract for Works of Civil Engineering Construction. One of the significant developments in the new conditions is the appointment of a dispute adjudication board ("DAB") for the purpose of dealing with disputes more efficiently. In the United States, a dispute review board ("DRB") is appointed in many construction projects. Onishi *et al.* [2] have shown that the efficiency of dispute resolution can be increased by the appointment of an impartial DAB/DRB, in the sense that the possibility of going to arbitration decreases. However, Onishi *et al.* [2] do not focus on the effect the DAB/DRB has on the possibility of dispute arising. It must be noted that 'dispute' in this article is defined as a controversial situation that the claim is referred to DAB/DRB. The term 'dispute' is variously defined and there is no definitive meaning of a dispute [3]. Therefore, our definition in a construction-specific context is not necessarily consistent with other definitions such as "any contract question or controversy that must be settled beyond the jobsite management" [4]. The DAB/DRB may give rise to a large number of disputes, not only because the cost of DAB/DRB is shared between the Employer and the Contractor, but also the marginal cost per claim is low. If the Contractor is excessively stimulated to refer claims to the DAB/DRB, the DAB/DRB will have a heavy workload dealing with too many claims. Therefore, the cost to be paid for the DAB/DRB can become very high and also the number of erroneous judgments of the DAB/DRB may increase. Consequently, construction cost may increase in the long run. In this sense, it is a very important issue to develop systems to control frivolous claims. This chapter, by developing a model using game theory in order to analyze the mechanism of dispute arising, focuses on the Contractor's decision-making process in referring a claim.

On the other hand, in Japan, Contractors have been entitled to make claims as a result of the revision of the Japanese standard conditions of construction contract for governmental works ("GCW") in 1995. So far, the contractual dispute issue does not seem to be significant in Japan. However, with the globalization of Japanese construction market, contractual disputes can become a significant issue. Therefore, we also analyze the mechanism of dispute arising under GCW contract. The models provide the comparative possibilities of dispute arising under the alternative dispute resolution procedures. Moreover, various procedural methods for reducing the incidence of frivolous claim, including cost-shifting rule for arbitration and DAB/DRB are examined. We conclude the chapter by indicating the several issues to be studied in the future.

Literature Review on Dispute Arising

A number of litigations are daily raised in the United States. Such an environment promotes studies on legal systems to control frivolous lawsuits effectively [5]-[7]. Literature on dispute arising in construction projects has already been accumulated [8], [9]. Love *et al.* [10] investigates the underlying pathogens that client and contractors perceive to contribute to dispute in construction projects. However, the meaning of dispute in the previous literature is loosely defined any type of contract question or controversy. Our interest in this article is the cost associated with a dispute resolution process when a dispute is referred to the third party such as DAB/DRB. In fact, the cost of dispute resolution process in a construction industry is not negligible [11]. Therefore, referring frivolous claims to the third party can be problematic in terms of cost efficiency of construction projects. The literature on frivolous lawsuits generally defines them to be those cases that would not succeed if they went to trial. According to a broader definition, it is possible that the case has some merit, but the plaintiff's cost of pursuing the case to trial exceeds its expected value [12]-[14]. In this chapter, this broader definition of the word "frivolous" is adopted. A number of frivolous lawsuits lead to an explosion of litigation costs. 'Law and economics' studies on frivolous lawsuits have answered for the following questions such as 'why the rational plaintiffs ever file frivolous suits' or 'when they do, why defendants ever agree to settle them'. Various models concerning dispute arising have been proposed, such as the "differing perception model" in which the plaintiff and the defendant have different perceptions for judgment at trial [15]-[17], "the legal error model" in which the plaintiff expects erroneous judgment at trial [18], [19], "the timing of litigation cost model" in which frivolous suits succeed as a result of the timing of litigation cost [20], and "the asymmetric information model" in which the plaintiff has private information about the merits of the case [18]. The effects of contingent legal fees on frivolous suits have also been analyzed [21]-[24].

Onishi *et al.* [2] proposed a model for the dispute resolution process under the FIDIC contract. They focused on the efficiency of dispute resolution procedures in terms of the possibility to go to arbitration, and pointed out that a DAB/DRB can reduce the possibility of going to arbitration. However, they have not discussed on the efficiency of dispute resolution procedure in terms of dispute arising. Even though a DAB/DRB can deal with disputes which have already arisen efficiently, they may bring about excessive disputes, that means claims with even less merit may be referred to the DAB/DRB. In this chapter, the Contractor's decision-making on referring a claim to the third party, that is defined dispute arising behaviour, is modelled by extending the dispute resolution model proposed by Onishi *et al.* [2]. Although the structure of the construction dispute resolution procedure is similar to that of civil lawsuits, the former includes the third parties' judgment before going to arbitration. This is the new aspect of our study compared to the existing studies on frivolous suits.

The Dispute Resolution Procedure of FIDIC

Figure 10.1: Dispute resolution procedure of the 1987 edition of FIDIC contract
Note C: Contractor, E: The Engineer (Employer)

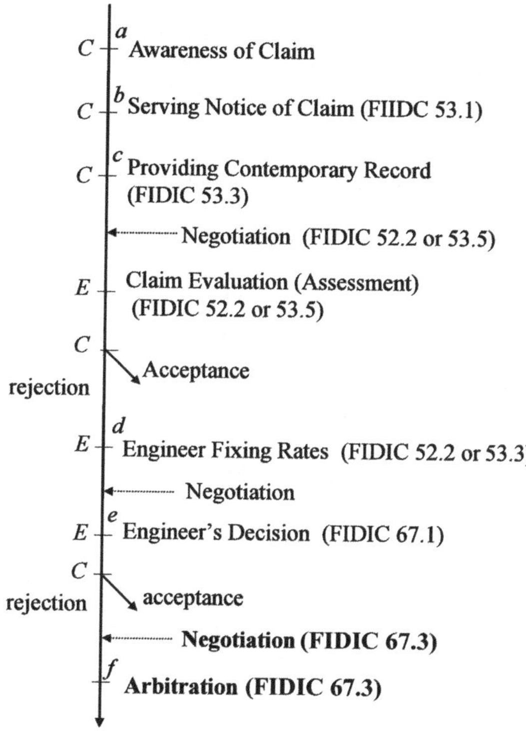

Figure 10.1 shows the dispute resolution procedure under the 1987 edition of FIDIC contract. When the Contractor recognizes an issue to be claimed (point *a*), it serves a notice of claim to the Engineer (point *b*). Subsequently, the Contractor has the contractual obligation to submit data and/or records to substantiate validity of the claim (point *c*). Although the Engineer evaluates the claim based on the data and/or records submitted by the Contractor, an amicable settlement does not always realize. If an amicable settlement does not realize, the Engineer evaluates the claim tentatively. Even if the Contractor is not satisfied with this tentative decision, the parties do not go to arbitration subsequently. Before going to arbitration, the FIDIC contract obliges the Engineer to make a decision again as a neutral professional (point *e*). The second decision by the Engineer is expected to reduce the possibility of going to costly arbitration. Nevertheless when the parties do not compromise, they go to arbitration (point *f*). One or three arbitrators is/are appointed by the parties to the contract as an arbitral tribunal. Appointed arbitrator(s) survey and examine the contract documents and drawings as well as the material relating to the dispute thoroughly, and make a decision which is final and legally binding.

Mechanism of dispute arising in construction contracts

The main issue of a dispute in construction contracts is the payment of additional costs. Real additional cost is fixed when a contractual variation is made. Therefore, in principle, a dispute in construction contracts is the matter of cost allocation between the parties, and additional cost itself does not change through the dispute process. However, dispute resolution requires transaction costs such as arbitration costs, and negotiation costs. In this sense, the number of dispute arising should be reduced to a minimum in order to cut down the costs of a construction project. But as long as construction contracts are incomplete, the Contractor is entitled to make claims, and disputes in construction contracts must be treated with fair procedures.

It is often the case with disputes in construction contracts that their generation is a result of the difference on interpretation between the parties. Therefore claims with no merit cannot be treated as disputes. Rather, the parties dispute about the validity of a claim. From the economic point of view, the problem of frivolous claims raises questions such as 'whether a claim should be treated as a dispute which requires transaction costs as mentioned.' In other words, we are wondering that even if a claim has some merit, whether it should be disputed spending negotiation costs or arbitration costs or not.

Figure 10.2: Dispute process model under the FIDIC 1987 edition

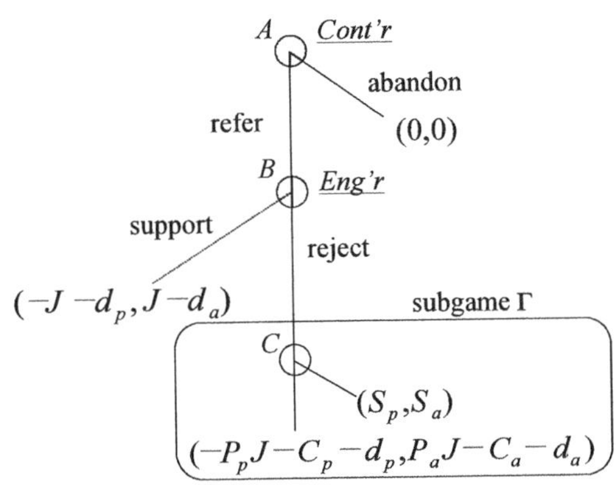

Especially when the costs for DAB/DRB are shared equally between the parties under the current system and costs for running DAB/DRB are fixed at the time of initial agreement, the Contractor can be motivated to refer to DAB/DRB excessively. In this chapter, the mechanism of dispute arising under the 1987 edition of FIDIC contract, revised FIDIC contract with DAB/DRB, and GCW contract are clarified by developing the theoretical models based on the game theory. Moreover, based on the clarified mechanism, several alternatives to control frivolous claims are examined.

The Mechanism of Dispute Arising in the FIDIC 1987 Edition

Assumption of the model

In this section, we analyze the Contractor's behaviour during the claim process and dispute resolution under the 1987 edition of FIDIC contract. In order to focus on the essential parts of the arising and resolution process of disputes in construction projects, we make the following assumptions:

(1) The cost of raising a claim is fixed for both parties; this cost does not fluctuate with the type of claim.

(2) The American / Japanese rule is adopted for sharing the cost of arbitration. In section 5, we analyze the case where the British rule is adopted.

(3) The Engineer and the Contractor do not make false allegations and reports when estimating the additional costs.

(4) All evidence to be in the verification of the claim is common knowledge.

In the process of a construction project, the Contractor serves a notice of claim to the Engineer when the Contractor recognizes or feels that it has or will suffer an impact on the program or costs. Negotiations are held by the parties, before the Engineer makes a judgment for the claimed issue. If the Engineer accepts the claim made by the Contractor, a dispute does not arise. However, if the Engineer rejects the claim, the Contractor decides whether to continue pursuing it or to abandon it as represented by the initial node *A* in Figure 10.2. If the Contractor rejects the Engineer's evaluation at the initial node *A* and proceeds to node *B*, the Employer and the Contractor will have claim costs d_p and d_a respectively. At node *B*, the Engineer has the opportunity to make a final decision from the neutral and impartial point of view. However, it is quite seldom in practice that the Engineer abandons the former decision which has been made once. Therefore, it is assumed that the Engineer always rejects the Contractor's substantiation of the claim. Afterwards, the Employer and the Contractor decide to come to a compromise or go to arbitration through negotiation at node *C*.

The parties cannot know the result of arbitration beforehand. They negotiate with a certain belief on the result of arbitration. Let $p_a = p_a(\gamma_a)$ $(0 < p_a < 1)$ be the Contractor's subjective assessment on the validity of its own claim. In the above, γ_a represents the Contractor's assessment of the capability of his/her evidence. The more complete the evidence is, the higher γ_a is assessed. Thus, $dp_a(\gamma_a)/d\gamma_a > 0$.

Considering the arbitral tribunal's error, η_1 is the Contractor's assessment on the probability that the arbitral tribunal does not support the claim, though the claim has the merit (type I error), and η_2 is the Contractor's assessment on the probability that the arbitral tribunal supports the claim, though the claim does not have the merit (type II error). Then, the Contractor expects to have a favourable award in arbitration with probability

$$P_a = p_a(1 - \eta_1) + (1 - p_a)\eta_2. \qquad (1)$$

Just like the Contractor, the Employer (the Engineer) has his own assessment $q_p = q_p(\gamma_p)$ $(0 < q_p < 1)$, where γ_p is the employer assessment of the capability of his evidence, and $dq_p(\gamma_p)/d\gamma_p > 0$. Assume the Employer (the Engineer) has the expectation to have the award in favor of him with a probability

$$Q_p = q_p(1 - \xi_1) + (1 - q_p)\xi_2, \qquad (2)$$

where ξ_1 and ξ_2 represent the Employer's assessments on the probability of the type I and II error respectively. Furthermore suppose the Employer's expectation on the probability that the claim will be accepted by the tribunal with probability $P_p = 1 - Q_p$. We assume that $P_a > P_p$ and $\eta_1 + \eta_2 < 1$, $\xi_1 + \xi_2 < 1$.

Payoffs in a contractual dispute

Suppose that the Contractor claims an additional payment J_a. The employer (or the Engineer) assesses the claim as J_p. The difference between the claimed amount by the Contractor and the assessment of the Employer $J = J_a - J_p$ comes from the dissimilarity of the interpretation of the variation or on other grounds of a claim in accordance to the conditions of the contract. This difference does not arise from an inflated or false claim by the Contractor, according to the premises of the model, but from different positions regarding which party should bear the additional costs. The payoffs of the Employer and the Contractor can be represented respectively as follows:

when the claim is approved $\begin{cases} \pi_p = -J - d_p \\ \pi_a = J - d_a \end{cases}$

when the claim is not approved $\begin{cases} \pi_p = -d_p \\ \pi_a = -d_a \end{cases}$

where d_p and d_a are the cost for claim arising at node A. If a dispute is referred to arbitration and its award supports the Contractor, the Employer has to make an additional payment J. If the award is in favor of the Employer, no additional payment is required. If a dispute is referred to arbitration, the Employer and the Contractor have to bear the arbitration cost C_p and C_a respectively. We assume that $C_p > d_p$ and $C_a > d_a$ (i.e. the claim costs less than the arbitration costs) and $J > C_p$ and $J > C_a$ (i.e. the additional payments exceed the arbitration costs).

Equilibrium of dispute arising game

We assume that the parties have enough evidence to support the validity of their view on the claim, that is they have a good understanding of the claim issues. In this model, we focus on the Contractor's decision-making at node A, i.e. whether to refer a claim to the Engineer's decision or to abandon it. Here, we define dispute arising when a claim is referred to the Engineer's decision. If the Contractor abandons a claim, the payoffs of the Employer and the Contractor are $(\pi_p, \pi_a) = (0, 0)$ respectively. If the Contractor does not abandon the claim, the parties play the subgame Γ. Omoto *et al.* [1] proposed the model of this subgame Γ in a construction contractual dispute context based on the Rubinstein [25] bargaining model with arbitration as a bilateral outside option. In the subgame Γ, the parties have the two alternative options; settlement and arbitration. If they go to arbitration, the arbitral tribunal makes a judgment on whether the claim has merit or not. At node C, the parties do not know the arbitral judgment, and they have their own subjective perception P_a, P_p. Therefore the expected payoffs of the parties to go to arbitration can be represented as $(w_p, w_a) = (-P_pJ - C_p - d_p, P_aJ + C_a - d_a)$. On the other hand, the parties gain the payoffs (S_p, S_a) respectively. Omoto *et al.* [1] pointed out that the condition that the parties compromise under the 1987 edition of FIDIC contract is

$$(P_a - P_p)\, J \leq C_p + C_a, \tag{3}$$

and the payoff to compromise(S_p, S_a) is represented as

$$(S_p, S_a) = \begin{cases} (s_p, s_a) & \text{if} \quad s_p \geq w_p,\ s_a \geq w_a \\ (w_p, -w_a) & \text{if} \quad s_p < w_p,\ s_a \geq w_a \\ (-w_a, w_a) & \text{if} \quad s_p \geq w_p,\ s_a < w_a \end{cases} \tag{4}$$

where

$$(s_p, s_a) = \left(\frac{r_a \Pi}{r_p + r_a} - P_aJ - d_p, \frac{r_p \Pi}{r_p + r_a} - P_pJ - d_a \right).$$

Here, r_p and r_a are the time discount rates of the Employer and the Contractor respectively, which represent thte bargaining power of the each party. And, $\Pi = (P_a - P_p)J$. On the contrary, if

$$(P_a - P_p)\, J > C_p + C_a \tag{5}$$

is satisfied, the parties go to arbitration. This subgame Γ represents the process of how arising disputes are resolved. However, to discuss the Contractor's dispute arising behavior, the Contractor's decision-making, whether to refer to the Engineer's decision or abandon the claim at node A must be analyzed.

1) If $(P_a - P_p)\, J \leq C_p + C_a$
   (a) if $s_p \geq w_p$ and $s_a \geq w_a$,
      a – 1) they compromise through the negotiation if $s_a > 0$
      a – 2) a claim is not referred to the Engineer if $s_a \leq 0$

(b) if $s_p < w_p$ and $s_a \geq w_a$, they compromise through the negotiation
(c) if $s_p \geq w_p$ and $s_a < w_a$,
 c – 1) they compromise through negotion if $P_a J > C_a + d_a$
 c – 2) a claim is not referred to the Engineer if $P_a J \leq C_a + d_a$
2) If $(P_a - P_p) J > C_p + C_a$, they go to arbitration.

The expected payoff of the Contractor for each case is described in **Appendix I.**

Proposition 1 The conditions that the Contractor refers a claim to the Engineer are
1) $P_a J - C_a > d_a$ or (6a)
2) $0 < P_a J - C_a \leq d_a$ (6b)

$$\text{and } t = \frac{r_a}{r_p + r_a} > \frac{-P_p J + d_a}{\Pi} = t_{cr}. \quad (6c)$$

The condition (6a) means that the Contractor's expected payoff to go to arbitration exceeds the arbitration cost and the claim costs. Note that the condition (6a) depends on only the parameters P_a, C_a, d_a related to the Contractor's property. In other words, the Contractor is stimulated to refer to the Engineer's decision regardless of the Employer's attributes. On the contrary, the condition (6b) and (6c) means that even in the case of the Contractor's expected payoff to go to arbitration does not exceed the arbitration and the claim costs, the claim can be referred to the Engineer's decision, depending on the Employer's attributes. In this case, even though the Contractor does not intend to refer a claim to arbitration, the contactor develops this claim to dispute, expecting the payoff to compromise, when the Contractor knows the Employer has no intension for arbitration. The payoff to compromise is determined by the bargaining power of the parties. The above case realizes when the Contractor's bargaining power exceeds a threshold represented as t_{cr}. However, when the Contractor's bargaining power is weak, such a case may not realize. Note that if $P_a J < C_a$, the condition (6b) is not satisfied regardless of the parameters r_p, P_p, C_p related to the Employer. In this case, the Contractor cannot gain the payoff to compromise because Contractor's thread to go to arbitration is not credible during the negotiation. Therefore the Contractor does not refer to the Engineer's decision, expecting that the Contractor cannot gain the payoff to compromise.

The Structure of Dispute Arising and the Third Parties' Adjudication

The third parties adjudication and dispute resolution process

FIDIC revised in 1999 makes it mandatory to appoint a DAB to adjudicate disputes between the Employer and the Contractor. An adjudication of the DAB is legally binding. If one of the parties is not satisfied with it, the dispute is referred to arbitration. In the United States, DRB is established instead of DAB. A judgment of the DRB is just a recommendation, compared to the one of the DAB which is legally binding. A recommendation made by DRB is expected to improve the possibility of compromise. We have already shown that the possibility of compromise in the case that a DRB is employed is more than that in the case that a DAB is employed. However, this result is focusing on how an existing dispute can be resolved. In this section, the effects of the third parties' adjudication on dispute arising are analyzed.

Revision of the objective perceptions by the third parties' judgment

The judgment of DAB/DRB plays a role in revising the parties' perceptions. First, we discuss the situation when DRB rejects the Contractor's claim. The Contractor's assessment of the merit of its claim will be revised by Bayesian' theorem, with a probability

$$\tilde{p}_a^r = \frac{p_a \eta_1}{p_a \eta_1 + (1 - p_a)(1 - \eta_2)} \quad (7)$$

under the condition that the claim has been rejected. Assume that the probabilities of errors by DRB are the same as those of arbitrators. The Contractor's expectation for a successful outcome of arbitration will be revised by Bayesian' theorem with a probability

Figure 10.3: Dispute process model under DAB

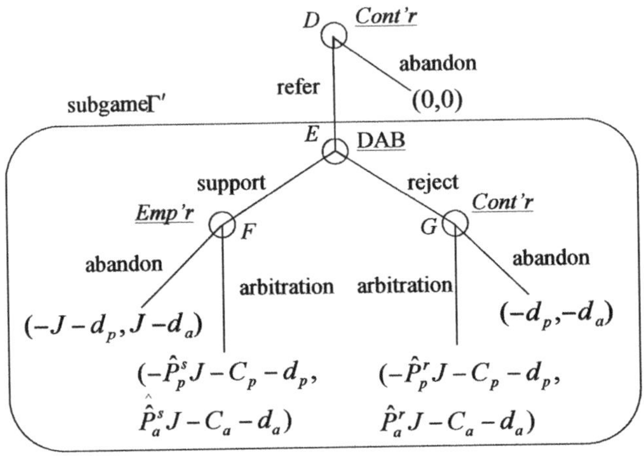

$$\tilde{p}^r_a = \tilde{p}^r_a(1-\eta_1) + (1-\tilde{p}^r_a)\eta_2 = \frac{p_a\eta_1(1-\eta_1) + (1-p_a)\eta_2(1-\eta_2)}{p_a\eta_1 + (1-p_a)(1-\eta_2)} \qquad (8)$$

when the dispute has been rejected. On the other hand, when the claim is rejected, i.e., the Employor's position is supported by DRB, his assertion for the Contractor's claim will be revised by Bayesian' theorem with a probability

$$\tilde{q}^r_p = \frac{q_p\zeta_1}{q_p(1-\zeta_1) + (1-q_p)\zeta_2} \qquad (9)$$

Then the employer's expectation for a successful outcome of arbitration will be revised by Bayes' theorem with a probability

$$\tilde{Q}^r_p = \tilde{q}^r_p(1-\zeta_1) + (1-\tilde{q}^r_p)\zeta_2 = \frac{q_p(1-\zeta_1)^2 + (1-q_p)\zeta_2^2}{q_p(1-\zeta_1) + (1-q_p)\zeta_2} \qquad (10)$$

when the Contractor's claim has been rejected. When the Contractor's claim is supported by DRB, the party's expectation for successful outcome of arbitration will be revised with probabilities

$$\hat{P}^s_a = \frac{p_a(1-\eta_1)^2 + (1-p_a)\eta_2^2}{p_a(1-\eta_1) + (1-p_a)\eta_2} \qquad (11)$$

$$\hat{Q}_p^s = \frac{q_p\xi_1(1-\xi_2)+(1-q_p)\xi_2(1-\xi_1)}{q_p\xi_1+(1-q_p)(1-\xi_2)} \tag{12}$$

respectively. Note that, $\hat{P}_p^s = 1-\hat{Q}_p^s, \hat{P}_p^r = 1-\hat{Q}_p^r$ denotes the Employer's perception on the probability that the claim is supported/rejected respectively.

DAB and structure of dispute arising

The dispute resolution process using DAB is modeled in Figure 10.3. The initial node D represents the Contractor's decision-making whether to refer a claim to DAB or abandon it. Node E is a chance move, which represents DAB's adjudication whether to support the claim or reject it. After DAB makes adjudication, if the Employer or the Contractor is dissatisfied with it, the claim is referred to arbitration.

If DAB supports the claim at node F, the Contractor accepts the DAB's decision as the final decision. The Employer makes a decision whether to accept the decision and pay the additional payment J or to reject the decision and go to arbitration. The Condition that the Employer accepts the decision is represented as follows:

$$(1-\hat{P}_p^s)J \le C_p \tag{13}$$

In case the Employer decides to go to arbitration, the Contractor's expected payoff is $\hat{P}_a^s J - C_a - d_a$. If the DAB rejects the claim at node G, the Employer accepts the DAB's decision as the final decision. The Contractor makes a decision whether to accept it, thus giving up the additional payment or to reject it and going to arbitration. Under the Contractor's revised perception, the condition that the Contractor accepts the decision is represented as follows:

$$\hat{P}_a^r J \le C_a \tag{14}$$

If the claim is referred to arbitration, the Contractor's expected payoff is $\hat{P}_a^r J - C_a - d_a$

If the Contractor abandons the claim at node D, the Contractor's payoff is zero. On the other hand, the Contractor refers the claim to DAB if the Contractor expects a positive payoff.

Here, we assume that η_1 and η_2 are sufficiently small so that the following condition is satisfied;

$$\hat{P}_a^r J > C_a \rightarrow P_a J - C_a > d_a \tag{15}$$

This condition means that a claim such as the Contractor is willing to bring to arbitration is always referred to DAB. Under this assumption, the condition that the Contractor refers the claim to DAB is represented as follows:

(1) If $\hat{P}_a^r J > C_a$ is satisfied, the claim is always referred to DAB.

(2) If $\hat{P}_a^r J \le C_a$ is satisfied

(a) If $P_a(P_a^s J - C_a) > d_a$ or (16a)

(b) If $P_a J > d_a$ and (16b)

$(1-\hat{P}_p^s)J \le C_p$ (16c)

are satisfied, the claim is referred to DAB (see **Appendix II**). If $\hat{P}_a^r J \leq C_a$ is satisfied, LHS of (16a) is always larger than LHS of (6a). Moreover, in the case that the Employer is not willing to go to arbitration, i.e. (16c) is satisfied, the range of P_a where the Contractor refers to the DAB become larger as shown by (16b). Here, we derive **Proposition 2** as follows:

Proposition 2 The possibility of a dispute arising becomes larger by introducing DAB.

Proposition 2 can be supported by the following two reasons. The first one is that DAB's adjudication is legally binding. If DAB supports the claim, the Employer decides whether to go to arbitration or not. If the Contractor expects that the Employer is not willing to go to arbitration when the claim is supported, the Contractor refers to DAB expecting DAB's favourable decision. The other reason is that the Contractor has flexibility of decision when introducing DAB. Even if DAB does not support the claim and the Contractor recognizes the negative expected payoff in the case of going to arbitration, the Contractor can abandon. In other words, the Contractor can get the option by paying the claim cost d_a to abandon the claim and the Contractor can get the positive payoff only if the claim is supported. Such flexibility endowed to the Contractor give rise to more disputes as compared to the 1987 edition of FIDIC contract.

Figure 10.4: Dispute process model under DRB

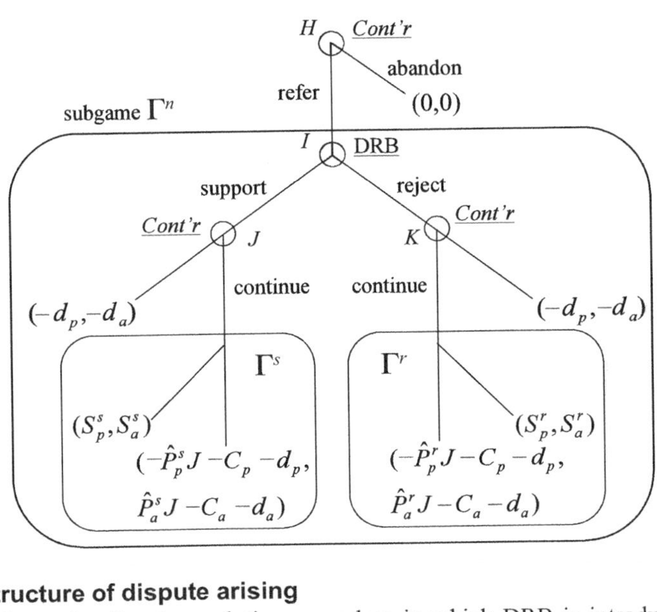

DRB and structure of dispute arising

Figure 10.4 shows the dispute resolution procedure in which DRB is introduced instead of DAB required in the 1999 edition of FIDIC contract. Recall that a judgment of DRB is just a recommendation as that of DAB is legally binding. DRB plays a role in revising the Contractor's perception to be supported by arbitral tribunal. The Contractor decides whether to continue to negotiate with the Employer or to abandon the claim. If the Contractor decides to continue negotiation, parties compromise or go to arbitration.

The condition that the Contractor refers the claim to DRB under the assumption (15) can be derived by the equilibrium of the DRB model as follows (see **Appendix III**):

(1) If If $\hat{P}_a^r J > C_a$ is satisfied, the claim is always referred to DRB.
(2) If $\hat{P}_a^r J \le C_a$ is satisfied,

(a) If $P_a(\hat{P}_a^s J - C_a) > d_a$ or (17a)

(b) If $0 < P_a(\hat{P}_a^s J - C_a) \le d_a$ and (17b)

$$(\hat{P}_a^s - \hat{P}_p^s)J \le C_a + C_p \text{ and} \tag{17c}$$

$$t > \frac{(d_a / P_a) - \hat{P}_p^s J}{(\hat{P}_a^s - \hat{P}_p^s)J} \tag{17d}$$

are satisfied, the claim is referred to DRB.
When $\hat{P}_a^r J \le C_a$ is satisfied, the LHS of (17a) is always larger than LHS of (6a). Moreover, the range of P_a which satisfies (17b) is wider than that of (6b). Here we can derive **Proposition 3**.

Proposition 3 The possibility of a dispute arising become larger by introducing DRB.

In the case that the Contractor can get the positive expected payoff of going to arbitration, even if the claim is rejected by DRB, the Contractor always refers to DRB. However, if the claim is rejected by DRB and the Contractor's expected payoff in the case of arbitration is negative, the Contractor can abandon it at that time. In other words, the Contractor has the option to abandon from the dispute resolution procedure when the continuing negotiation may not turn out to be profitable. Therefore, likewise DAB, DRB offers a sort of security to the Contractor in the sense that the Contractor can get the positive payoff only if the claim is supported. As a result, DRB attracts more disputes compared to the 1987 edition of FIDIC contract. Under DAB and DRB, the range of the Contractor's initial perception P_a that the Contractor refers to DAB/DRB varies depending not only on the additional cost and the claim/arbitration cost, but also on the Employer's perception. Thus, it is difficult to pinpoint which one causes more disputes, DAB or DRB.

GCW and structure of dispute arising

In the Japanese construction markets, the Employer evaluates the claim submitted by the Contractor, rather than negotiating on even grounds with the Contractor. Therefore, in the dispute resolution procedure of GCW, if the Contractor is not satisfied with the Employer's decision, the Contractor has no choice but to go to arbitration, established by Central Construction Industry Council of Japan, in order to reverse the Employer's decision. The Employer is always convinced of its own validity of judgment, and an amicable settlement is ideally infeasible. Such a situation is modelled as follows:

$$P_p = 0 \tag{18a}$$

$$P_a = p_a(1 - \eta_1) + (1 - p_a)\eta_2. \tag{18b}$$

The Contractor's expected payoff in case of arbitration is $P_a J - C_a - d_a$. If the Contractor does not go to arbitration, the Contractor's expected payoff is zero. Therefore, the condition that the Contractor goes to arbitration is represented as

$$P_a J - C_a \ge d_a. \tag{19}$$

Moreover, the Contractor has the belief that the Employer is always convinced of the validity of its own judgment and does not compromise through negotiation. Here, we derive **Proposition 4**.

Proposition 4 The condition that the Contractor goes to arbitration is represented as

$$P_a J - C_a \geq d_a. \tag{20}$$

(1) As far as the Employer has the ability to make always fair and neutral evaluation and
(2) as far as the Contractor trusts such Employer's ability, the dispute resolution procedure of GCW is efficient. However, if the Employer lacks the ability, inefficient dispute resolution brings about the costs, which leads to high costs in the long run.

Dispute resolution procedures and possibility of dispute arising

Figure 10.5: Alternative dispute resolutions and dispute arising

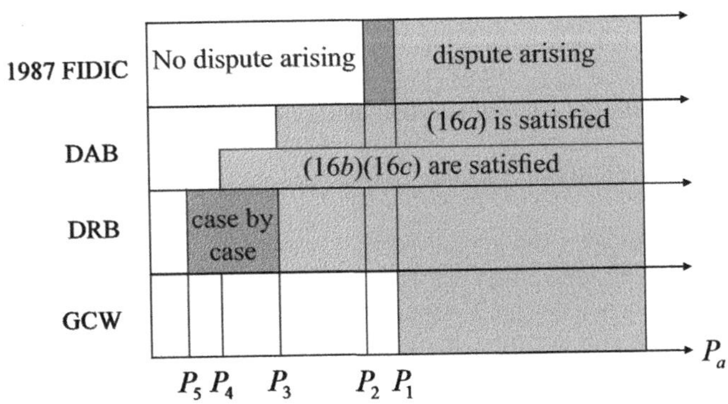

$$\text{If } \eta_1 = \eta_2 = \eta, \quad P_1 = \frac{C_a + d_a}{J}, \ P_2 = \frac{C_a}{J}, \ P_3 = \frac{\eta(1-\eta)J + d_a}{J - C_a}$$

$$P_4 = \frac{d_a}{J}, \ P_5 = \frac{\eta(1-\eta)J}{J - C_a}$$

Claim costs and arbitration costs may be different depending on the dispute resolution procedure. Therefore, we cannot simply compare the possibility of dispute arising among different dispute resolution procedures. To compare the efficiency of dispute resolution procedures, the data on the claim costs and the arbitration costs for each dispute resolution procedure are needed. However, such data are not available, because it has not been long since DAB/DRB were introduced. Here, we compare the efficiency of the dispute resolution procedures by analyzing the possibility of a claim developing to a dispute under the condition that the parameters of costs are uniform. Figure 10.5 shows the possibility of dispute arising and the initial Contractor's perception P_a under the dispute resolution procedure of old FIDIC, DAB, DRB, and GCW on the assumption that parameters (d_a, d_p, C_a, C_p, J, P_p, η_1, η_2 ζ_1, ζ_2) are constant. The bigger the Contractor's perception P_a is, the more the Contractor believes in the validity of the claim and the bigger the possibility of dispute arising becomes. As shown in Propositions, the possibility of dispute resolution is least in the case of GCW. In the case of GCW and DAB, the judgment of the Employer or the DAB is final and binding, and an amicable settlement cannot occur theoretically. Therefore, the thresholds under GCW and DAB do not depend on the Employer's bargaining power. However, the existence of DAB can cause the Contractor to generate disputes with less merit, expecting the Employer going to arbitration. Hence, the threshold under DAB depends on the Employer's perception. On the other hand, under the 1987 edition of FIDIC contract and under DRB, both parties negotiate to compromise with the option to go to arbitration. Therefore, the thresholds depend on the Employer's bargaining power. As mentioned above, it is difficult to compare the relative merits of the alternative dispute resolution methods, because necessary costs for dispute resolution vary by the methods. But we have pointed out that the introduction of the third parties such as DAB/DRB will cause more disputes as compared to the 1987 edition of FIDIC and GCW, ceteris paribus. Note that the order of P_4 and P_5 cannot be determined uniquely, because these depends on the value of η.

Mechanism of Dispute Arising and Control Measurement

The effect of the arbitration cost allocation rule

If we adopt the British rule for the arbitration cost allocation rule, the Contractor does not have to pay the arbitration costs in the case that the arbitral tribunal supports the claim. However, the Contractor has to pay the arbitration costs of both of Contractor and the Employer in case that the arbitral tribunal rejects the claim. Many papers have analyzed the effects of cost allocation rule on the frivolous litigation arising [26] - [32]. Earlier papers such as Landes [26], Gould [27], Bebchuk [28] pointed out that the British rule plays a role in controlling dispute arising. On the contrary, later papers pointed out that the British rule is not always superior to the American/Japanese rule in controlling the frivolous litigations. However, those papers examine subject to civil cases, and not dispute resolution with the third parties' adjudication.

Adopting the British rule, the Employer's and the Contractor's expected payoffs to go to arbitration in case that DAB supports the claim are $-\hat{P}_p^s(J + C_p + C_a) - d_p$ and $\hat{P}_a^s J - (1 - \hat{P}_a^s)(C_p + C_a) - d_a$ respectively. The Employer's and the Contractor's expected payoffs to go to arbitration in the case that DAB rejects the claim are $-\hat{P}_p^r(J + C_p + C_a) - d_p$ and $\hat{P}_a^r J - (1 - \hat{P}_a^r)(C_p + C_a) - d_a$. Also here, we assume that the probability of DAB's errors is sufficiently small so that the following condition is satisfied,

$$\hat{P}_a^r J > (1 - \hat{P}_a^r)(C_p + C_a)$$
$$\rightarrow P_a J - (1 - P_a)(C_p + C_a) > d_a \qquad (21)$$

Based on the above expected payoffs we can derive the following condition that the Contractor refers the claim to DAB.

1) If $\hat{P}_a^r J > (1 - \hat{P}_a^r)(C_p + C_a)$ is satisfied, the claim is always referred to DAB.
2) If $\hat{P}_a^r J \leq (1 - \hat{P}_a^r)(C_p + C_a)$ is satisfied
   (a) If $P_a\{\hat{P}_a^s J - (1 - \hat{P}_a^s)(C_p + C_a)\} > d_a$ or (22a)

   (b) If $P_a J > d_a$ and (22b)

   (c) $(1 - \hat{P}_p^s)J \leq \hat{P}_p^s(C_p + C_a)$ (22c)

are satisfied, the claim is referred to DAB (see **Appendix IV**). If the Employer's perception is less, i.e. the formula (22c) is satisfied, the condition of dispute arising (22b) under the British rule is same as to that under the American/Japanese rule (16b). Otherwise, by comparing (16a) and (22a), the British rule reduces the possibility of dispute arising more than the American/Japanese rule if $\hat{P}_a^s(C_p + C_a) < C_a$ is satisfied, and vice versa if $\hat{P}_a^s(C_p + C_a) > C_a$. Conclusively, we cannot derive a unique result as to which one has the best rule, the British or the American/Japanese.

The effect of the claim cost allocation rule

The problem of the arbitration cost allocation rule is that it affects not only the Contractor's dispute generating behaviour but also the dispute resolution behaviour. Therefore, the arbitration cost allocation rule affects various aspects of the dispute resolution procedure. Now, consider the claim cost allocation rule, which affects directly on the Contractor's dispute generating behaviour. For example, consider that the Contractor pays the claim cost D_a. If DAB supports the claim, the claim cost D_a is brought back to the Contractor. However, if DAB rejects the claim, the claim cost D_a is expropriated from the Contractor. This is sort of a penalty system. Under the assumption (15), the amount of D_a which reduces the possibility of dispute arising to the same level as the one under the GCW, is derived as follows (see **Appendix IV**);

$$D_a^* = \begin{cases} \dfrac{C_a J}{J - C_a - d_a} & \text{If } (1-\hat{P}_p^s)J \le C_p \\ C_a - \dfrac{\eta(1-\eta)J^2}{J - C_a - d_a} & \text{If } (1-\hat{P}_p^s)J > C_p \end{cases} \quad (23)$$

This result means that if $(1-\hat{P}_p^s)J \le C_p$, the claim cost D_a exceeds the arbitration cost C_a. Even if $(1-\hat{P}_p^s)J > C_p$, the value of D_a gets closer to the value of C_a as the possibility of errors gets smaller. We can reduce dispute arising under GCW, though an agreement on such a penalty system will make claim cost higher than the arbitration cost. Such an agreement may not be allowed legally.

The effect of the probability of DAB's error

The above discussion shows that it is difficult to reduce the possibility of dispute arising under the GCW or the 1987 edition of FIDIC contract when we introduce DAB. One of the primary purposes of the introduction of DAB is the reduction of dispute resolution costs through the reduction of the possibility of going to arbitration. If the possibility of DAB's error is sufficiently small, we can reduce the dispute resolution cost effectively. Now let us consider that the possibility of DAB's error is zero, i.e. $\eta_1 = \eta_2 = \zeta_1 = \zeta_2 = 0$, and this assumption is common knowledge shared by both parties. In this case, the initial perceptions of the parties are

$$P_a = p_a,\ P_p = 1 - q_p \quad (24)$$

The Objective perceptions after DAB's adjudication are

$$\hat{P}_p^r = 1,\ \hat{P}_a^r = 0 \quad (25a)$$

$$\hat{P}_p^s = 0,\ \hat{P}_a^s = 1 \quad (25b)$$

Even though DAB and arbitral tribunal make judgment independently, these judgments always coincide. Therefore, there is no possibility of arbitration. Hence, the condition that the Contractor refers the claim to DAB is represented as

$$p_a J > d_a \quad (26)$$

On the other hand, in case of $t > t_{cr}$ under the 1987 edition of FIDIC contract and in the case of GCW, the condition that the Contractor refers the claim to DAB is represented as

$$P_a J - C_a > d_a \tag{27}$$

This means that even if the possibility of DAB's error is zero, more disputes arise as compared to the case of the 1987 edition of FIDIC contract and GCW.

Towards the efficiency of dispute resolution procedure

As shown in **Proposition 2**, the possibility of dispute arising increases by introducing DAB. However, the more the reliability of DAB is, the less the possibility of dispute arising is. In order to evaluate improvement of dispute resolution due to the introduction of DAB, we must consider the total cost for dispute resolution in a project including the number of disputes, the cost for DAB's establishment and the arbitration costs. As mentioned above, a claim with no merit for the Contractor is unlikely to lead to a dispute as the Contractor and the Engineer discuss preliminary. However, it is practically not easy to assess the merit of a dispute definitely. When the Contractor strategically behaves, a claim with even less merit may lead to a dispute. It is proved that the possibility of such a dispute arising can be alleviated by decreasing the possibility of error in judgement. Because the arbitration cost is generally expensive, a claim with relatively large amount is likely to lead to a dispute under the GCW and FIDIC 1987.

On the other hand, the Contractor refers to DAB if the expected payoff exceeds the claim cost. The claim cost is not very high. Therefore, the Contractor is stimulated to refer even trivial claims such that the Contractor's perception is small, expecting the possibility of DAB's favourable adjudication. Moreover, if the possibility of DAB's error is zero, the arbitration cost is not necessary.

Based on the above analysis, in order to make dispute resolution more efficient, the improvement of reliability of DAB and arbitral tribunal is a better policy rather than the introduction of the claim cost allocation rule. Note that if $d_a = 0$, formula (26) is always satisfied for any p_a, that means all claims are referred to DAB. Therefore, appropriate claim cost d_a should be determined in order to prevent even trivial claims to be referred to DAB. This issue is not the scope of our study, and remains the subject of future research.

Conclusion

In this chapter, the mechanism of a claim developing to a dispute is analyzed. As a result of the analysis, we pointed out that the Contractor can be stimulated to dispute frivolous claims of less merit under the 1987 edition of FIDIC contract, expecting the payoff in the case of a settlement. The introduction of DAB/DRB can increase the number of disputes arising. Even adopting the British rule for arbitration cost allocation, the increase of dispute arising may not be controlled. Adopting the claim cost allocation rule is an effective measure to reduce the possibility of dispute arising, though the amount of penalty is too high.

In order to propose more efficient dispute resolution methods, a lot of issues remain to be studied. The number of dispute arising will affect the claim costs, the costs of establishing a DAB/DRB, and the arbitration costs in the long run. To compare the efficiency of dispute resolution procedures, total costs including those costs mentioned above must be calculated. So far, it is difficult to analyze the theoretical hypothesis proposed in this chapter using empirical data, because DAB/DRB was introduced not long ago. Moreover, in introducing DAB, the optimal claim costs rule must be designed to minimize the total costs of dealing with the claims.

Appendix

Appendix I Equilibrium of dispute resolution game under FIDIC 1987 edition

We show the subgame perfect equilibrium of the model shown in Figure 10.2. If $P_aJ \leq C_a$, the Contractor's threat to go to arbitration is not credible for the Employer. Therefore the Contractor cannot gain the payoff of compromise. At that time, the Contractor is not motivated to refer to the Engineer. Hereinafter, we focus on the case of $P_aJ > C_a$. 1). If $(P_a - P_p)\ J \leq C_p + C_a$, a) if $s_p \geq w_p$, and $s_a \geq w_a$, a – 1), in case of $s_a > 0$, a settlement realizes and the Contractor gains the payoff s_a.a-2) In case of $s_a \leq 0$, a dispute does not arise and the Contractor gains the payoff zero. b) If $s_p < w_p$ and $s_a \geq w_a$, a settlement realizes and the Contractor gains the expected payoff $PpJ + C_p - d_a$. c) If $s_p \geq w_p$ and $s_a < w_a$, c-1) in case of $P_aJ > C_a + d_a$, a settlement realizes and the Contractor gain the expected payoff $P_aJ - C_a - d_a$. c-2). In case of $P_aJ \leq C_a + d_a$, a dispute does not arise and the Contractor gains the payoff zero. 2) If $(P_a - P_p)\ J > C_p + C_a$, we derive $P_aJ > C_a + d_a$ from $P_aJ - C_a > P_pJ + C_p > d_a$. Therefore the Contractor gains the expected payoff $P_aJ - C_a - d_a$. Here, remind that $S_a = P_aJ - C_a - d_a$ if $t < t_1 = 1 - C_a/\prod$, $S_a = s_a = tP_aJ + (1-t)P_pJ - d_a$ if $t_1 \leq t \leq t_2 = C_a/\prod$, and $S_a = P_pJ + C_p - d_a$ if $t > t_2$, where $t = r_p/(r_p + r_a)$. Now consider the case of $(P_a - P_p)J \leq C_p + C_a$. The case of $t_1 \leq t \leq t_2$ corresponds to the case 1-a, the case of $t > t_2$corresponds to the case 1-b, and the case of $t < t_1$ corresponds to the case 1-c. Regarding s_a as the function of t, $s_a(t_1) = P_aJ - C_a - d_a (= w_a)$. In case of $t < t_1$, the Contractor is guaranteed to gain the expected payoff at least w_a. Therefore, if $s_a(t_1) = P_aJ - C_a - d_a > 0$ is satisfied, a dispute arises for any t. On the other hand, if $s_a(t_1) = P_aJ - C_a - d_a \leq 0$, there is the only $t = t_{cr} = (-P_pJ + d_a)/\prod$ which satisfies the equation, because $s_a(t)$ is increasing to $t \in [t_1, t_2]$ and $s_a(t_2) = P_pJ - C_p - d_a > 0$. Therefore if $t > t_{cr}$, a claim is escalated to a dispute.

Appendix II Equilibrium of the dispute resolution game under DAB

From $P_a - \hat{P}^r_a = (p_a - \hat{p}^r_a)(1 - \eta_1 \_ \eta_2)$ and

$$P_a - \hat{P}^r_a = \frac{p_a(1-p_a)(1-\eta_1-\eta_2)}{p_a\eta_1 + (1-p_a)(1-\eta_2)}$$

$\hat{P}^r_a < P_a$ is satisfied under the assumption $\eta_1 + \eta_2 < 1$. Likewise, we can derive $\hat{P}^s_a > P_a$. Here we derive the subgame perfect equilibrium of the game shown in Figure.10.3 by backward induction, and analyze the Contractor's behaviour at node D. If the DAB supports the claim, the Contractor accepts this judgment and the Employer decides whether to go to arbitration. If $(1 - \hat{P}^s_a)J \leq C_p$, the Employer abandons. And if $(1 - \hat{P}^s_p)J > C_p$, the Employer goes to arbitration. If the DAB rejects the claim, the Employer accepts this judgment and the Contractor decides whether to go to arbitration. The Contractor abandons if $\hat{P}^r_aJ \leq C_a$and goes to arbitration if $\hat{P}^r_aJ > C_a$. Considering the equilibrium of these subgames, the Contractor decides whether to go to the DAB or to abandon the claim. Therefore we derive the equilibrium as follows. 1) In case of $(1 - \hat{P}^s_p)J \leq C_p$and $\hat{P}^r_aJ \leq C_a$, a) if $P_aJ > d_a$, the Contractor goes to DAB and gains the expected payoff $P_aJ - d_a$. b) If $P_aJ \leq d_a$, the Contractors abandons and gains expected payoff zero at node D. 2) In case of $(1 - \hat{P}^s_p)J \leq C_p$and $\hat{P}^r_aJ > C_a$, a) if $P_aJ + (1 - P_a)(\hat{P}^r_aJ > C_a) > d_a$, the Contractor goes to the DAB and gains the expected payoff $P_aJ + (1 - P_a)(\hat{P}^r_aJ - C_a) - d_a$. b) If $P_aJ + (1 - P_a)(\hat{P}^r_aJ - C_a) \leq d_a$, the Contractor abandons the claim and gains the expected payoff zero. However, this 2-b case cannot realize, because $P_aJ + (1 - P_a)(\hat{P}^r_aJ - C_a) > P_aJ$ from $\hat{P}^r_aJ > C_a$ and $P_aJ - d_a > \hat{P}^r_aJ - C_a > 0$ from $P_a > \hat{P}^r_a$ and $C_a > d_a$. Therefore $P_aJ + (1 - P_a)(\hat{P}^r_aJ - C_a) > d_a$ is always satisfied. 3) In case of $(1 - \hat{P}^s_p)J > C_p$ and $\hat{P}^r_aJ \le C_a$, a) if $P_a(\hat{P}^r_aJ - C_a) > d_a$, the Contractor goes to DAB and gains the expected payoff $P_a(\hat{P}^r_aJ - C_a) - d_a$. b) If $P_a(\hat{P}^r_aJ - C_a) \le d_a$, the Contractor abandons and gains the expected payoff zero. 4) In case of $(1 - \hat{P}^s_p)J > C_p$ and $\hat{P}^r_aJ > C_a$, reminding that $P_a(\hat{P}^s_aJ - C_a) + (1 - P_a)(\hat{P}^r_aJ - C_a) = \{P_a\hat{P}^s_a + (1 - P_a)\hat{P}^r_a\}J - C_a = P_aJ - C_a$. a) If $P_aJ - C_a > d_a$, contactor goes to the DAB and gains the expected payoff $P_aJ - C_a - d_a$. b) If $P_aJ - C_a \le d_a$, the Contractor abandons and gains the expected payoff zero. Here under the assumption (15), $P_aJ - C_a > d_a$ is always satisfied if $\hat{P}^r_aJ > C_a$. Concluding the above equilibrium, if $\hat{P}^r_aJ \le C_a$, the conditions that the Contractor goes to DAB are 1) $P_a(\hat{P}^s_aJ - C_a) > d_a$ or 2) $P_aJ > d_a$ and $(1 - \hat{P}^s_p)J \le C_p$. If $\hat{P}^r_aJ > C_a$, the Contractor always goes to DAB.

Appendix III Equilibrium of the dispute resolution game under DRB

In Fig.4, 1) in case of $(\hat{P}^s_a - \hat{P}^s_p)J \le C_p + C_a$ and $(\hat{P}^r_a - \hat{P}^r_p)J \le C_p + C_a$, a) if $\hat{P}^s_aJ \le C_a$ and $\hat{P}^r_aJ \le C_a$, the Contractor cannot gain the payoff in the subgame Γ^s and Γ^r, because the threat that the Contractor goes to arbitration is not credible. Therefore, the Contractor abandons the claim at node H. b) If $\hat{P}^s_aJ > C_a$ and $\hat{P}^r_aJ \le C_a$ the Contractor can gain the positive payoff only in the subgame Γ^s, and the expected payoff to go to the DRB at node H, is $P_a\hat{P}^s_a - (1 - P_a)d_a\ (\ge P_a(\hat{P}^s_aJ - C_a) - d_a)$.

b-1) If $0 < P_a(\hat{P}^s_aJ - C_a) \le d_a$, the Contractor goes to the DRB if $t = \dfrac{(d_a/P_a) - \hat{P}^s_pJ}{(\hat{P}^s_a - \hat{P}^s_p)J}$ b-2) If $P_a(\hat{P}^s_aJ - C_a) > d_a$, the Contractor always goes to the DRB. c) If $\hat{P}^s_aJ > C_a$ and $\hat{P}^r_aJ > C_a$, the Contractor gains the positive payoff in both subgame Γ^s and Γ^r, the Contractor's expected payoff to go to the DRB at node H is represented as $P_aS^s_a - (1 - P_a)S^r_a\ (\ge P_a(\hat{P}^s_aJ - C_a - d_a) + (1 - P_a)(\hat{P}^r_aJ - C_a - d_a) = P_aJ - C_a - d_a)$. c-1) If $0 < P_aJ - C_a \le d_a$, the Contractor goes to the DRB if

$$t > \frac{d_a - P_a\hat{P}^s_pJ - (1 - P_a)\hat{P}^r_aJ}{\{P_a(\hat{P}^s_a - \hat{P}^s_p) + (1 - P_a)(\hat{P}^r_a - \hat{P}^r_p)\}J}$$

c-2) If $P_aJ - C_a > d_a$, the Contractor always goes to the DRB. 2) If $(\hat{P}^s_a - \hat{P}^s_p)J > C_p + C_a$ and $(\hat{P}^s_a - \hat{P}^s_p)J > C_p + C_a$, the parties go to arbitration in both subgame and Γ^s and Γ^r. The Contractor's expected payoff to go to the DRB at node H is represented as $P_aJ - C_a - d_a$. Because $P_a > \hat{P}^r_a$ and $\hat{P}^r_a - C_a > \hat{P}^r_p + C_p > d_a$, $P_aJ - C_p > d_a$. Therefore the Contractor always goes to the DRB in this case. 3) If $(\hat{P}^r_s - \hat{P}^s_a)J \le C_p + C_a$ and $(\hat{P}^r_a - \hat{P}^r_p)J > C_p + C_a$ the parties go to arbitration in case that the DRB rejects the claim and compromise in case the DRB supports the claim. The Contractor's expected payoff to go to the DRB is $P_aS^s_a - (1 - P_a)(\hat{P}^r_aJ - C_a - d_a)(\ge P_aJ - C_a - d_a)$.

Because $P_a > \hat{P}^r_a$ and $\hat{P}^r_a - C_a > \hat{P}^r_p + C_p > d_a$, $P_aJ - C_p > d_a$. Therefore the Contractor always goes to the DRB in this case. 4) If $(\hat{P}^s_a - \hat{P}^s_p)J > C_p + C_a$ and $(\hat{P}^r_a - \hat{P}^r_p)J \le C_p + C_a$, the parties go to arbitration in case that the DRB supports the claim and compromise in case the DRB rejects the claim. a) If $\hat{P}^r_aJ \le C_a$, the Contractor's expected payoff to goes to the DRB is $P_a(\hat{P}^s_aJ - C_a) - d_a$. The Contractor goes to the DRB if $P_a(\hat{P}^s_aJ - C_a) > d_a$. b) If $\hat{P}^r_aJ > C_a$, the

Contractor's expected payoff to goes to the DRB is $P_a(\hat{P}^s_a J - C_a - d_a) - (1 - P_a)S^r_a (\geq P_a J - C_a - d_a$.

b-1) If $0 < P_a J - C_a \leq d_a$, the Contractor goes to the DRB if

$$t > \frac{\dfrac{d_a - P_a(\hat{P}^s_a J - C_a)}{1 - P_a} - \hat{P}^r_p J}{(\hat{P}^r_a - \hat{P}^r_p)J}.$$

b-2) If $P_a J - C_a > d_a$, the Contractor always goes to the DRB. Here under the assumption (15), $P_a J - C_a > d_a$ is always satisfied if $\hat{P}^r_a J > C_a$. Concluding the above equilibrium, if $\hat{P}^r_a J \leq C_a$, the conditions that the Contractor goes to DAB are 1) $P_a(\hat{P}^s_a J - C_a) > d_a$ or 2) $0 < P_a(\hat{P}^s_a J - C_a) \leq d_a$ and $(\hat{P}^s_a - \hat{P}^s_p) J \leq C_p + C_p$, and $t = \dfrac{(d_a/P_a)\hat{P}^s_p J}{(\hat{P}^s_a - \hat{P}^s_p)J}$. If $\hat{P}^r_a J > C_a$, the Contractor always goes to DAB.

Appendix IV Effect of cost-shifting rule

Effect of arbitration cost-shifting rule

According to the same procedure in **Appendix II**, the subgame perfect equilibrium of the dispute resolution model under DAB applied the British rule. 1) If $(1 - \hat{P}^s_p) J - \hat{P}^s_p (C_p + C_a) \leq 0$ and $\hat{P}^r_a J \leq (1 - \hat{P}^r_a)(C_p + C_a)$, a) if $P_a J > d_a$, the Contractor goes to DAB and gains the expected payoff. b) $P_a J \leq d_a$, the Contractor abandons the claim and gains the payoff zero. By deriving the equilibrium strategies of the Contractor exhaustively as **Appendix II**, we can conclude the conditions that the Contractor goes to the DAB under British Rule are 1) $\hat{P}^r_a J > (1 - \hat{P}^r_a)(C_p + C_a)$ or 2-a) $\hat{P}^r_a J \leq (1 - \hat{P}^r_a)(C_p + C_a)$ and $P_a\{\hat{P}^s_a J - (1 - \hat{P}^s_a)(C_p + C_a)\} > d_a$, or 2-b) $\hat{P}^r_a J \leq (1 - \hat{P}^r_a)(C_p + C_a)$ and $(1 - \hat{P}^s_a) J \leq \hat{P}^s_a (C_p + C_a)$.

Effect of claim cost-shifting rule

If $\hat{P}^r_a J > C$, even under GCW contract, a claim is escalated to a dispute. Therefore, here we focus on the case of $\hat{P}^r_a J \leq C_a$. If $(1 - \hat{P}^s_p) J > C_p$, the Contractor's expected payoff to go to the DAB is $P_a J - (1 - P_a) D_a - d_a = P_a(J + d_a) - D_a - d_a$. Substituting the threshold $P^*_a = (C_a + d_a)/J$ of dispute arising under GCW contract for the above formula, we derive $D^*_a = C_a J/(J - C_a - d_a)(> C_a)$ that can be control dispute arising as the same level as that under GCW. If $(1 - \hat{P}^s_p) J \leq C_p$, the Contractor's expected payoff to go to the DAB is $P_a(\hat{P}^s_a J - C_a) - d_a$. In the same way, we can derive $D^*_a = C_a - \dfrac{\eta(1-\eta) J^2}{J - C_a - d_a}$.

References

[1] Omoto, T., Onishi, M., Kobayashi, K. (2002) Bargaining Model of Contractual Dispute Resolution. *Proceedings of IEEE Systems, Man, and Cybernetics Conference, IEEE*, (CD-ROM).

[2] Onishi, M., Omoto, T., Kobayashi, K. (2005) Introduction of a quasi-third party to dispute resolution in construction contracts. *Interdisciplinary Information Science*, Tohoku University, **32 (1)**: 105-116.

[3] Reid, A. and Ellis, R. (2007) Common sense applied to the definition of dispute. *Structural Survey* **25(3)**: 239-252.

[4] Diekmann, J. E., and Girard, M. J. (1995) Are contract disputes predictable. *ASCE Journal of Construction Engineering and Management* **121(4)**: 355-363.
[5] Posner, R (1992) *Economic Analysis of Law*, 4th *edition*. Little, Brown, Boston.
[6] Epstein, R (1995) *Simple Rules for a Complex World*. Harvard University Press, MA.
[7] Miceli, T. J (1997) *Economics of Law: Torts, Contracts, Property, Litigation*. Oxford University Press, New York.
[8] Kumaraswamy, M. (1997) Conflicts, claims and disputes. *Engineering, Construction and Architectural Management* **4(2)**: 95-111.
[9] Love, P. E. D., Davis, P. R., Ellis, J. M., Cheung, S. O. (2008) A systemic view of dispute causation. *International Journal of Managing Projects in Business* **3(4)**: 661-980.
[10] Love, P. E. D., Davis, P. R., Ellis, J. M., Cheung, S. O. (2010). Dispute causation: identification of pathogenic influences in construction. *Engineering, Construction and Architectural Management* **17(4)**: 404-423.
[11] Gebken, R. J. and Gibson, E. (2006). Quantification of costs for dispute resolution procedures in the construction industry. *ASCE Journal of Professional Issues in Engineering Education and Practice* **132 (3)**: 264-271.
[12] Huber, P (1990) *Liability: The Legal Revolution and its Consequences*. Basic Book, New York.
[13] Olson, W. (1991) *The Litigation Explosion*. Dutton, New York.
[14] Howard, P. (1994) *The Death of Common Sense*. Random House, New York.
[15] Png, I. (1983) Strategic behavior in suit, settlement, and trial. *Bell Journal of Economics* **14**: 539-550.
[16] Cooter, R., Ulen, T. (1988). *Law and Economics*. Harper Collins, New York.
[17] Cooter, R., Rubinfeld, D. (1989) Economic analysis of legal disputes and their resolution. *Journal of Economic Literature* **27(3)**: 1067-1110.
[18] Katz, A. (1990) The effect of frivolous litigation on the settlement of legal disputes. *International Review of Law and Economics* **10(1)**: 3-27.
[19] Hylton, K. (1990) Costly litigation and legal error under negligence. *Journal of Law, Economics & Organizations* **6(2)**: 433-452.
[20] Rosenberg, D. and Shavell, S. (1985) A model in which suits are brought for their nuisance value. *International Review of Law and Economics* **5**: 3-13.
[21] Danzon, P. (1983) Contingent fees for personal injury litigation. *Bell Journal of Economics* **14**: 213-224.
[22] Thomason, T. (1991) Are attorneys paid what they're worth? Contingent fees and the settlement process. *Journal of Legal Studies* **20(1)**: 187-223.
[23] Rubinfeld, D. and Scotmacher, S. (1993) Contingent fees for attorneys, An economic analysis. *Rand Journal of Economics* **24(3)**: 343-356.
[24] Miceli, T. J. (1994) Do contingent fees promote excessive litigation? *Journal of Legal Studies* **23**: 211-224.
[25] Rubinstein, A. (1982) Perfect equilibrium in a bargaining model. *Econometrica* **50(1)**: 97-110.
[26] Landes, W. (1971) An economic analysis of the courts. *Journal of Law and Economics* **14(1)**: 61-107.
[27] Gould, J. (1973). The economics of legal conflicts. *Journal of Legal Studies* **2(2)**: 279-300.
[28] Bebchuk, L. (1984) Litigation and settlement under imperfect information. *Rand Journal of Economics* **15(3)**: 404-415.

[29] Nalebuff, B. (1984) Litigation and settlement under imperfect information. *Rand Journal of Economics* **15(3)**: 198-210.
[30] Schweizer, U. (1984) Litigation and settlement under imperfect information. *Rand Journal of Economics* **15(3)**: 163-178.
[31] Spier, K. (1992) The dynamics of pretrial negotiation. *Review of Economic Studies* **59**: 93-108.
[32] Shavell, S. (1993) Suit versus settlement when parties seek nonmonetary judgments. *Journal of Legal Studies* **22(1)**: 1-13.

Joint Ventures in Construction 2
ISBN 978-0-7277-5783-8

doi: 10.1680/jvc2.57838.147

Chapter 11

Why Do Contractors Subcontract Work to Subcontractors?

Khairuddin Abdul Rashid, International Islamic University Malaysia

Introduction

This chapter reports on a study of the practice in the Malaysian construction industry of contractors subcontracting to others – otherwise known as subcontractors – work originally contracted to them by clients. Specifically, the objectives of the study were the following:

(1) To identify the presence and extent of contractors subcontracting work to sub-contractors, and
(2) To identify the motivation that drives contractors to subcontract work to subcontractors.

The motivation for the study stems from the general notion that in a typical construction project it is not uncommon for subcontractors to be the ones to perform almost all the construction work [1] - [3], but there is no credible literature that discusses the extent of the practice or the motivation that drives contractors to subcontract work to subcontractors.

The chapter is presented in 4 parts:

(1) Part 1 outlines the objectives of the chapter and the study to be reported herein.
(2) Part 2 provides an overview of the concept of subcontracting and discusses subcontracting as provided for under the relevant Contracts Act and the common standard forms of contract currently in use in Malaysia.
(3) Part 3 reports on a survey on the practice of subcontracting by contractors.
(4) Part 4 contains a discussion of the findings and the conclusion to the chapter.

Part 2

Subcontracting

The concept of subcontracting

There are two types of subcontractor, i.e. a subcontractor nominated by the client but who enters into a contract with the contractor; and a subcontractor chosen and appointed directly by the contractor. The former is referred to as a Nominated subcontractor (NSC) or Nominated supplier (N-Supplier), and the latter is referred to as a Domestic subcontractor (DSC) or Domestic supplier (D-Supplier) [i].

The subject of the study to be reported herein is the Domestic subcontractor, hereinafter referred to as subcontractor (Figure 11.1).

Figure 11.1: Client – Contractor – Subcontractor relationships

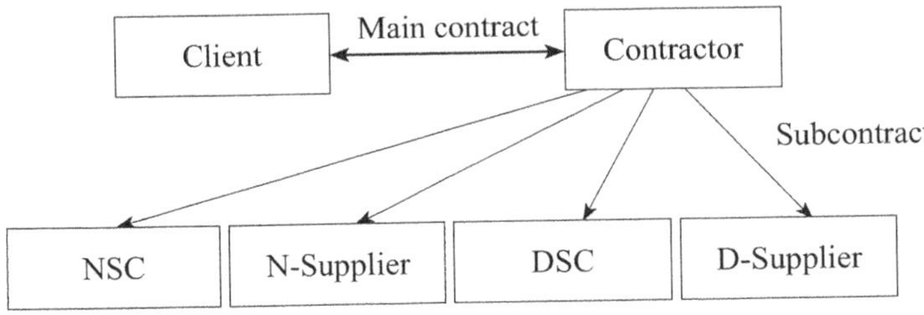

On the one hand, subcontracting 'helps the main contractor to lighten the load of the full construction work' and this in turn enables the contractor to address other issues such as 'long-term environmental uncertainties and to protect the technical core of the contractor's employee against short term contingencies' [3]. But on the other hand, subcontracting could be seen 'as a form of domination of large firms over small ones, where large firms (the contractors) benefited from the low wages and flexible work arrangements in small firms (the subcontractors)' [3].

The contractual position of subcontracting

The review of literature on the contractual position of subcontracting focuses on the Contracts Act of 1950 [4] and the two common standard forms of contract currently in use in Malaysia, namely the form PWD 203A (revised 2007) [5] and the forms PAM with Quantities (revision 1998 [6] and revision 2006 [7]. The form PWD 203A[5] is mandatory for all procurement of work projects-buildings and infrastructure-in the public sector, while the PAM forms are commonly used by private sector clients when procuring building work projects.

Section 41 of the Contracts Act of 1950 [4] states that:

"If it appears from the nature of the case that it was the intention of the parties to any contract that any promise contained in it should be performed by the promisor himself, such promise must be performed by the promisor. In other cases, the promisor or his representatives may employ a competent person to perform it."

Section 42 of the Contracts Act of 1950 [4] states that:

"... when a promisee accepts performance of the promise from a third person, he cannot afterwards enforce it against the promisor."

Clause 47.1 of the form PWD 203A (2007) [5] states that:

"The Contractor shall not without the prior written consent of the S.O.... subcontract the design for any portion of the works... Where the S.O. consents to any subcontract... such consent shall not in any way absolve the obligations of the Contractor...".

Clause 47.2 of the form PWD 203A [5] states that:

"The Contractor shall not subcontract the whole or any substantial parts of the Works without the prior written consent of the S.O. ... Any such consent, if given, shall not relieve the Contactor from any liability or obligation under this Contract and he shall be responsible for the due observance by such subcontractors, of all the terms, stipulations and conditions under this Contract...".

Clause 17.2 of PAM with Quantities (1998) [6] states that:

"The Contractor shall not without the written consent of the Architect ... sublet any portion or the whole of the Work except otherwise provided by the Contract. Such consent if given shall not relieve the Contractor from any liability or obligation under the Contract...".

Provisions under the Contracts Act of 1950 [4] and forms PWD 203A (2007) [5] and PAM with Quantities (1998) [6] therefore indicate that construction contracts do not always require personal performance.

The laws therefore permit vicarious performance of a contractual liability, which is nevertheless subject to Section 42 of the Contracts Act of 1950 [5] (Sundra Rajoo [8]; see also Powell-Smith in Lim Chong Fong [9]). Keating [10] points out that a contract binds the contractor to complete the work, not necessarily by his own efforts, but, if he prefers, by vicarious performance through a subcontract or otherwise.

However, not all standard forms of contract in use in Malaysia permit the practice of subcontracting. For example, the form PAM with Quantities [7] clause 17.3 states that:

"... the Contractor shall not wholly or substantially subcontract the Work ...".

But clause 17.3 also states that subcontracting is to be permitted if it has been provided for in the Contract.

According to Keating [10], whether or not a contractor has the right of securing vicarious performance depends on the way in which a contract was constructed. He argues that the contractor has the right where it is a matter of indifference whose hand should do the work, and that if the Client consents to vicarious performance he may lose the right to any objection he might otherwise possess. The contractor does not have the right to allow another to perform the work if there is prohibition in the contract or there is some personal element in his obligation [10].

Client – Contractor – Subcontractor relationships

In terms of the client – contractor – subcontractor relationships, the appointment of a subcontractor is such that the subcontractor enters into a direct contractual relationship with the contractor. The client, who has a direct contractual relationship with the contractor, is not a party to the subcontract (Figure 11.1). As such, the contract conditions that the client entered into with the contractor do not generally take into account the contractor's subcontractor. Consequently, the subcontractor has no cause of action against the client for the price of work done or goods supplied under the contract [10]

In addition, a subcontractor's employment is terminated immediately if the contractor's employment under the contract is terminated (PWD 203A 2007 [5], clause 47.4).

"It shall be a condition in any subcontract which has been consented to by the Government that upon termination of the Contractor's employment under the Contract, the employment of the subcontractor under the subcontract shall terminate immediately. No claim whatsoever shall be made by the Contractor done or materials or goods supplied and/or subcontractor against the Government for any work done or materials or goods supplied".

The absence of a contractual relationship between the client and the subcontractor places the subcontractor somewhat at the 'mercy' of the contractor. One such situation occurs when the contractor pays his subcontractor only when he himself receives payment from the client. This 'pay when paid' practice has prevented subcontractors from receiving money for work they have completed on time. The problem of delayed payment or non-payment to subcontractors is serious and has contributed somewhat to a 'bad local culture' [11].

The construction industry and the government, for their parts, are taking steps to combat the various problems faced by subcontractors. For example, to promote a more equitable contract between contractors and their subcontractors, the Construction Industry Development Board Malaysia encourages the use of the subcontract form it published in 2006, entitled Model Terms of Construction Contracts for Subcontract Work [2].

In addition, a Malaysian Construction Industry Payment & Adjudication Act has also been proposed. The government believes that the Act (the Bill is yet to be tabled in Parliament) would be able to mitigate non-payment issues and help ensure a well-planned process of construction payment to subcontractors.

Part 3
Survey on the practice of subcontracting by contractors

In an effort to unveil the presence and extent of contractors subcontracting work to subcontractors, a questionnaire survey among contractors [ii] was carried out between December 2007 and March 2008 [12].

Specifically, the survey endeavours to:
(1) Identify the presence and extent of contractors subcontracting works to subcontractors, and
(2) Identify the motivation that drives contractors to subcontract work to subcontractors.

The questionnaire was structured in 3 parts. Part 1 relates to the contracting firms' background, Part 2 focuses on identifying the presence and extent of subcontracting by contractors, and Part 3 focuses on identifying the contractors' motivation to subcontract work to others (i.e. to subcontractors).

The responding contractors were selected through a simple stratified random sampling among Kuala Lumpur based contracting firms. All firms are registered with the Construction Industry Development Board Malaysia (CIDB) and the Contractors' Service Centre (CSC). The CIDB is a government agency with statutory powers under the Construction Industry Development Board Act (Act 520) of 1994 to register qualified contractors, and the CSC is a government agency entrusted to assist in the development of indigenous contractors.

A total of 78 contracting firms–registered in Class A to F and permitted to undertake work in fields ranging from civil and building to specialized engineering, mechanical, sanitary and water and telecommunications–were identified, and a set of questionnaires was sent to each firm by post and e-mail, followed up by telephone contact. A total of 28 responses (35.90%) were received. Table 11.1 shows a breakdown of respondents to the survey.

Table 11.1: Respondents' background (in terms of their registration with the CSC)

Class of Registration with CSC[iii]	No. of firms contacted	No. of firms responding
A (no financial limit on projects to be tendered)	13	6 (21.40%)
B	13	6 (21.40%)
C	13	7 (25.00%)
D	13	4 (14.40%)
E	13	2 (7.10%)
F (financial limit up to RM200,000.00)	13	3 (10.70%)
Total	78	28 (100%)

Data on the background of the respondents shows that the contracting firms have experience ranging from 5 to 20 years, and the persons providing the responses are those that hold positions in the firms ranging from managers to directors.

Given that the contracting firms are registered with the CIDB and CSC and that the persons providing the responses are people holding senior positions in their respective firms, it is therefore considered that the data they provided are reliable and appropriate for the purpose of the study.

The presence and extent of contractors subcontracting works to subcontractors

In the questionnaire, the contractors were asked about the practice of subcontracting, and to rate the extent of the practice. In response to the former question, the contractors were asked to state their answers on a 5-point scale ranging from 1 (strongly disagree) to 5 (strongly agree). For the latter question, only those who answered 4 (agree) and 5 (strongly agree) to the former question were asked to rate the extent of subcontracting as low, medium, or high. The questions were presented in accordance with the following 7 categories of work projects: civil engineering, building, mechanical, sanitary and water engineering, other specialist engineering, quarrying, etc., forest and land development, and telecommunications.

Table 11.2 shows the results of the survey in terms of the practice of contractors subcontracting work to subcontractors, and Table 11.3 shows the results in terms of the extent of subcontracting. The results, based on the high numbers and percentages of respondents scoring 4 (agree) and 5 (strongly agree), as shown in Table 11.2, suggest that the practice of contractors subcontracting work projects to others exists and that it exists in all seven categories of work.

Table 11.2: Results on the presence of the practice of subcontracting work according to categories of work

Categories of work	1 Strongly disagree		2 Disagree		3 Neutral		4 Agree		5 Strongly agree		Did not answer		Total	
	No	%	No	%	No	%	No	%	No	%	No	%	No	%
Civil	0	0	1	3.6	0	0	17	60.7	7	25	3	10.7	28	100
Building	0	0	1	3.6	0	0	21	75	6	21.4	0	0	28	100
Mechanical, sanitary and water	0	0	0	0	6	21.4	11	39.3	4	14.3	7	25	28	100
Other civil	0	0	0	0	5	17.9	7	25	5	17.9	11	39.3	28	100
Quarrying, etc.	1	3.6	1	3.6	5	17.9	7	25	5	17.9	9	32.1	28	100
Forest and land	1	3.6	1	3.6	6	21.4	8	28.6	2	7.1	10	35.7	28	100
Telecommunications	1	3.6	0	0	5	17.9	8	28.6	3	10.7	11	39.3	28	100

Table 11.3: Results on the extent of subcontracting according to categories of work

Categories of work	Low		Medium		High		Not relevant / not sure		Total	
	No	%	No	%	No	%	No	%	No	%
Civil	1	3.57	15	53.6	8	28.6	4	14.3	28	100
Building	4	14.3	10	35.7	14	50.0	0	0	28	100
Mechanical, sanitary and water	2	39.3	7	25	6	21.4	13	46.0	28	100
Other civil	0	0	6	21.4	6	21.4	16	57.1	28	100
Quarrying, etc.	0	0	6	21.4	6	21.4	16	57.4	28	100
Forest and land	0	0	6	21.4	4	14.3	18	64.3	28	100
Telecommunications	0	0	7	25	4	14.3	17	60.7	28	100

In addition, the results, based on the high numbers and percentages of respondents rating the extent of the subcontracting to be medium and high, as shown in Table 11.3, suggest that subcontracting is highest in building work projects, followed by civil engineering and mechanical, sanitary and water engineering work projects.

The results in terms of the high numbers and percentages of respondents scoring 4 (*agree*) and 5 (*highly agree*) on the presence of the practice of subcontracting in building work projects (96.4%), followed by civil engineering (85.7%) and mechanical, sanitary and water engineering work projects (53.6%); and in terms of the high numbers and percentages of respondents indicating medium and high on the practice of subcontracting in building work projects (85.7%), followed by civil engineering (82.2%) and mechanical, sanitary and water engineering work projects (46.4%), respectively, suggest that the data obtained is highly consistent in identifying the top 3 areas where subcontracting is the most prevalent, i.e. (i) building, (ii) civil engineering, and (iii) mechanical, sanitary and water engineering work projects.

Why do contractors subcontract work to subcontractors?

Respondents who indicated the presence of subcontracting were asked to state the motivation behind the practice. In the questionnaire, the motivations considered relevant by the author were classified into 2 broad categories, namely (i) internal – sub-categorized into factors within the control of the contractors themselves, such as attitude, financial constraints, and lack of knowledge and skills; and (ii) external – sub-categorized into factors outside the control of the contractors, such as procedural matters, constraint in the availability of key construction materials, constraint in the availability of adequate supply, properly maintained and efficient plant and equipment, and constraint in the availability of skilled and semi-skilled labour. The results (due to space constraints, only the top two in each sub-category, where applicable, are provided) are as shown in Tables 11.4 and 11.5, respectively.

Table 11.4: Results on 'internal reasons' why contractors subcontract work to subcontractors

Internal reasons for subcontracting	1 Strongly disagree		2 Disagree		3 Neutral		4 Agree		5 Strongly agree		Did not answer		Total	
	No	%	No	%	No	%	No	%	No	%	No	%	No	%
Attitude														
Lack of professionalism	3	10.7	5	17.9	7	25	12	42.9	1	3.6	0	0	28	100
Subsidy mentality	1	3.6	5	17.9	11	39.3	10	35.7	1	3.6	0	0	28	100
Financial constraint														
Cash flow problem	1	3.6	1	3.6	3	10.7	15	53.6	8	28.6	0	0	28	100
Could not provide collateral	1	3.6	2	7.1	3	10.7	16	57.1	6	21.4	0	0	28	100
Knowledge and skills														
Lack technical skills	1	3.6	10	35.6	5	17.9	12	42.9	0	0	0	0	28	100
Lack management skills	2	7.14	10	35.6	5	17.9	10	35.6	1	3.6	0	0	28	100

Table 11.5: Results on 'external reasons' why contractors subcontract works to subcontractors

External reasons for subcontracting	1 Strongly disagree		2 Disagree		3 Neutral		4 Agree		5 Strongly agree		Did not answer		Total	
	No	%	No	%	No	%	No	%	No	%	No	%	No	%
Procedures														
Late payments	2	7.1	0	0	4	14.3	16	57.1	6	21.4	0	0	28	100
Delay in getting loan approval	1	3.6	3	10.7	4	14.3	15	53.6	5	17.9	0	0	28	100
Materials constraint														
Price uncertainty	1	3.6	2	7.1	4	14.3	16	57.1	5	17.9	0	0	28	100
Monopoly by suppliers	1	3.6	4	14.3	3	10.7	7	25	13	46.4	0	0	28	100
Plant and equipment constraint *High rental charges*	1	3.6	7	25	6	21.4	13	46.4	1	3.6	0	0	28	100
Labour constraint *Over-dependence on foreign labour*	1	3.6	4	14.3	9	32.1	12	42.9	7.1	2	0	0	28	100

Part 4
Discussion and conclusion
Discussion of the findings

The literature review presented at the beginning of this chapter indicates that the law permits the practice of subcontracting and that the two major standard forms of contracts for construction work currently in use in Malaysia [5], [6] provided the pathway for contractors to subcontract work to subcontractors.

Contractors, subject to obtaining permission from the Client's representative, are therefore entitled to secure the vicarious performance of their obligations by means of subcontractors. Notwithstanding any subcontract permitted by the Client's representative and made by the contractor, the contractor shall remain liable for non-performance himself (Powell-Smith, in Lim Chong Fong [9], Sundra Rajoo [13], Keating [10], forms PWD 203A[5] and PAM with Quantities [6]).

In the context of the Malaysian construction industry, the findings on the presence of the practice of subcontracting confirm the general notion that contractors are subcontracting their own contracted work to others, otherwise known as subcontractors.

The findings have also suggested that the practice of subcontracting is widespread; i.e., subcontracting is present in almost all types of work: building, civil engineering, mechanical, sanitary and water engineering, specialist civil engineering, quarrying, forest and land development, and telecommunications work. Subcontracting is most prevalent in building work projects, followed by civil engineering, mechanical, sanitary, and water engineering work.

The presence of provisions relating to subcontracting in the major standard forms of contract in use in Malaysia suggests that the practice of subcontracting in Malaysia is not new. In addition, the widespread practice of subcontracting in Malaysia as suggested by the research findings is not unique to Malaysia. According to Ashworth [1] and Ofori and Lim [3], subcontracting has always existed, occurring in many different areas of life and in many industries. These authors observe that the practice of subcontracting in the construction industry is also widespread in the UK, in Singapore, and throughout the developed world.

Ashworth [1] contends that the widespread use of subcontractors illustrates the rapidly changing nature of the construction industry worldwide during the later part of the 20th century, as efforts are made to increase efficiency, effectiveness, and economic performance:

"...the scope of change and growth amongst subcontracting firms has been considerable. There are now relatively few main contractors who carry out a comprehensive range of the different construction trades. The process of change is of course evolutionary in order to meet the demands placed upon the industry by employers and the need for contractors to manage their work in the best possible way" [1].

The findings also indicate the motivations behind the widespread practice of contractors subcontracting works to subcontractors, namely the following:

(1) Internal factors, i.e. factors arising from within the contractors themselves
- Attitude – the contractors' lack of professionalism towards work and the presence of a subsidy mentality;
- Financial constraint – cash flow problems and the inability to provide collateral for the purpose of contract funding; and
- Knowledge and skills – lack of technical and management skills.

(2) External factors, i.e. factors outside the control of the contractors
- Procedural matters – delays in getting payment and delays in getting loan approvals;
- Materials constraint – price uncertainty and monopoly by suppliers;
- Plant and equipment constraint – high rental charges; and
- Labour constraints – over-dependence on foreign labour whose skills and commitment to the work are difficult to ascertain.

The above mentioned motivations to subcontract appear to be related to solving operational problems prevailing in the Malaysian construction industry. Khairuddin [14], [15] has pointed out that negative attitudes among contractors (reflected in such behavior as 'selling' contracts awarded to them by the government to others for a quick profit, subletting work for a fee, etc.), bureaucratic constraints, and constraints in the availability, price and timely supply of materials, plant and equipment, and indigenous construction by skilled and semi-skilled workers are not new to Malaysia. His research findings [14], [15] point out that those constraints have prevailed in the Malaysian construction industry since the mid-1990s and have had a significant impact on the construction industry's rate of growth and on the quality and timely delivery of infrastructure and other constructed facilities.

In contrast, in Singapore, Ofori and Lim [3] observe that subcontracting is used by contractors, among others, as a business strategy. Subcontracting works 'to help the contractor to lighten their load of the full construction works, and this in turn enables the contractor to deal with organizational related problems such as long-term environmental uncertainties and protecting the technical core of the contractors' employees against short-term contingencies.'

Conclusion

Conclusions drawn from the study are as follows:

(1) Subcontracting is recognized by law, is contractually permissible, and is akin to outsourcing.

(2) The practice of subcontracting in Malaysia can be found across all major sectors of the construction industry but is most highly concentrated in building, civil engineering, mechanical, sanitary and water engineering work projects.

(3) Seven key reasons or motivations driving contractors to subcontract work to subcontractors have been identified as follows:
   - Attitude – lack of professionalism and the presence of a subsidy mentality;
   - Financial constraint – cash flow problems and inability to provide collateral for the purpose of contract funding;
   - Knowledge and skills – lack of technical and management skills;
   - Procedural matters – delays in getting payment and delays in getting loan approvals;
   - Materials constraint – price uncertainty and monopoly by suppliers;
   - Plant and equipment constraint – high rental charges; and
   - Labour constraints – over-dependence on foreign labour whose skills and commitment to the work are difficult to ascertain.

(4) The practice of subcontracting by contractors appears to be primarily a way of addressing recurring operational problems rather than a carefully crafted strategy to achieve efficiency and to combat the effect of the cyclical nature of the construction industry.

The research findings are considered significant in terms of contributing data on subcontracting and on contractors' motivations in engaging in the practice.

However, the research and therefore its findings are not without limitations. These limitations are related to the methodology used for the research, especially the small sample size of only 28 contractors based in Kuala Lumpur, and the lack of further in-depth analysis and detailed discussions of the data. Consequently, it is important to carry out in the future a wider, more in-depth study on the subject of contractors subcontracting their work to subcontractors.

References

[1] Ashworth, A. (2001) *Contractual Procedures in the Construction Industry*. Pearson Education Limited Essex.

[2] CIDB (2006) *Model Terms of Construction Contract for Subcontract Work*. CIDB Malaysia, Kuala Lumpur.

[3] Ofori, G. and Lim, K.H. (2009). Multi-layered Sub-contracting System in Singapore's Construction Industry. In Kobayashi, K., Khairuddin Abdul Rashid, Ofori, G. and Ogunlana, S. (2009). *Joint Ventures in Construction. London*: Thomas Telford London, pp. 216-226.

[4] Laws of Malaysia, *Act 136*, Contracts Act 1950.

[5] PWD Form 203A (Revision 2007).

[6] PAM Contract 1998 (with Quantities).

[7] PAM Contract 2006 (with Quantities).

[8] Sundra Rajoo (1999) *The Malaysian Standard Form of Building Contract (The PAM 1998 Form)*. Malayan Law Journal, Kuala Lumpur.

[9] Lim Chong Fong (2004) *The Malaysian PWD Form of Construction Contract*. Sweet & Maxwell, Asia Petaling Jaya.

[10] Keating, D (1978) *Building Contracts* 4th *Edition*. Sweet & Maxwell, London.

[11] Master Builders Association Malaysia (2008) *Master Builders Association of Malaysia (MBAM) Response* on *Sub-contractors to be Paid Directly by WOR*. Online athttp://www.mbam.org.my/mbam/index.php?option=com_content&task=view&id=1004&Itemid=331 accessed 4th July 2010.

[12] Mazni Sahabuddin (2008). *Leakages Among Bumiputra Contractors*. Unpublished Undergraduate Dissertation. International Islamic University Malaysia.

[13] Sundra Rajoo, Dato' W.S.W. Davidson, Ir. Harbans Singh K.S. (2006) *The PAM 2006 Standard Form of Building Contract*. Lexis Nexis, Kuala Lumpur.

[14] Khairuddin Abdul Rashid (1998) *The Process of Construction Procurement in Malaysia: Identification of Constraints and Development of Proposed Strategies in the Context of Vision 2020*. Unpublished PhD Thesis. Nottingham Trent University.

[15] Khairuddin Abdul Rashid (2002) *Construction Procurement in Malaysia: Processes and Systems, Constraints and Strategies*. International Islamic University Malaysia, Kuala Lumpur.

Endnotes

(i) Nominated Supplier/Domestic supplier refer to supply only subcontracting; nominated subcontractor/domestic subcontractor refer to supply, install and complete subcontracting.

(ii) The survey was part of a data collection exercise for an undergraduate dissertation by Mazni Sahabuddin (2008), *Leakage among Bumiputra Contractors*, unpublished, International Islamic University Malaysia. The author is the grant holder and supervised the survey, FRGS Grant; Ref 0207-85.

(iii) All contractors are registered with the CIDB.

Joint Ventures in Construction 2
ISBN 978-0-7277-5783-8

doi: 10.1680/jvc2.57838.159

Chapter 12

Corruption in Construction Projects and Deterrence Policies in Developing Countries

Lei Shi, Dalian University of Technology
Taisuke Miyao, Asian Development Bank
Kiyoshi Kobayashi, Kyoto University

Introduction

Construction contracts for public works can be interpreted to mean that governments entrust construction work to the private sector. The relationship between the government and the private sector can be viewed as a principal–agent relationship, in which the adverse selection problem due to asymmetric information becomes the main focus. According to traditional agency theory, the adverse selection problem occurs if the principals cannot know exactly what type of agents they are dealing with before entering into a contract. Similarly, the adverse selection problem occurs in a construction contract, since the owner (principal) cannot know the type of contractor (agent) before signing the contract. It is therefore important to improve the information asymmetry that exists between the contractor and the government in public construction work.

Numerous cases of bribery and other irregularities have been reported to occur in construction projects in developing countries [1], [2]. In particular, due to their incompleteness, construction contracts for public works are prone to cause fraud, bribery, and collusion between local governments and contractors [3]. Corruption is generally defined as the misuse of public power for private benefit [4]. An accumulation of research based on development economics exists covering the effects of corruption on the economic growth or development of developing countries [5]-[8]. Literature in the field of public procurement points out that corruption may have a negative effect on the efficiency of public procurement [9]-[11]. On the other hand, many studies with a principal-agent approach have analyzed the design of rules to deter corruption, focusing on incentive mechanisms such as bonuses and penalties [12]-[16]. The point is to find ways to trade off the quasi-rent paid to the person who misuses power and the loss arising from corruption [13]. In this sense, social efficiency may not be accomplished even when corruption is completely eliminated. In addition, Rose-Ackerman found that penalties do not suffice to deter and may in fact trigger the occurrence of corruption [9]. Therefore, a penalty system should be implemented with careful deliberation. However, all of the research cited above focuses on the general impact of corruption on the efficiency of social welfare or organizations, and there is no theoretical analysis regarding construction project corruption.

Considering the parties concerned, corruption can be classified into two types. One is the collusion between the agent entrusted with supervisory power over the project and clients (hereafter external corruption); the other type is corruption occurring inside the organization responsible for the project (hereafter internal corruption). This chapter analyzes two types of collusion occurring at the implementation phase of a construction project, one between the contractor and officers entrusted with power, the other between an upper level officer and a subordinate officer inside the ordering organization.

On the other hand, corruption occurring in the implementation phase also affects the contractor's bidding strategy. We classify construction contractors into two types, low compliant contractors (hereafter LC contractor) and high compliant contractors (hereafter HC contractor). LC contractors are those who tend to engage in strategic behavior (bribery) which harms the project. On the other hand, HC contractors are compliant with the owner. Since corruption can benefit the LC contractor ex pose, the LC contractor can ex ante submit an undervalued price in the tender to defeat the HC contractor; that is, an adverse selection problem occurs. Many cases have been reported of adverse selection problems happening at the tender stage [17]. The government should take steps to prevent such problems by properly designing the tender system and incorporating a qualifying examination. However, the tender system is not effective if a contractor intends to go bankrupt or breach the contract [18].

When international donors finance projects in developing countries, they can clarify contractual clauses and bidding rules. However, it is difficult to deter collusion between the contractors and officers as well as collusion between upper level and subordinate officers. Therefore, we need to design a system, from the standpoint of international donors, to deal with these two types of collusion occuring in construction projects. In this chapter, we assume two types of bureaucratic structure: centralized organization and decentralized organization. In the former case, decisions concerning the implementation of the project are made through the whole organization, and the subordinate officer is not granted decision rights. In the latter case, the subordinate officer can carry out personnel decisions related to the project. So far, there has been little research on the impact of organizational stucture on collusion in the construction contract.

In this chapter, we investigate the mechanism of the adverse selection problem and corruption occurring in construction projects in centralized and decentralized organizations. Moreover, we analyze the impact of penalties on collusion, focusing on the allocation of authority in two different organizations. The remainder of the chapter is organized as follows. Section 2 builds a benchmarking model. The centralized organization model is formalized in section 3. Section 4 investigates the two types of collusion in the decentralized stucture. Concluding remarks follow in section 5.

Basic Model

Assumptions

The construction project is interpreted as a 2-period game. The main players in this game are an upper level officer, a subordinate officer, and a contractor. At the beginning of period 1, a tender for determining the contract price R is implemented. The upper level officer responsible for public works then enters into the contract with the winning bidder. Information about the type of contractor is asymmetric between the upper level (or subordinate) officer and the contractor. In other words, an adverse selection problem occurs. The contractor can be either the HC contractor denoted by θ_g or the LC contractor denoted by θ_b. Suppose that the HC contractor possesses excellent technology and the completion risk is small. But the cost overrun risk is large if the low compliant contractor is selected. In the first period, cost related to contractor type is defined as

$$F = \begin{cases} C_H & (\text{if } \theta_i = \theta_g). \\ C_L = 0 & (\text{if } \theta_i = \theta_b) \end{cases}$$

Suppose C_i $(i = g, b)$ satisfies

$$C_H = 0. \tag{1}$$

Additional cost K, which depends on contractor type, may occur at period 2. Define

$$K = \begin{cases} 0 & (\text{if } \theta_i = \theta_g) \\ \left.\begin{array}{l} 0 \text{ (with probility } p) \\ M \text{ (with probility } 1-p) \end{array}\right\} & (\text{if } \theta_i = \theta_b). \end{cases} \tag{2}$$

The additional cost is observable by both the upper level (or subordinate) officer and the contractor. According to (2), the additional cost is zero if the HC contractor is selected at period 1, whereas the additional cost is 0 or M with probability p or 1-p, respectively, if the first-period contractor is not efficient. The contractor then requires extra payment when the additional cost M is determined. From a cost minimizing perspective, the upper level (or subordinate) officer will reject the requirement in the absence of collusion between the upper level (or subordinate) officer and the contractor. However, the upper level (or subordinate) officer may take bribes from the contractor. Collusion between the contractor and upper level (or subordinate) officers is analyzed in sections 3 and 4. In the basic model, we assume that collusion is forbidden. In addition, we assume

$$R < M. \tag{3}$$

If (3) is not satisfied, the private contract never obtains negative profit. There is nothing but an income transfer from the government to the private contractor.

The contractor is selected through a tender process in this model. Adverse selection occurs if the LC contractor is selected. The last assumption is the premise condition for the adverse selection problem:

$$C_H \leq (1-p)M. \tag{4}$$

Define $\Delta = (1-p)M - C_H$

Equilibriums

The project cost is C_H when the HC contractor is selected, whereas the project cost covers 0 and the additional cost M which occurs with probability 1-p when the LC contractor is selected.

Therefore, the expected payoff $\prod_i^E (i = g, b)$ obtained by the contractor θ $(i = g, b)$ is written as

$$\prod_g^E = R = C_H \tag{5a}$$

$$\prod_b^E R - (1-p)M. \tag{5b}$$

According to (4), the condition ensuring the compliant contractor's participation in the project is

$$R \geq v_1. \tag{6}$$

Define $v_1 = C_H$ similarly, the participation condition for the low compliant contractor is written as

$$R \geq v_2. \tag{7}$$

Define $v_2 = (1-p)M.$

In the tender process at the beginning of the first period, the bidder offering the minimum price will ultimately be selected as the contractor. The equilibriums obtained by this tender process are referred to as competitive tender equilibriums. Suppose that a perfect competition is accomplished during the tender process. In the real world, considerable transaction costs occur, such as costs for participating in the tender, etc. Because in many cases the number of bidders who participate in the tender is restricted for cost reasons, perfect competition might not be achieved. There is extensive literature [19][20] investigating the corruption occurring at the tender stage. As we assume that perfect competition is accomplished in the tender process, the mechanism by which corruption occurs at this stage is beyond the scope of this chapter.

The minimum contract prices (hereafter marginal contract prices) satisfying (6),(7) by $R^{\theta_g}_{\min}$ and $R^{\theta_b}_{\min}$ respectively, can be written as

$$R^{\theta_g}_{\min} = v_1 \quad (8)$$
$$R^{\theta_b}_{\min} = v_2. \quad (9)$$

According to assumption (3), $v_1 \leq v_2$ holds. Therefore, the competitive tender equilibrium is

[competitive tender equilibrium A $\theta^* = \theta_g$, $R^* = v_1$.

In competitive tender equilibrium A, the compliant contractor is selected; thus the socially optimal contract is achieved.

Proposition 1 Under the assumption that collusion between the contractor and the upper level (or subordinate) officer is forbidden, adverse selection can be deterred by competitive tender.

Figure 12.1: Centralized bureaucratic structure

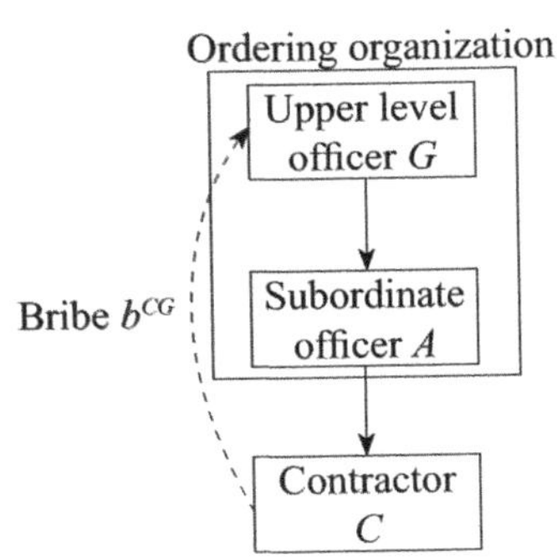

Centralized Bureaucratic Structure

Bargaining process

In the basic game, collusion is considered forbidden. However, the contractor might bribe the officer when the additional cost M occurs. As shown in Figure 12.1, in the centralized bureaucratic structure, the subordinate officer is not granted decision rights, which means that no collusion occurs between the upper level and the subordinate officers. On the other hand, we assume that collusion between the international donor and the officer is forbidden and that the donor's monitoring of misbehavior is not perfect, indicating that collusion can be detected by the donor at probability $r < 1$. In a case where collusion is exposed, the donor imposes a penalty H on the officer. Because it is enforced by the external organization, we call penalty H an external penalty comprised of not only monetary but also social and personal punishment. In order to obtain the sub-game perfect equilibriums, first consider the bargaining process regarding collusion between the upper level officer and the contractor. Consider a case in which the LC contractor has already signed a contract with the government. Given the contract price $\hat{R}$ and assumption (3), $\hat{R} < M$. The LC contractor pays a bribe to the upper level officer in order to obtain additional payment. The amount of the bribe is determined through bargaining between the LC contractor and the upper level officer. The status quo of the LC contractor and upper level officer is represented as $\hat{R} - M$ and 0, respectively. Meanwhile the gross payoff is $M - \hat{R} - rH$ when the bargaining is successful. If the officer has all the bargaining power, then he or she can demand a bribe of $M - \hat{R}$. In such an extreme case, bribery does not occur because the contractor has no incentive to bribe the officer. On the other hand, if the contractor has all the bargaining power, the contractor can get away with paying no bribe at all, and deterrence is zero. Many authors have taken the middle ground, whereby bribes are determined by a Nash bargaining solution with equal bargaining powers. The exact formulation does not matter significantly, but results must be interpreted accordingly. Using such a bribe determination process, it can be shown that the contractor would have to pay a bribe b^{CG}, which is

$$b^{CG} = (M - \hat{R} - rH)/2. \tag{10}$$

The upper level officer's expected payoff thus is

$$\prod^{G} = b^{CG} - rH = (M - \hat{R} - 3rH)/2. \tag{11}$$

Therefore, if and only if

$$rH \leq (M - \hat{R})/3, \tag{12}$$

$\prod^{G} \geq 0$, which means that the upper level officer's participation condition is satisfied. For convenience, we assume (12) holds in the rest of the chapter.

Equilibriums

The expected payoff $\prod_i^E (i = g, b)$ obtained by the contractor $\theta_i\ (i = g, b)$ is written as

$$\prod_g^E = R - C_H \tag{13a}$$

$$\prod_b^E = R - (1 - p)b^{CG}. \tag{13b}$$

The condition that ensures the HC contractor's participation in the project thus can be written as

$$M > R \geq v_1. \tag{14}$$

Meanwhile, the participation condition for the LC contractor is written as

$$M > R \geq v_3. \tag{15}$$

Define $v_3 = \{(1 - p)(M - rH)\}/(3 - p)$.

Now, consider the tender process. The bidder offering the minimum price will ultimately be selected as the contractor. The minimum contract prices (hereafter marginal contract prices) satisfying (13), (14) by $R_{\min}^{\theta_g}$ and $R_{\min}^{\theta_b}$, respectively, can be written as

$$R_{\min}^{\dot{\theta}_g} = v_1 \tag{16a}$$

$$R_{\min}^{\dot{\theta}_b} = v_3. \tag{16b}$$

Therefore, the competitive tender equilibrium is

$$\begin{cases} \text{competitive tender equilibrium } B\ \theta^{**} = \theta_g,\ R^{**} = v_1 \\ \text{competitive tender equilibrium } C\ \theta^{**} = \theta_b,\ R^{**} = v_3 \end{cases}$$

The necessary and sufficient condition (proof omitted) for the competitive tender equilibriums can be written as

$$\text{(competitive tender equilibrium } B) \begin{cases} \Delta \geq \tau_1 \\ \text{or} \\ rH > M/(4-p) \end{cases}$$

$$\text{(competitive tender equilibrium } C) \begin{cases} \Delta \geq \tau_1 \\ rH > M/(4-p). \end{cases}$$

Define $\tau_1 = (1-p)(C_H + M + rH)/2$ which is the threshold for comparing v_1 and v_3. In addition, $rH \leq M/(4-p)$ satisfies (12). In competitive tender equilibrium B, the HC contractor is selected, and collusion does not occur. However, in tender equilibrium C, the LC contactor is selected. As a result, collusion between the contractor and the upper level officer takes place.

External penalty and corruption

We took penalty H as a given in the above analysis. In order to analyze the impact of the penalty on collusion, however, we need to regard the penalty as a policy variable. From the foundation condition of equilibrium B, $\Delta \geq \tau_1$ ensures that the HC contractor is selected and no collusion occurs. However, if $\Delta > \tau_1$, the LC contractor is selected in the tender. In addition, $rH \leq M/(4-p)$ satisfies (12), which means that the LC contractor has the incentive to pay a bribe to the upper level officer. Therefore, the minimum penalty to prevent bribery is

$$H^* = \frac{1}{r}\frac{M}{4-p} \tag{17}$$

From the analysis above, we obtain the following proposition.

Proposition 2 In a case where the ordering organization has a centralized bureaucratic structure, implementing an external penalty satisfying (17) makes it possible to stop collusion between the contractor and the upper level officer.

Figure 12.2: Decentralized bureaucratic structure

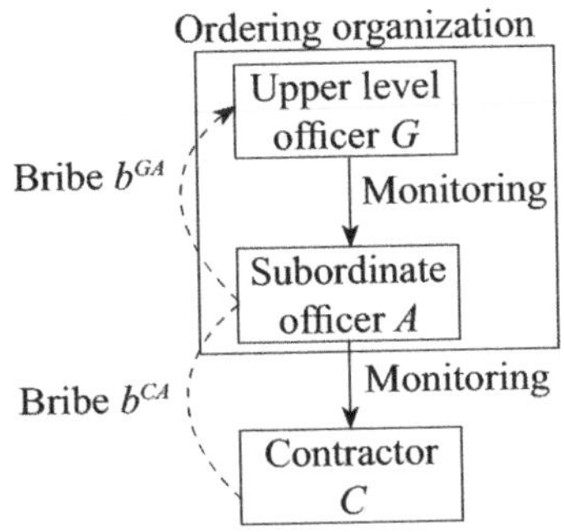

Decentralized Bureaucratic Structure

Bargaining process

In the decentralized structure, the subordinate officer is assumed to have decision rights regarding the project (Figure 12.2). Assume that the upper level officer can detect bribe taking by the subordinate officer with some probability q. Now, the subordinate officer can be subject to penalty G for bribery and possibly confiscation of the bribe. Because penalty G is implemented inside the ordering organization, we call penalty G an internal penalty. Following the detection of bribery, it is possible that the contractor also faces the same penalty and possibly an additional penalty for bribe giving. To keep matters simple, assume that it is not possible to punish the contractor (evidence might have been destroyed by the subordinate officer), and the upper level officer does not confiscate the bribe paid to the subordinate officer. However, there is nothing to guarantee that the upper level officer will take honest action. The upper level officer can also take a bribe from the subordinate officer and decide not to punish him/her.

In order to obtain the sub-game perfect equilibriums, first consider bribery-related bargaining between the contractor and the subordinate officer. The status quo of the upper level officer and the subordinate officer is represented as 0 and $-G$, respectively. Meanwhile the gross payoff is G when the bargaining is successful. The Nash bargaining solution to bribe negotiations between the upper level officer and the subordinate officer will simply be $G/2$, that is

$$b^{GA} = G/2 \tag{18}$$

But this also has important implications for bribe negotiations between the subordinate officer and the contractor. The status quo of the LC contractor and the subordinate officer is represented as $\hat{\hat{R}} - M$ and 0, respectively. Meanwhile the gross payoff is $M - \hat{\hat{R}} - qG$ when the bargaining is successful. The bribe b^{EA} to be paid by the contractor is given by

$$b^{EA} = (M - \hat{\hat{R}})/2 - qG/4, \tag{19}$$

where $\hat{\hat{R}}$ is the contract price which is taken as a given. The subordinate officer's expected payoff thus is

$$\Pi^{A} = b^{EA} = qG/2 = (M - \hat{\hat{R}})/2 - 3qG/4. \tag{20}$$

According to (20), if and only if

$$qG \leq 2(M - \hat{\hat{R}})/3, \tag{21}$$

$\Pi^{A} \geq 0$, which means that the subordinate officer's participation condition is satisfied. For convenience, we assume (21) holds in the rest of section 4.

Equilibriums

The expected payoff Π_i^E $(i = g, b)$ obtained by the contractor θ_i $(i = g, b)$ is written as

$$\Pi_g^E = R - C_H \tag{22a}$$

$$\Pi_b^E = R - (1 - p)b^{CA}. \tag{22b}$$

The condition that ensures the HC contractor's participation in the project thus can be written as

$$M > R > v_1. \tag{23}$$

Meanwhile, the participation condition for the LC contractor is written as

$$M > R > v_4. \tag{24}$$

Define $v_4 = \{(1-p)(M - qG/2)\}/(3-p)$.

Now, consider the tender process. The bidder offering the minimum price will ultimately be selected as the contractor. The minimum contract prices (hereafter marginal contract prices) in the equilibriums A and B by $R^E{}_{\min}$ and $R^F{}_{\min}$, respectively, can be written as

$$R^E{}_{\min} = v_1 \tag{25a}$$

$$R^F{}_{\min} = v_4 \tag{25b}$$

Therefore, the competitive tender equilibrium is

[competitive tender equilibrium D $\quad \theta^{**} = \theta_g,\ R^{**} = v_1$

[competitive tender equilibrium E $\quad \theta^{**} = \theta_b,\ R^{**} = v_5$

The necessary and sufficient condition (proof omitted) for the competitive tender equilibriums can be written as

$$\text{(competitive tender equilibrium } D) \begin{cases} \tau_2 \leq \Delta \\ \text{or} \\ qG > \dfrac{M}{5-2p} \end{cases}$$

$$\text{(competitive tender equilibrium } E) \begin{cases} \Delta \geq \tau_2 \\ qG > \dfrac{M}{5-2p} \end{cases}$$

Define $\tau_2 = (1-p)(C_H + M + qG/2)/2$. In competitive tender equilibrium D, the HC contractor is selected, and collusion does not occur. However, in tender equilibrium E, the inefficient contactor is selected. As a result, collusion between the contractor and the upper level officer takes place. On the other hand, the expected payoffs obtained by the upper level officer in competitive equilibrium D and E are

$$\text{[competitive tender equilibrium } D)\ \prod^G (D) = 0 \tag{26a}$$

$$\text{[competitive tender equilibrium } E)\ \prod^G (E) = qG/2. \tag{26b}$$

Internal penalty and corruption

The above analysis showed that if $\tau_2 \leq \Delta$, any penalty G is effective to stop both the external corruption occurring between the upper level officer and contractor and the internal corruption occurring between the upper level officer and subordinate officer. However, collusion between the upper level officer and contractor and that between the upper level officer and subordinate officer happens in the case where $\Delta < \tau_2$. According to the foundation condition of competitive tender equilibrium E, if the expected penalty qG satisfies

$$qG > \frac{M}{5-2p}, \tag{27}$$

the upper level officer rejects the bribe from the subordinate officer. As a result, collusion between the upper level officer and subordinate officer never happens. However, (27) may not hold in the case where the upper level officer determines the amount of penalty G. We next analyze the impact on corruption of decision making related to penalty G.

(26a) and (26b) show that the expected payoff obtained by the upper level officer receiving the bribe is larger, which means that the larger penalty G is, the larger the upper level officer's expected payoff will be. Therefore, the maximum penalty G that the upper level officer determines is

$$G^* = \frac{1}{q}\frac{M}{5-2p}. \quad (28)$$

Competitive tender equilibrium E therefore is achieved. According to (20), the expected payoff obtained by the subordinate officer is $\prod^A = 0$, which means that the quasi-rent (bribe) is squeezed by the upper level officer who determines penalty G. As a result, an internal penalty cannot stop corruption from occurring in the decentralized organization.

Responsibility of supervision

The donor cannot penalize the subordinate officer who receives the bribe because she cannot find collusion inside the organization. However, the donor can imperfectly monitor collusion between the subordinate officer and the contractor. When finding any irregularity, suppose the donor imposes a penalty T on the upper level officer for neglect of supervisory responsibility. Assuming the probability the donor finds the irregularity is s, the expected payoff obtained by the upper level officer is written as

$$\prod^{G'} = \frac{qG^*}{2} - sT. \quad (29)$$

Therefore, the minimum penalty T^* to deter corruption from occurring in the decentralized organization is

$$T^* = \frac{1}{s}\frac{M}{2(5-2p)}. \quad (30)$$

If we let $r = s$, then compare the minimum penalty H^* and T^* according to (28) and (30), we get

$$H^* > T^*. \quad (31)$$

(30) shows that the optimal external penalty T^* to stop corruption in the decentralized organization is smaller than H^* in the centralized organization. Therefore, the following proposition is obtained

Proposition 3 In a case where the ordering organization has a decentralized bureaucratic structure, implementing an external penalty satisfying (30) can deter both collusion between upper level officers and that between the subordinate officer and the contractor. In addition, the optimal external penalty in the decentralized organization is smaller than that in the centralized organization.

Conclusion

In this chapter, we analyzed the mechanism of collusion occurring in centralized and decentralized bureaucratic structures, focusing mainly on the two types of collusion common to construction projects in developing countries. One is collusion between the officer and the contractor, and the other is that between the upper level officer and subordinate officer. An external penalty is effective in preventing collusion from occurring in the centralized organization, while an internal penalty introduced in the decentralized organization triggers collusion between the upper level officer and the subordinate officer. We conclude that 1) implementing an external penalty makes it possible to stop collusion between the officers and the contractor; and 2) the optimal external penalty introduced in the decentralized organization is smaller than that in the centralized organization. However, there are a few remaining issues. First, this chapter shed light on the mechanism of corruption occurring in the implementation phase of construction projects, but there have also been many cases reported of corruption occurring in the bidding phase. Research is needed on the impact of corruption on the efficiency of public procurement. Secondly, the feasibility of implementing an external penalty has not yet been discussed. The issue of how penalties should continue to be implemented has become important. Thirdly, this chapter analyzes corruption occurring in a solo project. Corruption occurring in many projects may play a role in resource reallocation if externalities exist among the projects. Meanwhile, monitoring costs become huge if the number of officers is large. In this case, other incentive policies are necessary, such as external review and internal audit policies.

Acknowledgements

This work is supported by Science Foundation of Ministry of Education of China (08JC790012); Scientific Research Foundation for Returned Scholars, Ministry of Education of China; the Fundamental Research Funds for the Central Universities (DUT11RW415).

References

[1] Rose-Ackerman, S. (1978) *Corruption: A Study in Political Economy*. Academic Press, New York.

[2] Abdiweli, A.M. and Hodan, S.I. (2003) Determinants of Economic Corruption: A Cross-country Comparison. *Cato Journal* **22(3)**: 449-466.

[3] Dreher, A., Christos, K.and Steve, M. (2007) Corruption around the World: 13 Evidence from a Structural Model. *Journal of Comparative Economics* **35(3)**: 446-466.

[4] Lambsdorfe, J.G. (2007) *The Institutional Economics of Corruption and Reform*. Cambridge Press, New York.

[5] Williams, R. (ed.) (2000) *The Politicals of Corruption, Vol.1, Explaining Corruption*. Edward Elgar, Cheltenham UK.

[6] Williams, R. and Theobald, R., (ed.) (2000) *The Politicals of Corruption, Vol.2, Corruption in the Developing World*. EdwardElgar, Cheltenham UK.

[7] Williams, R., Moran, J. and Flanary, R. (ed.) (2000) *The Politicals of Corruption, Vol.3, Corruption in the Developed World*. Edward Elgar, Cheltenham UK.

[8] Williams, R. and Doig, A., (ed.) (2000) *The Politicals of Corruption, Vol.4, Controlling Corruption*. Edward Elgar, Cheltenham UK.

[9] Rose-Ackerman, S. (1975) The Economics of Corruption. *Journal of Public Economics* **4**: 187-203.

[10] Shleifer, A.andVishny, R.W.(1993) Corruption. *Quarterly Journal of Economics* **108**: 599-618.

[11] Aidt, T.S. (2003) Economic Analysis of Corruption: A Survey. *The Economic Journal* **113**: F632-F652.

[12] Becker, G. and Stigler, G.(1974) Law enformance, Malfeasance and the Compensation of Enforcers. *Journal of Legal Studies* **3**: 1-19.

[13] Tirole, J. (1994) The Internal Organization of Government. *Oxford Economic Papers* **46**: 1-29, 1994.

[14] Tirole, J.(1992) Collusion and theTheory of Organization. In Tirole, J.et al.(eds.)(1992). *Advance sin Economic Theory Sixth World Congress.* Cambridge University Press, Cambridge.

[15] Dhami, S. (2007) Corruption and the Provision of Public Output in a Hierarchical Asymmetric Information Relationship. *Journal of Public Economic Theory* **9(4)**: 727-755.

[16] Acemoglu, D. and Verdier, T. (2000)The Choice between Market Failure and Corruption. *American Economic Review* **90(1)**: 194-211, 2000.

[17] Campos, J.E. and Pradhan, S. (2007) *The Many Faces of Corruption.* Washington: The World Bank.

[18] Shi,L., Onishi, M., and Kobayashi, K. (2008) Pay off Externality and Contract Efficiency in BOT Projects. *Proceedings of the16th Seminar of JSPS-MOE Core University Program on Urban Environment.* pp.53-61.

[19] Compte, O., Lambert-Mogiliansky, A. and Verdier, T.(2005) Corruption in Procurement Auctions. *Journal of Economics* **36(1)**: 1-15.

[20] Menezes, F.M. and Monteiro, P.K. (2006) Corruption and Auctions. *Journal of Mathematical Economics* **42**: 97-108.

SECTION 3:
PERFORMANCE AND RISK

Joint Ventures in Construction 2
ISBN 978-0-7277-5783-8

doi: 10.1680/jvc2.57838.173

ice
Institution of Civil Engineers
publishing

Chapter 13

Key Factors of Successful Joint Ventures in Korea – Two Different Case Scenarios

Myungsik Do, Hanbat National University
Sang Hyuk Lee, Hanbat National University

Introduction

PPP (Public Private Partnership) has become a common way to implement SOC (Social Overhead Capital) investments in infrastructure in developed countries like the U.S., the U.K., etc., since the 1970s, due to an increase in government budget deficits. Since the 1990s, as industrialization and the capacity of the Korean economy has dramatically increased, infrastructure works such as roads, railways, schools, amenities, and environmental facilities have needed to be enlarged.

However, the government budget deficit has been deemed an impediment to carrying out SOC. PPP was recognized as a way to solve the government's budget constraints regarding economic and social infrastructure. A law supporting PPP for infrastructure works was passed in 1994 as "The Enactment of the Promotion of Private Capital into Social Overhead Capital Investment Act." In 1997, with the Korean economy in the midst of the Asian Financial Crisis, the Korean government decided to expand and improve SOC. For this reason, it needed foreign capital and foreign companies' advanced management systems, and sought to acquire these by means of joint ventures.

In its attempt to overcome budget constraints, lighten the financial burden, and help the private sector surmount financial difficulties, in 1998 the Korean government amended the above-mentioned act to the "Act on Private Partnership in Infrastructure (PPI)." This was followed by the government's official announcement of MRG (Minimum Revenue Guarantee), whereby contracted private sector organizations could receive subsidies from the government for up to 80~90% of expected annual operational revenue of infrastructure [1].

PPP can be implemented in different ways. One way is through a joint venture, whereby two or more groups who are interested in government issued projects come together as one group to carry out projects and to operate facilities for a certain period of time after projects' implementation [2]. Joint ventures are now a common way to fulfill infrastructure projects for reasons such as (1) minimal financial risk can be distributed to each member of a joint venture group, (2) companies lacking the qualifications and capability of completing projects on their own can be supported by companies which specialize in those areas, enabling joint ventures to promote small and medium sized businesses by transferring technologies from advanced groups and vice versa, and (3) rational construction management and optimal facilities operation management can be applied, therefore shifting from inefficient government systems to superior systems of the private sector.

Even though joint ventures have advantages for the public and private sectors, joint ventures can fail without government support. Furthermore, in order for infrastructure projects to succeed through joint ventures, we realized that both legal and financial support measures from the government are key factors. Legal support can provide private sector companies with stability and sustainability, and financial support can allow these companies to reinvest their money in infrastructure projects. Therefore, appropriate legal and financial assistance can stimulate and encourage the private sector to invest resources in the construction and operation of infrastructure.

In this study, the history of Korean joint ventures will be introduced and the key factors of successful joint ventures will be defined through case studies. The development process of Korean joint ventures is separated into four stages, namely the beginning, enhancement, transition, and maturity periods, and legislation based on the private sector's participation in infrastructure projects is introduced.

Furthermore, two different cases are introduced as cooperative and uncooperative cases within the public and private sectors. In this chapter we focus on joint venture projects within transportation related government and local government PPP projects. The cooperative case is a project of successful SOC implementation with the government's willingness to provide legal support and financial aid, and the uncooperative case is a project of local government based-SOC investment faced with a shortage of legal and financial support from the local government. In other words, the cooperative case was completed through the construction and operation stage by joint venture and had government support, and the uncooperative project confronted barriers preventing its start due to a lack of legal assistance. The case study shows that the cooperative case with public and private rewards conducted joint venture projects successfully. Thus, with both legislative and financial government support, joint ventures lighten the burden of budget constraints and increase the efficiency and effectiveness of infrastructure construction and operation.

Advancement and Current Status of Joint Ventures in Korea

The development process of Korean joint ventures can be classified into four phases based on legislation regarding private sector participation in infrastructure projects. Before the start of joint ventures in Korea, there were single acts of legislation such as the Road Act, the Port Act, etc. At that time, most infrastructure investments were pushed ahead with the government budget as government-sponsored development. Due to budget constraints, these acts of legislation were designated for permitting the private sector to participate in infrastructure development. Private investors were interested only in specific infrastructure like roads, ports, and railway stations.

The insufficiently detailed institutional strategies of individual acts of legislation limited private sector investment. A new law was needed to stimulate investment in infrastructure projects, so "The Enactment of the Promotion of Private Capital into Social Overhead Capital Investment Act" was legislated in August 1994. Accordingly, the period from August 1994 to March 1999 can be defined as the first phase of joint ventures in Korea. This act helps clarify the procedures for private investment in the targeted infrastructure projects and can define the criteria for the operating period of private sector participation and government support, etc. Even though institutional strategies were reinforced through legislation, only 10 infrastructure projects were bid upon out of 45 due to undeveloped financial markets, lack of experience, insufficient capabilities of public and private sectors, and the Asian Financial Crisis [3]. So in order to stimulate investment and overcome the Asian Financial Crisis, the act was amended to the "Act on Private Partnership in Infrastructure (PPI)" in December 1998.

The second phase, the enhancement period of joint ventures from April 1999 to December 2005, began when the MRG (Minimum Revenue Guarantee) went into effect in April 1999. MRG provided government revenue to the private sector when revenue could not otherwise be raised. MRG relieved the revenue risk and future uncertainties for private sector participants operating infrastructure. Beginning in 1999, MRG encouraged investment in infrastructure by foreign and domestic private investors.

The benefits of MRG were reduced due to erroneous travel demand forecasting and an increased government budget deficit [1]. The third phase of joint ventures occurred from January 2006 to October 2009 and was called the transition period. During this period, the benefits of MRG were reduced in public-proposed infrastructure projects and abolished in private-proposed infrastructure projects. Therefore, both the private sector and the government realized that an appropriate estimate of demand and cooperation between the two sectors were key factors for successful infrastructure projects.

The fourth phase was the maturity period of joint ventures and has taken place from November 2009 to present. At the beginning of this period, MRG was abolished in all joint venture projects to prevent an inappropriate estimate of demand and to relieve the financial burden on the government. Through their accumulated experience with PPP projects, both the government and the private sector are constantly seeking the best solutions to construction and operation problems. Even though the benefits of MRG have been abolished for infrastructure projects, private investments have been gradually increasing. Table 13.1 shows a summary of the development process of Korean joint ventures.

Table 13.1: The development process of Korean joint ventures

Phase	Period	Characteristics
First Phase	Aug. 1994 ~ Mar. 1999	– "The Enactment of the Promotion of Private Capital into Social Overhead Capital Investment Act" was legislated – Asian Financial Crisis occurred – "Act on Private Partnership in Infrastructure (PPI)" was amended
Second Phase	Apr. 1999 ~ Dec. 2005	– MRG was introduced – Joint venture was widespread in SOC projects
Third Phase	Jan. 2006 ~ Oct. 2009	– MRG benefits for private investors were reduced and abolished – Lead appropriate estimate of demand and prevent moral hazards
Fourth Phase	Nov. 2009 ~ Present	– MRG benefits for private and government proposed infrastructure projects were abolished – Appropriate demand estimation and legal and financial supports from the government became the key factors

Transportation related assets such as roads, railways, and air and marine ports were valued at approximately 215 trillion KRW (197 billion USD) in 2004 [4]. In spite of the increasing amount of investment, it is insufficient to satisfy all demand for economic and social infrastructure [5]. Therefore, private investment has been increasing for public-proposed and private-proposed projects to overcome the budget constraints of the government, after the legislation of PPI Act.

To complement insufficient public investment, private investment has come to play a key role in providing infrastructure. The proportion of private investment to public investment in SOC has increased from 6.6% in 1998 to 17.3 % in 2008 [6]. Table 13.2 shows detailed information regarding the proportion of private investment in infrastructure.

Table 13.2: The proportion of private investment (Unit: Billion USD, %)
(Source: Do and Park [1])

	2000	2001	2002	2003	2004	2005	2006	2007	2008
Private	0.91	0.50	1.09	1.09	1.54	2.36	2.73	2.73	3.09
Government	13.82	14.54	14.54	16.73	15.82	16.64	16.73	16.73	17.82
Ratio (%)	6.6	3.4	7.5	6.6	9.8	14.2	16.3	16.3	17.3

Until September 2009, the total number of PPP projects, including joint ventures contracted and awarded, was 461. 106 BTO (Build Transfer Operate) and 145 BTL (Build Transfer Lease) projects have been completed to provide service to the public [6]. BTO projects have been a common method for constructing infrastructure in transportation related facilities such as roads, railways, ports, etc.; road projects especially account for more than half of all investment [7]. BTL has different aspects of conditions compared with BTO. Only BTL projects permitted by the government, for instance, educational facilities, cultural facilities and welfare facilities – not profitable businesses – can be implemented and operated by private investors.

Cooperative Case Scenario of a Joint Venture

The Incheon International Airport Expressway

In the late 1980s, highway traffic demand had been increasing dramatically with economic growth, but due to budget constraints of the government, the construction of highway facilities had failed to keep up with demand. Traffic congestion on the highways and social costs had risen considerably. For this reason, the Korean government decided to construct a new highway connecting Seoul to Incheon International Airport. Both the basic and final design were completed from November 1990 to February 1994, with construction of the first sector (Yeonyouk Bridge: 3.6 km) beginning in December 1993 as a result of an increase in the government budget [8].

With the legislation of "The Enactment of the Promotion of Private Capital into Social Overhead Capital Investment Act," private investors-several construction companies-took over the construction of the remaining sectors (36.6 km) in November 1994. The Incheon International Airport Expressway was the first infrastructure facility completed through a joint venture in Korea after the act was legislated in 1994.

For the Incheon International Airport Expressway project, 11 companies established a consortium called the "New Airport Highway Corporation" and made agreements with the Korean government regarding construction costs, budget procurement methods, and operation periods. The first construction project by means of joint venture began in November 1995 and was completed in November 2000. The total cost of the project was 1.4 billion USD, with 0.21 billion USD coming from the government budget. Upon project completion, then, the government became the owner of the Incheon International Airport Expressway, with the right of highway operation belonging to the New Airport Highway Corporation for the subsequent 30 years for the recovery of investment and operating costs, in accordance with the BTO method [1].

Figure 13.1: Location of Incheon International Airport Expressway

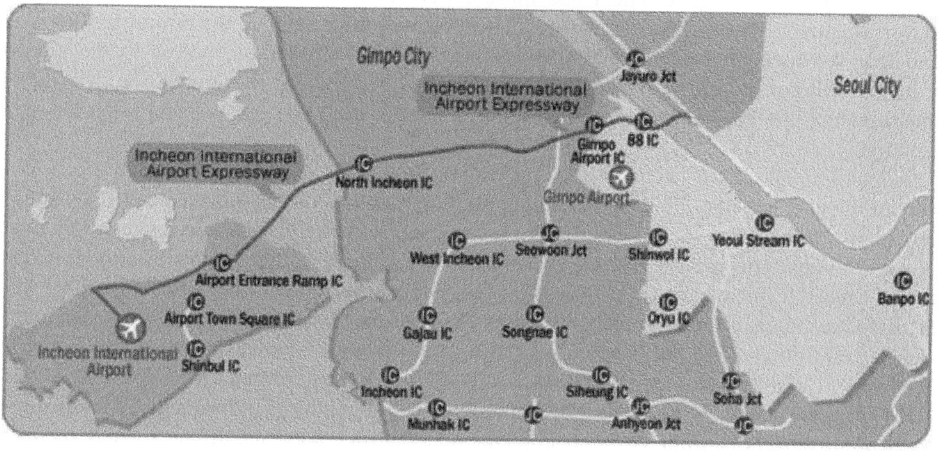

Later the consortium sold their stakes to financial companies such as the Macquarie Group. Table 13.3 provides a list of joint venture companies, ratios of investment in the Incheon International Airport Expressway, and current stockholders of the New Airport Highway Corporation.

Table13. 3: List of joint ventures in the consortium and current stock holder of company (Source:https://www.hiway21.com/index.html [9])

Investors (Shareholders)	Ratio of Investment (%)	Stockholders of Company	Ratio of Stock held (%)
Samsung C & T Corporation	29.50	Korea Teachers' Credit Union	45.07
Hanjin Heavy Ind. & Con.	23.43	Macquarie Korea Infrastructure Fund	24.10
Dong Ah Construction	10.50	Kyobo Life Insurance Co., LTD	15.00
Posco E & C	6.93	Samsung Life Insurance Co., LTD	8.85
Kolon E & C	6.00	Korea Life Insurance Co., LTD	3.50
LG E & C	5.00	Woori Bank Co., LTD	2.10
Kumho E & C	4.86	Samsung Fire & Marine Co., LTD	1.38
Kukdong E & C	3.45	–	–
Lotte E & C	3.45	–	–
Poong Lim Construction	3.45	–	–
Doosan E & C	3.43	–	–

As the Incheon International Airport Expressway project was the first private investment project in Korea, the Korean government had more experience with development than the private sector. Realizing that Incheon International Airport would be seen as the airport hub of Asia and that the Incheon International Airport Expressway was the sole access to the airport from other domestic cities at that time, the Korean government recognized the priority of this infrastructure project and supported it through legislation and guaranteed profits for joint ventures.

One of the guaranteed policies for encouraging joint ventures in this project was MRG, which provided revenue from the government to private sector participants when revenue fell short of expectations. This project collected revenue from tollbooths along the expressway. Initially, the forecasted traffic demand was overestimated. For example, the traffic demand for 2001 was predicted as being 110,622 vehicles per day. This estimate was made before the government had agreed to undertake the development as a joint venture [1]. However, the actual traffic in 2001 was 51,936 vehicles per day. This was only 46.9 % of the expected traffic demand because other projects such as the airport support complex, the international business district, and other residential projects had not yet been implemented. As time went by, circumstances changed and other facilities such as the Incheon Bridge and the Airport Railroad were implemented to connect with Incheon International Airport.

Prediction of future demand is very difficult but very important. Even though the expected traffic demand did not reach the actual estimate, MRG was able to make this project a profitable joint venture investment within 20 years of its completion. In 2004, MRG policies were revised to prevent future overestimated traffic demand and to reduce government spending. Detailed MRG information can be found in Table 13.4.

Table 13.4: MRG comparison of the Incheon International Airport Expressway

	Original MRG (December 2000)	**Revised MRG (April 2004)**
Guaranteed Revenue	– 20 years after implementation – Up to 90% of expected revenue	– 20 years after implementation – Up to 80% of expected revenue
Redemption Revenue	– 20 years after implementation – Over 110% of expected revenue	– 20 years after implementation – Constant ratio revenue over 80% of expected revenue

Also, Table 13.5 shows the status of traffic demand differences and guaranteed operational revenue of the Incheon International Airport Expressway project.

Table 13.5: MRG status of guaranteed operational revenue of the Incheon International Airport Expressway
(Source:http://www.mltm.go.kr/ [10])

Year	Expected Traffic Demand (Vehicles/ day) (A)	Observed Traffic Demand (Vehicles/ day) (B)	Ratio(A/B)	Received Revenue from the Government by MRG (Millions, USD)
2001	110,622	51,936	2.13	56.3
2002	121,496	54,244	2.24	65.0
2003	133,438	55,323	2.41	90.8
2004	146,554	59,780	2.45	96.1
2005	119,026	62,831	1.89	62.9
2006	125,322	65,571	1.91	67.6
2007	131,965	68,711	1.92	72.7
2008	138,930	64,956	2.14	85.7
2009	146,282	62,165	2.35	90.5
2010	93,094	53,490	1.74	65.7

The Incheon International Airport Expressway project is an example of a successful joint venture because the rate of return on the investment surpassed the rate of interest. It is also an example of government success because the necessary infrastructure was carried out without putting forth an enormous amount of budgetary money all at once. As a result, government support of and legislation regarding infrastructure development stimulated the private sector to invest in both public-proposed and private-proposed projects.

Uncooperative Case Scenario of a Joint Venture

Yongin Light Electric Railway

After the Asian Financial Crisis came to an end, population growth accelerated in the Seoul area. Urban sprawl affected Seoul satellite cities such as Gwangmyeong, Anyang, Goyang, and Yongin, causing unsustainable development and traffic congestion in suburban areas. The nation's largest amusement park and numerous manufacturing plants in Yongin also resulted in increased traffic demand between Seoul and Yongin. In order to deal with increasing future traffic demand and to establish an efficient city transportation system in Yongin, the City of Yongin decided to construct a light electric rail connecting residential districts and the CBD (Central Business District), with an 18.8km overall length of railway. Light railway is a complementary public transportation system using electricity instead of the less energy efficient systems of heavy rail and bus. The light rail system has lower capacity and speed than heavy rail, but higher capacity and speed than the public bus system.

The City of Yongin decided to push ahead with a light electric railway in December 1996, but implementation of the plan was postponed due to budget constraints. Yongin Light Electric Railway decided to invest in this project by way of a joint venture. Therefore, in July 2004, the City of Yongin made an agreement through a consortium formed by Bombardier Transportation of Canada, Hanjin Heavy Industries and Construction, and Iljin Electric with respect to the Yongin Light Electric Railway project. According to the agreement, the City of Yongin has the right of ownership of Yongin Light Electric Railway, and the consortium has the right to operate Yongin Light Electric Railway for 30 years, in return for investing 59% of the total budget of this project using the BTO method.

The agreement included MRG of up to 90% of the expected annual operational revenue for 20 years. When the Yongin Light Electric Railway agreement was prepared in 2001, the forecasted future demand (number of daily passengers) was estimated to be 146,000 persons per day. However, demand decreased not only because other public transportation systems were improved due to the integrated fare system of metropolitan public transit and the execution of weekday exclusive bus lanes, but also because the development of eastern Yongin and the extension of the Bundang line were delayed [6]. According to the MRG agreement, the City of Yongin was responsible for compensating Yongin Light Electric Railway for the revenue shortfall, and a revised agreement was negotiated to lower the MRG rate from 90% to 79.9%. Table 13.6 denotes the progress of construction of Yongin Light Electric Railway.

Figure 13.2: Location of Yongin Light Electric Railway

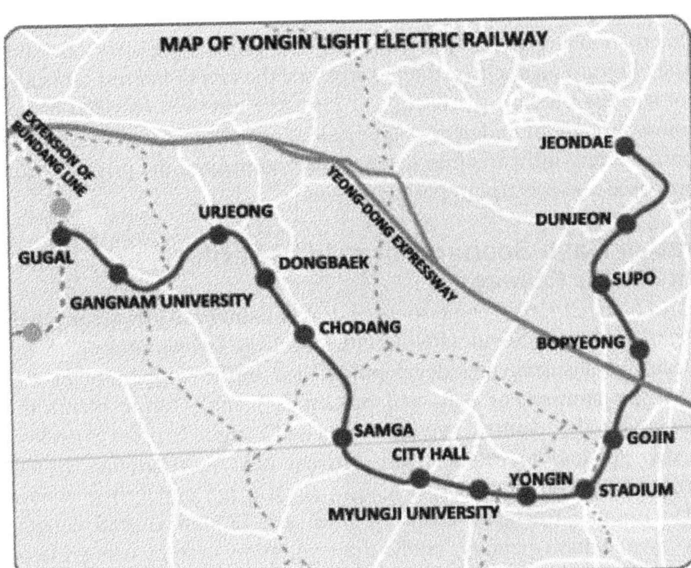

Table 13.6: Construction progress of Yongin Light Electric Railway (Source:Yong-in Light Electric Railway Co. [6], [11])

	Contents
Dec 1996	– Establishing construction and general operation plans of Yongin Light Electric Railway
Dec 2001	– Notifying of general plans for PPP projects (Notification of Yongin city No. 2001-295)
Sep 2002	– Designating the preferred bidder
Mar 2004	– Passing deliberation by deliberation committee of PPP projects (Ministry of Planning and Budget)
Apr 2004	– Establishing Yongin Light Electric Railway Co., Ltd.
Jul 2004	– Designating a business executor, Entering into an enforcing agreement, Entering into a financial agreement, Entering into a construction contract and a management contract.
May 2005	– Applying for approval to enforce plans
Oct 2005	– Completing deliberation of city management plans (Approval of plans for a city train project) by Gyeonggi province city planning committee (Ministry of Construction and Transportation)
Dec 2005	– Starting construction
Dec 2010	– Filing provisional disposition to cancel the refusal to confirm completion of Light Electric Railway
Jan 2011	– Notifying of project cancellation of Yongin Light Electric Railway Co., Ltd.
Feb 2011	– Filing arbitration to International Committee of Arbitration (ICA)
Oct 2011	– First verdict as City of Yongin pays project expenses to Yongin Light Electric Railway by ICA

The Yongin Light Electric Railway construction costs thus far have been approximately 693.1 million USD, but the City of Yongin has delayed completion through administrative procedures related to the lack of noise reduction, boarding systems, inappropriate components in the trains, lack of financing plans, etc. Yongin Light Electric Railway claims that these are spurious reasons for delays in completion, that in fact the railway was planned by the former mayor of Yongin, that there was failure to take into consideration other related transportation projects and the delayed extension of the Bundang line and metropolitan subway, that demand was overestimated, and that the rate of MRG would therefore become an excessive burden on the City of Yongin. After a new mayor was elected, the project was reviewed and the City of Yongin has found ways to reduce the budget by cutting back on excessive expenditures.

The Yongin Light Electric Railway has a shortage of operation funds to run the system; in particular they claim that the project is in limbo because they will soon default as a result of not being able to pay the interest on the 466.7 million USD bank loan [6]. The Yongin Light Electric Railway filed for arbitration calling for the payment of 723.8 million USD including payment according to the cancellation of an enforcing agreement with Yongin Light Electric Railway and compensation for damages, etc., against the City of Yongin, to the International Committee of Arbitration (ICA) through the International Chamber of Commerce (ICC) [6].

The first verdict, issued by the ICA in October 2011, requires that the City of Yongin pay 491.3 million USD to Yongin Light Electric Railway. In addition, the City of Yongin charged Yongin Light Electric Railway with alleged corruption and overestimation of forecasted demand, etc. Consequently, this pork-barrel project has led to the misuse and waste of taxpayers' money.

Key Factors of Successful Joint Venture Projects

As these two cases demonstrate, government support through legislation and unwavering commitment are key factors for conducting infrastructure projects carried out by joint venture. Additionally, in order for projects to be successful, joint ventures need to estimate more accurately project expenditures and future demand.

Key factors for joint venture projects to succeed are (1) the selection of proper infrastructure projects for joint ventures, (2) the selection of a qualified private partner, (3) government financial support, and (4) mutual cooperation between the government and joint ventures.

When joint venture projects are selected, certain criteria should be considered, such as project profitability, expected financial support capability, and project area potential. For analyzing such criteria, appropriate future demand estimation is important for joint venture projects. However, both completed and current joint venture projects have not correctly estimated future demand according to systematic and scientific methods, thus causing some projects to fail. Due to inaccurate estimates of future demand and unethical and fraudulent actions of some private investors who intentionally overestimate future demand, the benefits of MRG have been reduced for government-proposed projects and abolished for private-proposed projects.

Selecting joint ventures for certain infrastructure projects also has procedural problems. In the negotiation phase, the preferred bidder is screened only once, so it is not easy to select the best proposal for quality and cost. In addition, even though the winning bidder may not meet certain criteria, the bidder can be successful in obtaining the project because of the composite score system. For a more detailed description of the selection procedures of joint ventures, it is suggested that readers refer to the *Current Status and Perspectives of Public Private Partnership for Infrastructure Projects in South Korea* [1].

Infrastructure facilities are public goods, constructed under budget, so user costs are often low because the government bears the responsibility for the deficit created by operational expenditures. If joint ventures did not receive government financial support, they would raise fees to cover operational expenditures. In other words, either the government or users must pay for utilizing infrastructure to defray the costs. A key factor in carrying out infrastructure projects successfully, government financial support can prevent user resistance and aid the supervision of joint ventures. Proper financial support, such as compensation for foreign exchange loss, MRG, and other financial support legislation, can stimulate potential joint venture participants to invest in infrastructure projects. Therefore, the government should create a better investment environment to lessen the burden of operational expenditures on investors, in particular by providing financial support for large-scale SOC projects which inevitably must be implemented regardless of low economic or feasibility demands.

As the two different cases of this study demonstrate, mutual cooperation between the government and joint ventures is one of the key factors in carrying out SOC projects successfully in Korea. SOC projects require huge amounts of capital for construction and operation. For successful SOC projects, joint ventures raise the necessary funds and look for technical partners for infrastructure projects, while the government provides support for joint ventures through legal and administrative means. Hence, the government and joint ventures should recognize that mutual cooperation helps guarantee the goals of successful implementation of SOC projects for both citizens and companies alike.

Conclusion

As Korean society and its economy developed in the early 1990s, it was necessary to expand infrastructure, but the government faced difficulty due to budget constraints. New methods were needed to replace the traditional involvement of only the government in infrastructure construction and operation. With its advantages of risk diversification regarding costs and benefits and experience sharing with partners, etc. [12], the joint venture has emerged as a new paradigm in social and economic infrastructure.

Joint ventures never invest in projects without profitable guarantees, especially infrastructure investments requiring huge amounts of capital and long-term commitment. SOC is not typically a business area because certain benefits and profit cannot be quantified in currency value, such as serviceability, accessibility, mobility, etc. Thus the government must guarantee a certain amount of profit to private investors for infrastructure projects through particular legislation and administrative support, both of which are very important to maintaining the vital role of joint ventures in expanding and improving infrastructure in Korea.

After the first legislated law "The Enactment of the Promotion of Private Capital into Social Overhead Capital Investment Act" in 1994, joint ventures were expected to alleviate the budget constraints of the government for infrastructure projects. However, overestimation of future demand and the lack of the government experience made it necessary for the government to spend a huge amount of money to compensate for the shortage in operating expenditures, according to the MRG agreement. Accordingly, the government had to reduce the MRG benefits after 2006, although MRG remains a helpful system for encouraging investment in government-proposed projects.

As the above cases of joint ventures in Korea have shown, the mutual cooperation of government with the private sector is the key factor to successful joint venture projects and maintaining a level of investment conducive for expanding and improving infrastructure. The government must be willing to support joint ventures by eliminating any barriers to investing actively in infrastructure development. These days, joint ventures are absolutely necessary to the expansion and improvement of infrastructure facilities in Korea.

References

[1] Do, Myungsik and Park, Hyeon (2011) Current Status and Perspectives of Public Private Partnership for Infrastructure Projects in South Korea. *Proceedings of international workshop on taking the stock of public-private partnerships and private finance initiatives around the world.* 151-168.

[2] Farooqui, U. Rizwan, Ahmed, M. Syed and Umer Dilshad (2009) Best Practices for Joint Ventures: Issues and Critical Success Factors. *ICCEM-ICCPM.*369-374.

[3] Wang, Se Jong (2005) *A Study on the Role of PPI Project on National Economy*. Construction & Economy Research Institute of Korea.
[4] Ahn, H. and Kim, M. (2006) *Reginal Allocation of Transportation Infrastructure Investment and Development of Local Economy*. Korea Research Institute for Human Settlements.
[5] Do, Myungsik and Kwon, Sooahn (2009) Current status and perspectives of asset management in South Korea. *Proceedings of international workshop on asset management implementation in Asian countries*. 17-35.
[6] Do, Myungsik and Choi, Keechoo (2011) Current Status of Private Transportation Investments and the Perspective of Yong-In Light Rail Transit. *Transportation Technology and Policy* **8(2)**: 25-35.
[7] Korea Development Institute (KDI) (2010) *Public-Private Partnership in Infrastructure in Korea Brochure*. http://www.kdi.re.krl [Accessed in 2011].
[8] Kim, Sigon, Oh, Jaehak and Kang, Seungpil (2004) An Experience of Construction and Operation of the Incheon Airport Exclusive Expressway by Private Capital. *Journal of Korea Society of Civil Engineers* **23(2D)**: 143-148.
[9] Incheon International Airport Expressway. https://www.hiway21.com/index.html [Accessed in 2011].
[10] Ministry of Land, Transportation and Maritime Affairs. http://www.mltm.go.kr.
[11] Yong-in Light Electric Railway Co. http://www.ever-line.co.kr/ [Accessed in 2010].
[12] Suhr, Myong Suk and Kim, Moo Han (2002) A Study on the Effective Operation model of Joint-Venture Contract-With a Focus on the Example of Domestic Construction Industry. *Korea Journal of Construction Engineering and Management* **3(3)**:103-111.

Joint Ventures in Construction 2
ISBN 978-0-7277-5783-8

doi: 10.1680/jvc2.57838.185

Chapter 14

Application of Target Cost Contracts and Guaranteed Maximum Price Contracts in the Construction Industry of Hong Kong

Joseph H.L.Chan, The University of Hong Kong
Daniel W.M.Chan, The Hong Kong Polytechnic University
Patrick T.I.Lam, The Hong Kong Polytechnic University

Introduction

The construction industry has long been fraught with a lack of co-operation, limited trust between the employers and contractors and misalignments of common objectives, often leading to an adversarial working relationship amongst members within the project team and eventually impairing overall project performance [1]. A multitude of industry review reports have also identified that employers in both the public and private sectors were not satisfied with the traditional procurement approach (e.g. Latham [2], Egan [3], Construction Industry Review Committee [4]).

There have been strong calls to adopt alternative integrated procurement strategies in order to overcome the problems with fragmented working relationship and lack of incentives for project team members to contribute more than just fulfilling the minimum contractual requirements under the traditional design-bid-build approach[2],[4]. Guaranteed Maximum Price (GMP) and Target Cost Contracts (TCC), which align the individual objectives of the employer and contractor, have been applied in the construction industry of the United Kingdom, Australia and the United States to achieve more satisfactory project outcomes [5] - [7].

Both TCC and GMP (TCC/GMP) are relatively novel in Hong Kong. There have been some examples of application of TCC in the private sector and quasi-government sector in Hong Kong over the recent years [8], reflecting that its application is still at a germinating stage of development. In the public sector, the Government of the Hong Kong Special Administrative Region has introduced the first New Engineering Contract Version 3 (NEC3) Option C (target cost with activity schedule) on a pilot basis to an open nullah improvement works project in Sai Kung commissioned by the Drainage Services Department in 2009 [9]. TCC has been practised in the infrastructure sector projects of Hong Kong such as the Tseung Kwan O Railway Extension, Tung Chung Cable Car Project, Tsim Sha Tsui Metro Station Modification Works [10], West and South Island Railway Lines. On the other hand, GMP has been applied in some private residential buildings and office developments in Hong Kong [8]. However, not every project procured with TCC/GMP is equally successful in terms of time, cost and quality. For instance, Bogus *et al.* [7] reported that contracts with GMP provisions were more likely to have no schedule or cost overruns when compared with contracts with lump-sum provisions in municipal water and wastewater facilities projects in the United States. Chan *et al.* [10] reported a case study on an underground railway station modification works project with TCC in Hong Kong. This case study concludes that both time and cost savings can be achieved with TCC contractual arrangement. In contrast, an earlier research by Rojas and Kell [11] suggested that only 6 of the 24 public school projects investigated were completed at or below the contract GMP value. This finding did not support the expectation that GMP was really "guaranteed".

It is thus valuable and indispensable to explore the underlying motives as well as the potential pitfalls of adopting TCC/GMP procurement strategies. The objectives of this chapter are to review the development and application of TCC/GMP in the construction industry and discuss the key motives of application and potential pitfalls associated with these procurement approaches. This chapter is envisaged to assist the decision-makers in providing some useful pointers into applying TCC/GMP at an early stage of project delivery and capture the lessons learnt from previous cases of TCC/GMP in both the local and international context.

Concepts of TCC and GMP

Definitions of TCC and GMP

Target Cost Contract (TCC)

A target cost contract is considered as a means of sharing risks between the employer and contractor with the latter being encouraged to maximise his performance [12]. Broome and Perry [13] suggested that a target cost is introduced in this kind of project and any cost under spent or overrun against the target cost is divided with pre-agreed and specified portions. Wong [14] regarded that the contractor was paid the actual cost for the works done during the construction stage. When the final construction cost, which is termed as the final total cost, differed from the initial contract target cost, the difference would be shared between the employer and the contractor based on a pre-determined gain-share/pain-share ratio stated in the contract. The gain-share/pain-share operational mechanism of TCC is portrayed in Figure 14.1.

Figure 14.1: Gain-share/pain-share mechanism of TCC procurement strategy in construction
(Adapted from the Construction Industry Council [15])

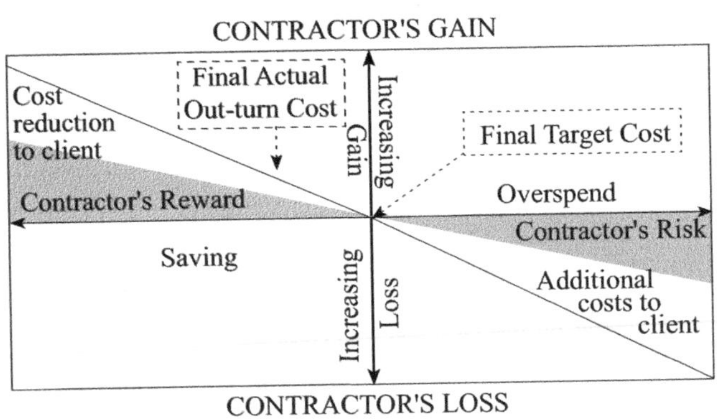

Guaranteed Maximum Price (GMP) contract

GMP is defined as "a sum established in an agreement between owner and contractor as the maximum compensation to be paid by the owner to the contractor for performing specified works on the basis of the cost of labour and materials plus overhead expenses and profits" [16]. GMP can be regarded as a lump sum price for a project in which the amount of money which the employer pays is the maximum price under the contract [17]. It has been postulated that GMP is not a form of contract, but it is a condition which can be applied to any form of contract [17], [18].

Masterman [19] viewed GMP as an agreement which will reward the contractor for any savings made against the GMP and penalise him when this sum is exceeded as a result of his own mismanagement or negligence. The gain-share operational mechanism of GMP is illustrated in Figure 14.2.

Figure 14.2: Gain-share mechanism of GMP procurement strategy in construction (Adapted from Cheng [20])

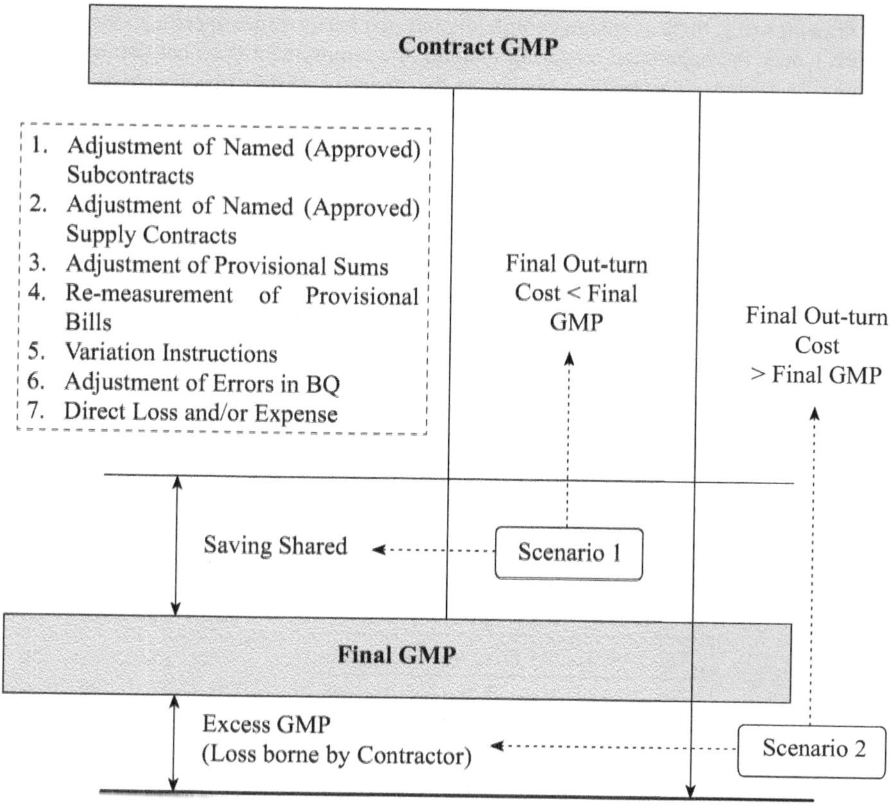

Pricing Mechanism of TCC/GMP

If a TCC/GMP project is procured under negotiated tendering, the preferred contractor has been identified by way of a long-term corporate business relationship [8]. In particular in Hong Kong, a large portion of GMP contracts are procured with a preferred contractor through negotiation in the private sector. In the case of selective tendering, tenderers are invited to participate in a pre-qualification exercise in the normal way by submitting an initial proposal containing the details of corporate strength, relevant work experience, past track record, expertise in alternative procurement methods, technical competence and the like to the client. The proposals will then be evaluated by the employer together with his team of consultants. A group of pre-qualified tenderers will be shortlisted and invited to submit a tender after meticulous assessment. Figure 14.3 illustrates the typical procurement route for launching a new TCC/GMP construction project.

If procured by the two-stage tendering method, pre-qualified tenderers will be invited to submit their tenders based on the following preliminary materials supplied by the employer and his consultants:

(1) Cost plan
(2) Base schematic/outline design drawings (e.g. 20% of design complete)
(3) Performance specifications for works packages
(4) Other available information

After tender evaluation, the shortlisted tenderers are then requested during the second stage to submit more detailed proposals based on: (i) Bill of Quantities; (ii) More complete set of design drawings (e.g. 80% of design complete); and (iii) Performance specifications for works packages. Under the negotiated tendering method, the requirement does not detract from the objective of assuring a competitive tender, as the majority of the subcontract packages are ultimately tendered on an 'open-book' competitive tender basis [8]. It should be stressed that this exchange of information requires a high level of mutual trust within the project team, especially the main contractor. The quantum of the subcontract packages competitively tendered may represent a range of 60-80% of the contract value.

Figure 14.3: A typical procurement route of TCC/GMP contracts (Adapted from the Hong Kong Housing Authority [21])

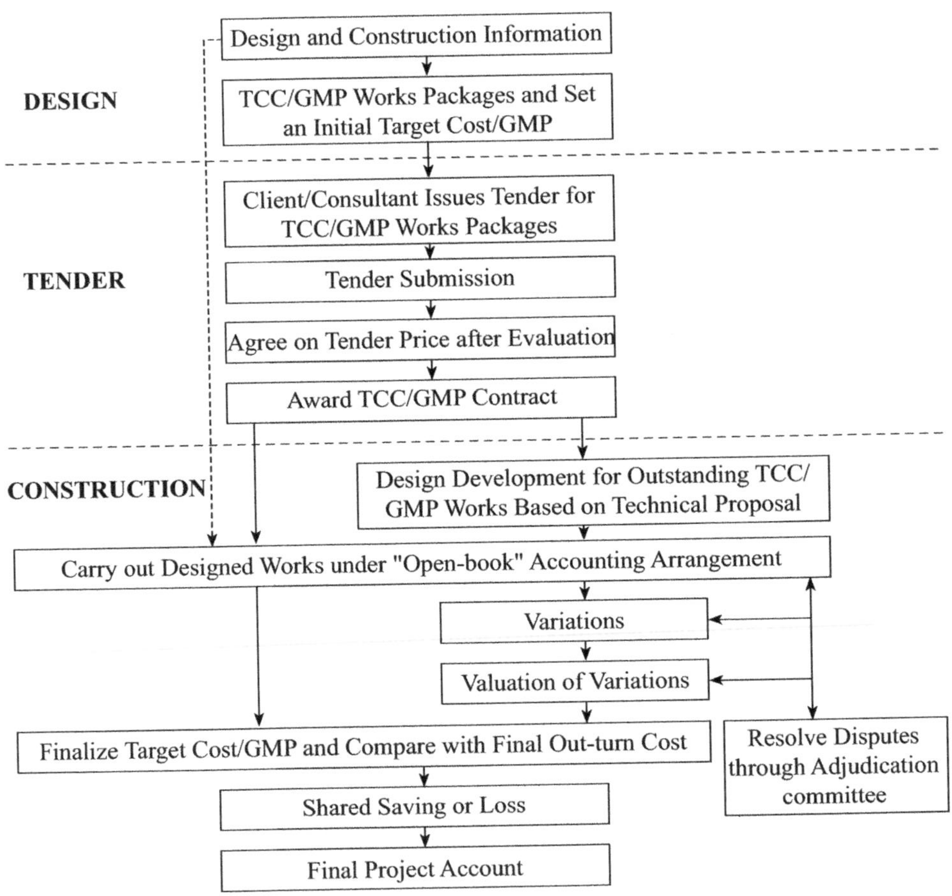

Both the guaranteed maximum price and target cost are estimated based on preliminary design documentation provided by the client and his team of consultants. Tender documents for TCC/GMP contracts are usually composed of: (1) cost for main contractor's direct works (e.g. substructure works, reinforced concrete superstructure works, finishing works, etc); (2) domestic subcontractor's works packages (e.g. electrical and mechanical installation, plumbing and drainage, fire services installation, lift installation, etc); (3) provisional quantities; (4) provisional sums; and (5) design development allowance [21]. The information provided in the tender documents is not sufficient for construction and completion of the works. The contractor is thus allowed in his tender pricing for design development. This arrangement provides an opportunity for contractors to input their innovations and experience to further improve the project design and buildability [22]. Further design information will be provided by the client and his team of consultants after the guaranteed maximum price or target cost is agreed upon and issued to the main contractor under Architect's Instructions.

Previous Research Studies on TCC/GMP

Nicolini *et al.* [23] studied two successful pilot projects procured with TCC. They opined that target costing may be one of the ways to support supply chain integration and improve both profitability and quality of the construction industry within the United Kingdom.

Perry and Barnes [24] proposed methods of tender evaluation of TCC and suggested that the contractor's share of cost overrun and under spent should not be less than 50%. Both Broome and Perry [13] and Badenfelt [25] explored how the gain-share/pain-share ratio in TCC should be determined in the British and Swedish perspectives respectively.

Bower *et al.* [26] examined three projects under different contractual arrangements, including one with TCC, to illustrate the effective use of incentive mechanisms. They concluded that contractual incentive structures should provide appropriate incentives for contractors to meet the targets of cost, schedule and quality; correctly allocate risks and allow a suitable level of client's involvement in the projects.

Both Walker *et al.* [27] and Hauck *et al.* [28] investigated the case of the Australian National Museum procured with TCC arrangement. Bubshait [29] conducted a questionnaire survey on incentive/disincentive contracting with the clients and contractors of industrial projects from Saudi Arabia, and his findings supported the use of this kind of contract. Tang *et al.* [30] conducted a similar research on incentive contracts in Mainland China through an empirical survey and a case study of the Three Gorges Project.

Wong [14] introduced a computerised system for cost management in a cable car project in Hong Kong to enable a more efficient management of documentation in TCC projects. Chan *et al.* [6] reported on the findings of 8 structured interviews in respect of motives, benefits, difficulties, risks and success factors of TCC/GMP and suitability to adopt those procurement strategies. Moreover, Chan *et al.* [31] identified some key risk factors and developed some effective risk mitigation measures for guaranteed maximum price and target cost contracts in construction based on a series of in-depth interviews with some senior industrial practitioners having abundant direct hands-on experience with TCC/GMP in Hong Kong.

Furthermore, Chan *et al.* [32] also evaluated the effectiveness of partnering on an underground railway extension project with TCC arrangement in Hong Kong via another study, suggesting that partnering and TCC may be keys to success of this infrastructure project. The practice of pre-project peer reviews in GMP contracts of contractors in the United States was reviewed in

Kaplanogu and Arditi's [33] study. This study concluded that the cost of pre-project peer review was justified since its benefit included minimising the risk of underestimating the cost of a project, evaluating the appropriateness of schedule, reviewing contract conditions and the like.

A recent study undertaken by Badenfelt [34] advocated a plethora of informal control mechanisms performed by project participants involved in a TCC project in Sweden. This study revealed that mutual trust is fragile which must be continuously preserved under TCC, and a high level of mutual trust between the client and contractor is essential to project success. This study indicated that on one hand clients try to maintain the level of mutual trust by letting contractors beware that they are knowledgeable and not easy to mislead. On the other hand, the contractors attempt to communicate that they cherish the existence of a long-term working relationship and a goodwill reputation.

Rose and Manley [35] launched 4 case studies applying financial incentives (including TCC arrangement) in Australia and concluded that financial incentives could incorporate rewards across all key organisations contributing to team performance and a reward amount should be sufficient to be valued by potential receipts. They further perceived that the benefits derived from financial incentive mechanisms could be maximised with equitable risk allocation, early contractor's involvement in design, value-driven tender selection, relationship workshop and future job opportunities. Chan *et al.* [10] reported on a case study of an underground railway station modification works procured with TCC in Hong Kong. This project was completed ahead of schedule by 7 months and achieved a cost saving of 5%.

Lahdenpera [36] proposed a two-stage target cost arrangement model to encourage early contractor's involvement in design under TCC. This model is claimed to be a means for owners to enter into a co-operative relationship which is of value for projects involving special challenges and risks. Senam *et al.* [37] carried out a questionnaire survey to evaluate the suitability of applying the GMP approach as an alternative procurement method for public sector projects in Malaysia. It was indicated that industrial practitioners had limited experience or low awareness of the concepts of GMP but would welcome the introduction of GMP to the construction industry in Malaysia, in order to ease the problems inherent with the traditional procurement system such as delay of project completion.

Perkin and Ma [38] highlighted the perceived benefits, difficulties and key risk factors associated with GMP construction projects in South Australia by means of an empirical questionnaire survey. It is followed by two GMP case studies gleaned through face-to-face interviews which are used to illustrate the opinions of various project stakeholders on the reasons for adopting GMP form of procurement and inherent risk factors throughout the course of the projects.

Table 14.1: Some previous research studies on TCC/GMP between 2000 and 2010 (Adapted from Chan *et al.* [31])

Authors	Year	Location	Focus
Nicolini *et al.*	2000	United Kingdom	Two case studies of TCC in the United Kingdom
Perry and Barnes	2000	United Kingdom	Tender evaluation of TCC
Bower *et al.*	2002	United States	Comparison of incentive features of 3 case studies
Broome and Perry	2002	United Kingdom	Determination of share ratios of TCC with utility theory
Walker *et al.*	2002	Australia	Case study of the Australian National Museum Project procured with TCC arrangement
Bubshait	2003	Saudi Arabia	Perceptions of owners and contractors on incentive/ disincentive contracting in industrial projects
Hauck *et al.*	2004	Australia	Case study of the Australian National Museum Project procured with TCC arrangement
Wong	2006	Hong Kong	Study on a computer system for cost monitoring in cable car project with TCC
Chan *et al.*	2007b	Hong Kong	Report of interview findings on motives, benefits, difficulties, risks, success factors and suitability of adopting TCC/GMP
Chan *et al.*	2008	Hong Kong	Case study of an underground railway extension project with TCC arrangement
Tang *et al.*	2008	China	Perceptions of stakeholders on incentives in the Chinese Mainland construction industry
Badenfelt	2008	Sweden	Derivation of share ratio in TCC
Kaplanogu and Arditi	2009	United States	Practice of pre-project peer reviews in GMP and lump sum contracts
Badenfelt	2010a	Sweden	Informal and formal control mechanisms performed by the employer in a TCC project with a 3-year longitudinal study
Rose and Manley	2010	Australia	Four case studies applying financial incentives (including TCC arrangement)
Chan *et al.*	2010b	Hong Kong	Case study of a underground railway station modification and extension works project procured with TCC arrangement
Lahdenpera	2010	Finland	Proposing a two-stage target cost arrangement for competitive co-operation
Senam *et al.*	2010	Malaysia	Evaluation of the suitability of applying GMP in Malaysia
Perkin and Ma	2010	Australia	Report of survey findings on benefits, difficulties and risks associated with GMP projects in South Australia, followed by two GMP case studies
Chan *et al.*	2011	Hong Kong	Report of survey findings on motives and benefits of applying TCC/GMP in Hong Kong

Application of TCC in Other Industries

The concepts of TCC were pioneered by the international oil industry during the 1950s and 1960s when they built the huge refinery and petrochemical complexes around the world [39]. Sakal [40] reported on the concepts of target costing by British Petroleum in the early 1990s with open-book accounting regime. It was also advocated by Sakal [40] that the projects with target cost contracting arrangement of British Petroleum were successful with a significant cost saving and duration of 6 months ahead the schedule. Not only were the concepts of target cost contracting adopted in the manufacturing sector of Japan and oil industry of the United Kingdom, it was applied to the information technology (IT) industry as well [41] with a view to fostering harmonious, productive client relationships. It was suggested that TCC can be as complex as traditional contracts with regard to change in project scope. The two complexities presented in their case study project included: (1) deciding whether a change was an unanticipated complexity of a within-scope change feature or a new feature which could lead to change in target cost; and (2) if the change was determined to be a new feature, figuring out the quantum of increase in fixed profit of the service provider in the IT industry [41]. Bourn [42] discovered that about 30% of the 115 projects surveyed in a study in the defence industry of the United Kingdom applied the concepts of target costing.

Application of TCC/GMP in Construction

Target cost contracting was widely adopted in the construction of health services sector premises by the Department of Health of the United Kingdom as well [43]. Up to June 2010, there were 317 schemes completed under this procurement framework applying the New Engineering Contract Version 2 (NEC2) Option C (target cost with activity schedule). It is reported that there are savings in both time and cost in the projects procured with this framework, when compared with those under the traditional procurement approach in the United Kingdom [43].

As mentioned in the previous sections, there have been some examples of application of TCC in Hong Kong over recent years, especially in the infrastructure sector. For example, TCC was applied to the Tsim Sha Tsui Metro Station Modification Works. It was found that the project was completed within budget and with a shorter project duration than expected [10]. The same procurement strategy was applied to the civil engineering works contracts of the South Island Railway Line in Hong Kong commenced in late 2010 [44]. Interestingly, all these projects procured with TCC are civil engineering works projects. In the private sector, developers intend to apply GMP contracts [8], which cap the expenditure of developers if the GMP value is not adjusted, for their residential and office buildings. Perhaps, if the GMP approach is applied to civil works projects, contractors would tend to price a higher risk premium, which may not be acceptable to the employer, to cover their potential risks such as unforeseen ground conditions and change in scope of works.

Perceived Motives Behind Introducing TCC/GMP

Chan *et al.* [22] reviewed the prevailing practices of TCC/GMP and reported on the key findings derived from an empirical questionnaire survey on the perceived motives and benefits of adopting TCC/GMP schemes in the construction industry of Hong Kong. It was found that the three most important motives of clients behind their decision of applying TCC/GMP include: (1) To generate an incentive for contractor to achieve cost saving; (2) To develop a more harmonious working relationship within the project team; and (3) To tap in contractor's expertise in both design and innovation.

To generate an incentive for contractor to achieve cost saving

Walker *et al.* [27] opined that the gain-share/pain-share mechanism encourages a teamwork approach to create innovative ideas in problem solving via the case study of the Australian

National Museum. Bower *et al.* [26] advocated that the TCC/GMP contracting method can be an effective means of motivating contractors to achieve better value for money and more satisfactory project performance by linking their own individual financial objectives to the overall objectives of the project together. The gain-share/pain-share mechanism indeed generates an enormous impetus for effective collaboration between client and contractor in order to minimise the final out-turn cost of a project [45], [46]. Pre-construction planning for design development which involves all relevant key project stakeholders can reduce the potential conflicts and intractable disputes at later time. This contracting approach also allows the contractor and client to determine the appropriate ownership of risks and encourages various contracting parties to agree on an equitable allocation of risks, which is in the client's long-term interest [46]. In addition, the contractors are usually involved in an earlier phase of project development to tap in their innovations in design for the sake of achieving cost savings under the TCC/GMP methodology. Integration of different types of competence is most beneficial in the early phase of the project, since the potential for improving the quality and buildability of project design is the greatest at this stage [36]. An early involvement of contractor in the TCC/GMP procurement process would probably offer an opportunity for contractors to input their experience and innovations in both design and construction, as well as achieve better value for money for the employer.

To develop a more harmonious working relationship within the project team

Turner [47] suggested that the performance of a construction project cannot be accurately predicted and the contracts for construction projects are considered to be incomplete in most cases. Such incompleteness of contracts is caused by the reality of transaction cost, bounded rationality and information asymmetries which make both the employer and the contractor design a complete contract [48]. Badenfelt [34] also opined that the extent of contract completeness is significant in determining which kinds of incentives to be adopted in a contract. Traditional fixed-price contracts have been restricted to projects with fewer uncertainties on technology and economics. In practice, due to information asymmetries, even a risk-neutral contractor may not be willing to sign on a fixed-price contract without offering a higher price. However, the higher cost of identifying unforeseen events renders it difficult to draft a very elaborate contract to deal with all kinds of uncertainties at the post contract stage [49].

It is recommended by Badenfelt [50] that cost-plus contracts may minimise the problem of overpayment, provided that the contract is well documented, but the client may expose himself to the problem of cost padding. In order to mitigate the negative effect of cost-plus contracts, it is common to replace standard cost-plus contracts with target cost contracts which are believed to reinforce the collaboration between the client and the contractor [51]. Anvuur and Kumaraswamy [52] presented their findings based on two case studies of building projects with a GMP contractual arrangement. Their findings indicated that the project performance of the two GMP cases was satisfactory in terms of time and cost, and concluded that the GMP form of contract can be a useful tool for work group integration which helped to achieve collaborative working relationships within an integrated working team. In addition, Bayliss *et al.* [53] reported on a successful case of applying the TCC arrangement and opined that both the periodic partnering review workshops and the use of an incentivisation agreement improved the working relationship amongst the project team members.

Partnering is often implemented in parallel to TCC/GMP schemes in the construction industry of Hong Kong. The application of gain-share/pain-share arrangement helped to tie the individual goals of the employer and contractor together and establish a more harmonious working relationship [46]. In particular, contractual incentives are usually included under the gain-share/pain-share mechanism for deeper collaboration between client and contractor with a view to lowering the out-turn cost of a project [45], [46]. By involving all of the relevant major project stakeholders, the pre-construction planning for the design development can avoid contentious conflicts and intractable disputes often plaguing contracts at later time. This arrangement also urges different contracting parties to agree on an equitable allocation of potential risks [46]. Furthermore, a fair and effective dispute resolution mechanism and prompt communication channels are provided via adjudication meetings, which are not only conducive to reduction in claim / dispute occurrence, but also to improved partners' working relationship as a result of inter-disciplinary efforts [54].

Chan *et al.* [10] shared a similar view that TCC together with partnering initiatives could motivate the contractor to work with the employer and shared any gains or pains at project completion. In this case study of the underground railway station modification works, project participants expressed that more opportunities were available for them to voice out their opinions and concerns openly and freely under the umbrella of TCC. The gain-share/pain-share mechanism built in the TCC/GMP methodology and partnering arrangement introduced in several construction projects indeed contributed to the harmonious, co-operative working relationship. The cost incentives generated from TCC/GMP schemes could serve as a vehicle to align individual objectives of different contracting parties. Ting [54] held a similar notion that the incentivisation approach can create a more proactive co-operation and initiate the cultural shift away from the traditional adversarial approach to new collaborative contracting.

To tap in contractor's expertise in both design and innovation

Import of contractor's expertise in both design and innovation is considered as one of the most significant motives behind applying TCC/GMP. Under the TCC/GMP procurement arrangement, the capability of tapping in contractor's expertise and innovative ideas in both design and construction can largely enhance the constructability of project. This procurement option allows the contractor to be brought in at the early design stage to offer technical advice on construction costs, building design, project programming, construction materials, alternative construction techniques and other constructability and environmental issues to be integrated into the design, so as to mitigate the construction risk [8]. Lahdenpera [36] opined that under TCC, incentives would be created for contractors to lower the target cost in order to achieve cost saving. The contractor would probably tap in his expertise in buildability and innovation through the submission of alternative design proposals to strive for possible savings, since the contractor himself would directly benefit from any cost savings generated. What is more, TCC/GMP can facilitate the commencement of construction activities before the design is fully completed [22]. The substantial overlapping of design with construction, analogous to the design-and-build contract, would shorten the duration required for overall project development.

Other key benefits of applying TCC/GMP include setting an agreed ceiling price of the project at main contract award for the client in case of GMP, increased client's involvement in problem solving and subcontractor selection, an open-book accounting regime to enable better accountability and quantification of the costs of risk, early settlement of final project account, as well as higher quality of constructed facilities [22].

Potential Pitfalls of Adopting TCC/GMP

In addition to the above identified motives and benefits, a review of the literature indicates several potential pitfalls and difficulties in implementing TCC/GMP schemes, which are highlighted as follows [55].

Setting an inappropriate sharing ratio

Despite there being a number of benefits of applying TCC/GMP in construction as stated in previous literature, there could be potential pitfalls in implementing TCC/GMP procurement strategy in construction. One of the potential pitfalls may be associated with the selection of sharing ratio on gain and pain between the employer and contractor [13]. In the United Kingdom, Perry and Barnes [24] analysed the methods of tender evaluation of TCC and proposed one to increase the likelihood of the contract being awarded to the tenderers whose final price will be the lowest. Their study found that a low contractor's share would decrease his motivation to reduce the actual cost. It was strongly suggested that the contractor's share should not be less than 50%. In Sweden, Badenfelt [25] launched a series of interviews with eight clients and eight contractors, followed by a case study, to study the factors affecting the selection of sharing ratio. It was found that perception of fairness, knowledge of target cost contracts and long-term relationship were key factors influencing the selection of sharing ratio. It is further recommended that a long-term relationship was important to the success of TCC since such relationship could reduce the risks associated with asymmetric information [25].

Change in scope of works

In order to maintain the incentives for contractor to reduce the actual cost, the target cost should be adjusted for changes in scope of works and matters outside the control of the contractor [12]. Change in scope of works is more common with TCC/GMP as the scope is not totally defined in many cases of TCC projects in Hong Kong [8]. A recent study by Olawale and Sun [56] also reflected that design changes are considered as the most important factor inhibiting the ability of industrial practitioners in both time and cost controls in their questionnaire survey. Another earlier research launched by Cox *et al.* [57] in the United Kingdom also revealed that change in employers' requirements is one of the most frequently cited reasons for design changes in their cases explored. If the circumstances under which the target cost could be adjusted are not clearly defined in the contracts, there can be room for disputes as to whether a variation is a change in scope of works (which is a ground to change the target cost) or a design development item (which does not change the target cost). This view is supported by Fan and Greenwood [18] that a change in the scope of works could be a major source of disputes in GMP contracts within the construction industry.

Lack of familiarity with and understanding of TCC/GMP methodology

As Chan *et al.* [8] report, unfamiliarity with or misunderstanding of TCC/GMP concepts by senior management is one of the difficulties in implementing TCC/GMP in Hong Kong. It is reported that obtaining endorsement from senior management to apply the concepts of TCC in Hong Kong was a challenge [58]. Such difficulty was compounded because there was no precedent of applying a fully cost reimbursable target cost contract with the gain-share/pain-share arrangement. Another study by Chan *et al.* [55] suggested that some consultants and subcontractors might not fully understand the concepts of TCC, and this could adversely affect the proper implementation of the procurement strategies and lead to failure of the whole project.

Lack of mutual trust between employer and contractor

Target cost contracts are considered to serve as a means for building mutual trust between the employer and contractor by aligning the goals of both parties [26]. The cost reimbursement nature of TCC requires a transparency of cost information because the visualization of cost is necessary in this kind of project. Thus, open-book accounting is often employed in TCC/GMP, giving the client to access of project cost [8]. Badenfelt [59] conducted the case study of a laboratory project with TCC, gathering data through interviews and non-participant observations. The study suggested that the mutual trust between client and contractor is fragile and must be continuously preserved. It is recommended by both Badenfelt [59] and Chan *et al.*[10] that the partnering arrangement can be implemented in TCC with the purpose of fostering mutual trust and mitigating potential disputes between employer and contractor.

Figure 14.4: Summary of the perceived motives and potential pitfalls of applying TCC/GMP schemes

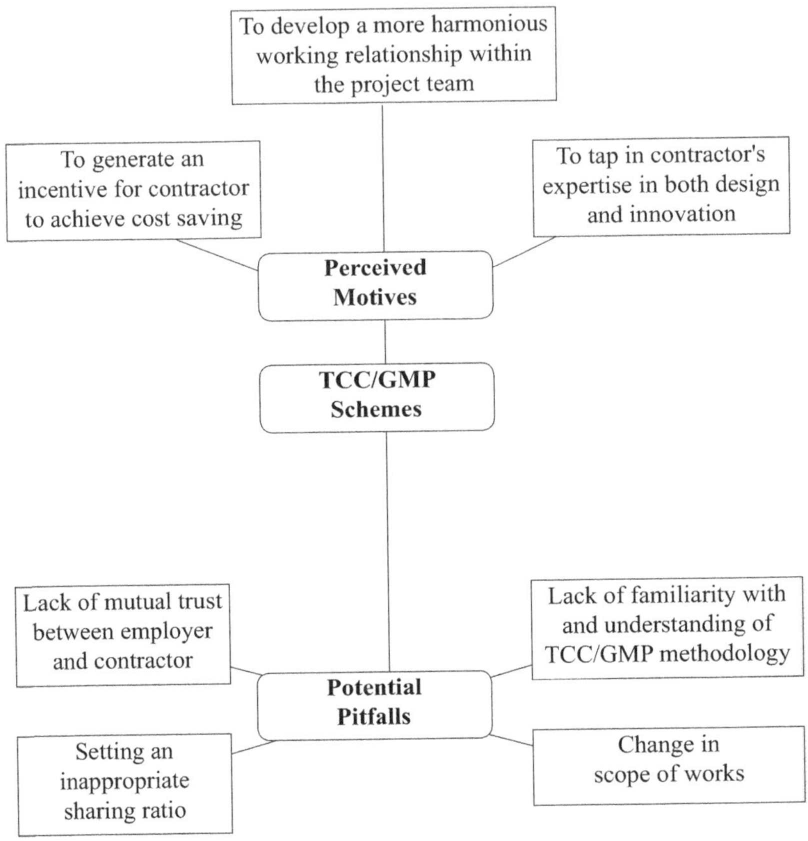

Conclusion

Figure 14.4 gives a summary of the major perceived motives and potential pitfalls of applying TCC/GMP schemes in construction. To improve the traditional adversarial working relationship amongst project team members, employers are encouraged to apply TCC/GMP styles of procurement which tie the individual objectives of employer and contractor together. The motives behind the increasing application of TCC/GMP in Hong Kong are discussed in this chapter. The nature of TCC/GMP of generating incentives for contractor to achieve cost saving, providing a platform to develop a more harmonious working relationship and importing contractor's expertise in both design and construction appear to be the attraction of these procurement strategies. However, the application of these contractual arrangements is still at a geminating stage of development in Hong Kong. This may be due to some potential pitfalls such as unfamiliarity with the concepts of TCC/GMP by project team members, setting an unreasonable sharing ratios, etc. This chapter has presented a concise review of international and local literature of TCC/GMP and has discussed some essential perceived motives and potential pitfalls of applying TCC/GMP in the construction industry of Hong Kong. It is expected to benefit the industrial practitioners on important issues for implementation to be carefully examined before introducing the TCC/GMP methodology. This chapter also forms a solid foundation for further researches on these contractual arrangements in both the local and international context.

Acknowledgements

The authors would like to thank The Hong Kong Polytechnic University for providing financial support to this research endeavour (HK PolyU General Research Grants Allocation 2008/09 with Project Account Code: BRE-A-SA42). The work described in this chapter was further supported by a grant from the Research Grants Council of the Hong Kong Special Administrative Region, China (RGC Project No. PolyU 5107/07E). This chapter forms part of a RGC funded research project entitled "Evaluating the Key Risk Factors and Risk Sharing Mechanisms for Target Cost Contracting (TCC) Schemes in Construction", from which other deliverables have been produced with several objectives but sharing common background of study and research methodology.

References

[1] Kaka, A., Wong, C. and Fortune C. (2008) Culture change through the use of appropriate pricing systems. *Engineering, Construction and Architectural Management* **15(1)**: 66-77.

[2] Latham, M. (1994) *Constructing the Team.* Final Report of the Joint Government/Industry Review of Procurement and Contractual Arrangements in the United Kingdom Construction Industry. HMSO, London.

[3] Egan, J. (1998) *Rethinking Construction.* Department of Environment, Transport and the Regions, London.

[4] Construction Industry Review Committee (2001) *Construct for Excellence.* Report of the Construction Industry Review Committee, Hong Kong SAR, 207 pages.

[5] Trench, D. (1991) *On Target – A Design and Manage Target Cost Procurement System.* Thomas Telford, London.

[6] Chan, D.W.M., Chan, A.P.C., Lam, P.T.I., Lam, E.W.M. and Wong, J.M.W. (2007b) Evaluating guaranteed maximum price and target cost contracting strategies in Hong Kong construction industry. *Journal of Financial Management of Property and Construction,* **12(3):** 139-149.

[7] Bogus, S.M., Shane, J.S. and Molenaar, K.R. (2010) Contract payment provisions and project performance: An analysis of municipal water and wastewater facilities. *Public Works Management and Policy*, **15(1)**, 20-31.

[8] Chan, D.W.M., Chan, A.P.C., Lam, P.T.I., Lam, E.W.M. and Wong, J.M.W. (2007a) *An Investigation of Guaranteed Maximum Price (GMP) and Target Cost Contracting (TCC) Procurement Strategies in Hong Kong Construction Industry.* Research Monograph. Department of Building and Real Estate, The Hong Kong Polytechnic University, 152 pages, ISBN 978-962-367-593-2, October 2007. Available at URL: http://repository.lib.polyu.edu.hk/jspui/handle/10397/2376.

[9] Cheung, I. (2008) New Engineering Contract (NEC). *EC Harris Asia Commentary*, July, 2008.

[10] Chan, D.W.M., Lam, P.T.I., Chan, A.P.C. and Wong, J.M.W. (2010b) Achieving better performance through target cost contracts: The tale of an underground railway station modification project. *Facilities – Special Issue on Performance Measurement and Management in Facilities Management* **28(5/6)**: 261-277.

[11] Rojas, E.M. and Kell, I. (2008) Comparative Analysis of Project Delivery Systems Cost Performance in Pacific Northwest Public Schools. *Journal of Construction Engineering and Management, ASCE* **134(6)**: 387-397.

[12] Potts, K. (2008) *Construction Cost Management.* Taylor & Francis Group, New York.

[13] Broome, J. and Perry, J. (2002) How practitioners set share fractions in target cost contracts. *International Journal of Project Management* **20(1)**: 59-66.

[14] Wong, A.K.D. (2006) The application of a computerised financial control system for the decision support of target cost contracts. *Journal of Information Technology in Construction (ITcon)* **11**: 257-268 (Special Issue on Decision Support Systems for Infrastructure Management).

[15] Construction Industry Council (2010) *Guidelines on Partnering, Version 1*, August 2010, 62 pages. Available at URL: http://www.hkcic.org/WorkArea/linkit.aspx?LinkIdentifier=id&ItemID=6529&libID=6578 (accessed on 16 November 2010).

[16] American Institute of Architects (2008) *Architect's Handbook of Professional Practice, 14th edition*. John Wiley & Son, Hoboken, New Jersey.

[17] Davis Langdon and Seah (DLS) (2003) *Research Study on Alternative Procurement Approaches for Public Construction Works Projects – Final Report*. Research Study for the Environment, Transport and Works Bureau of Hong Kong SAR Government. October 2003.

[18] Fan, Avan C.W. and Greenwood, David (2004) Guaranteed maximum price for the project? *Surveyors Times*. The Hong Kong Institute of Surveyors, March, 20-21.

[19] Masterman, J.W.E. (2002) *Introduction to Building Procurement System, 2nd Edition*. London New York Spon Press.

[20] Cheng, R.L.L. (2004) *Investigation of the Application of Guaranteed Maximum Price in the Hong Kong construction industry*. Unpublished BSc(Hons) Dissertation in Construction Economics and Management. Department of Building and Real Estate, The Hong Kong Polytechnic University, Hong Kong, 58 pages.

[21] Hong Kong Housing Authority (2006) *Internal Guidelines for Guaranteed Maximum Price Contract Procurement Based on Private Sector Model*. The Hong Kong Housing Authority, Hong Kong SAR Government, 19 pages.

[22] Chan, D.W.M., Chan, A.P.C., Lam, P.T.I., Lam, E.W.M. and Wong, J.M.W. (2011) An empirical survey of the motives and benefits of adopting guaranteed maximum price and target cost contracts in construction. *International Journal of Project Management* **29(5)**: 577-590.

[23] Nicolini, D, Tomkins, C, Holti, R and Oldman, A. (2000) Can target costing and whole life costing be applied in the construction industry? Evidence from two case studies. *British Journal of Management* **11(4)**:303-324.

[24] Perry, J.G. and Barnes, M. (2000). Target cost contracts: An analysis of the interplay between fee, target, share and price. *Engineering, Construction and Architectural Management* **7(2)**: 202-208.

[25] Badenfelt, U. (2008) The selection of sharing ratios in target cost contracts. *Engineering, Construction and Architectural Management* **15(1)**: 54-65.

[26] Bower, D., Ashby, G., Gerald, K. and Smyk, M. (2002) Incentive mechanism for project success. *Journal of Management in Engineering, ASCE* **18(1)**: 37-43.

[27] Walker, D.H.T., Hampson, K.D. and Peters, R. (2002) Project alliancing vs project partnering: A case study of the Australian National Museum Project. *Supply Chain Management: An International Journal* **7(2)**: 83-91.

[28] Hauck, A.J., Walker, D.H.T., Hampson K.D. and Peters, R.J. (2004) Project alliancing at National Museum of Australia – Collaborative process. *Journal of Construction Engineering and Management, ASCE* **130(2)**: 143-152.

[29] Bubshait, A.A. (2003) Incentive/disincentive contracts and its effects on industrial projects. *International Journal of Project Management* **21(1)**: 63-70.

[30] Tang, W., Qiang, M., Duffield, C.F., Young, D.M. and Lu Y. (2008) Incentives in the Chinese construction industry. *Journal of Construction Engineering and Management, ASCE* **134(7)**: 457-467.

[31] Chan, D.W.M., Chan, A.P.C., Lam, P.T.I. and Chan, J.H.L. (2010a) Exploring the key risks and risk mitigation measures for guaranteed maximum price and target cost contracts in construction. *Construction Law Journal* **26(5)**: 364-378.

[32] Chan, A.P.C., Chan, D.W.M., Fan, L.C.N., Lam, P.T.I. and Yeung, J.F.Y. (2008) Achieving partnering success through an incentive agreement: Lessons learned from an underground railway extension project in Hong Kong. *Journal of Management in Engineering, ASCE* **24(7)**: 128-137.

[33] Kaplanogu, S.B. and Arditi, D. (2009). Pre-project peer reviews in GMP/lump sum contracts. *Engineering, Construction and Architectural Management* **16(2)**:175-185.

[34] Badenfelt, U. (2010a) Fixing the contract after the contract is fixed: a study of incomplete contracts in IT and construction projects. *International Journal of Project Management* **29(5)**: 568-576.

[35] Rose, T. and Manley, K. (2010) Client recommendations for financial incentives on construction projects. *Engineering, Construction and Architectural Management* **17(3)**: 252-267.

[36] Lahdenpera, P. (2010) Conceptualizing a two-stage target-cost arrangement for competitive cooperation. *Construction Management and Economics* **28(7)**: 783-796.

[37] Senam, M.R., Ehrhardt, C. and Zaini, R.M. (2010) An evaluation of applying Guaranteed Maximum Price (GMP) procurement method for public construction projects in Malaysia. In: Kobayashi, K., Rashid, K.A., Onshi, M. and Lei, S. (Editors): Towards New Paradigm of Partnership for the Increasingly Global Construction Markets, *Proceedings of the 6th International Conference on Multi-National Joint Venture for Construction Works*, 22-23 September, Kyoto, Japan, 63-80.

[38] Perkin, T. and Ma, T. (2010). Guaranteed Maximum Price (GMP) contracting in the South Australian construction industry. *Proceedings of the 8th International Conference on Construction and Real Estate Management (ICCREM 2010) - Leading Sustainable Development through Construction and Real Estate Management.* 1-3 December 2010, Queensland University of Technology, Brisbane, Australia, Volume 1, 178-181.

[39] Ritz, G.J. (1994) *Total Construction Project Management*. McGraw-Hill, United States of America.

[40] Sakal, M.W. (2005). Project alliancing: A relational contracting mechanism for dynamic projects. *Lean Construction Journal* **2(1)**: 67-79.

[41] Eckfeldt, B., Madden, R. and Horowitz, J. (2005) Selling Agile: Target-Cost Contracts. *Proceedings of the Agile Development Conference.*

[42] Bourn, J. (2006) *Using the Contract to Maximise the Likelihood of Successful Project Outcome.* Report by the Comptroller and Auditor General, 7 June 2006.

[43] ProCure 21 Guide (2010). *Overview of ProCure 21.* National Health Service. Available at URL: http://www.nhs-procure21.gov.uk/content/overview.asp (accessed on 30 June 2010).

[44] Mass Transit Railway Corporation (2010). *Notice of Pre-qualification for Tenderers for Contract No. 901* – Admiralty Integrated Station and SCL Enabling Works and Contract No. 903 – Ocean Park Station, Wong Chuk Hang Station, Viaduct and Aberdeen Channel Bridge. Available at URL: http://www.mtr.com.hk/eng/tenders/SIL_E_901_903.html (accessed on 16 July 2010).

[45] Chevin, D. (1996) The max factor. *Building*, 17 May 1996.

[46] Sadler, M.C. (2004). *The Use of Alternative Integrated Procurement Approaches in the Construction Industry*, Unpublished MBA Dissertation in Construction and Real Estate. Department of Construction Management and Engineering, University of Reading, UK, 132 pages.

[47] Turner, R.J. (2004). Farsighted project contract management: Incomplete in its entirety. *Construction Management and Economics* **22(1)**: 75-83.

[48] Spier, K.E. (1992). Incomplete contracts and signalling. *RAND Journal of Economics* **23(3)**: 432-443.

[49] Crocker, K.J. and Reynolds, K.J. (1993). The efficiency of incomplete contracts: An empirical analysis of Air Force engine procurement. *RAND Journal of Economics* **24(1)**: 126-146.

[50] Badenfelt, U. (2010b) I trust you, I trust you not: A longitudinal study on control mechanisms in incentive contracts. *Construction Management and Economics* **24(1)**: 301-310.

[51] Bresnen, M. and Marshall, N. (2000). Building partnerships: Case studies of client–contractor collaboration in the UK construction industry. *Construction Management and Economics* **18(7)**: 819-832.

[52] Anvuur, A.M. and Kumarawamy, M.M. (2010). Promises, pitfalls and shortfalls for the guaranteed maximum price approach: A comparative case study. In: Egbu, C. (Editor), *Proceedings of the 26th Annual ARCOM Conference*, 6-8 September 2010, Leeds, United Kingdom, Association of Researchers in Construction Management, 1079-1088.

[53] Bayliss, R., Cheung, S.O., Suen, H.C.H. and Wong, S.P. (2004). Effective partnering tools in construction: A case study on MTRC TKE Contract 604 in Hong Kong. *International Journal of Project Management* **22(3)**: 253-263.

[54] Ting, W. (2006). The Impact of the Interdisciplinary Efforts on the Receptivity of Guaranteed Maximum Price (GMP) Project. *Unpublished MSc in International disciplinary Design Management Dissertation.* Department of Real Estate and Construction, The University of Hong Kong, Hong Kong, 97 pages.

[55] Chan, D.W.M., Chan, A.P.C., Lam, P.T.I. and Wong, J.M.W. (2010c). Empirical study of the risks and difficulties in implementing guaranteed maximum price and target cost contracts in construction. *Journal of Construction Engineering and Management, ASCE* **136(5)**: 495-507.

[56] Olawale, Y.A. and Sun, M. (2010). Cost and time control of construction projects: Inhibiting factors and mitigating measures in practice. *Construction Management and Economics* **28(5)**: 509-526.

[57] Cox, I.D., Morris, J.H.R. and Jared, G.E. (1999) A quantitative study of post contract design changes in construction. *Construction Management and Economics* **17(4)**: 427-439.

[58] Avery, D. (2006) How collaborative commercial strategies give certainty to the delivery of major railway infrastructure projects. *Proceedings of the PMICOS 2006 Annual Conference.* 23-26 April 2006, Orlando, Florida, United States.

[59] Badenfelt, U. (2007) Trust and control in the early phases of target cost contracts. In: Boyd, D. (Editor), *Proceedings of the 23rd Annual ARCOM Conference*, 3-5 September 2007. Belfast, United Kingdom, Association of Researchers in Construction Management, 23-32.

Joint Ventures in Construction 2
ISBN 978-0-7277-5783-8

doi: 10.1680/jvc2.57838.201

Chapter 15

The Effect of Performance-based Contracting in Public-Private Partnerships – The Japanese Case

Masamitsu Onishi, Kyoto University
Yasuhiro Shoji, Ministry of Land, Infrastructure and Transportations, Japan
Kiyoshi Kobayashi, Kyoto University

Introduction

Public-Private Partnership (PPP) is increasingly pervasive around the world as a method of public acquisition. PPP requires the formation of a corporation called a 'Special Purpose Company (SPC)' to undertake a specific project from design to operation and maintenance. The SPC is responsible for any decisions concerning the method of work to meet the required level of performance stipulated in a contract. In contrast to conventional public procurement, whereby the government specifies the method of work, a contract for a PPP project is inevitably performance-based, with the government specifying not a particular method of work but rather measures of performance. Hereinafter, a contracting scheme with conventional procurement is called method-based contracting. With PPP projects, the contractor is allowed to select a method of work at the contractor's own discretion on the condition that the completed work satisfies the required performance level. Contractors' discretion in selecting a method of work can incentivise them to make technological innovations to gain a competitive advantage over competitors in terms of cost and quality. It is claimed that the rationale for applying PPP to public procurement is to employ the private sector's know-how to achieve higher value-for-money on the project. Performance-based contracting as a feature of PPP may change the outcome of public procurement, as it would seem to have an impact on the contractors' incentive for technological innovation and bidding strategy. However, it has been pointed out that private firms' motive for innovation is dependent on the market environments they face [1]. In various market and technological environments, it may not always be guaranteed that the performance-based contracting characteristic of PPP contributes to motivating contractors to make appropriate technological innovation. Therefore, it is meaningful to analyse the effect of performance-based contracting theoretically by developing a model that explicitly considers behavioural changes in contractors when such contracting is introduced in various market and technological environments.

There are papers that share our interest, i.e. the effects of market conditions on private firms' motivation to engage in technological development [1]–[4]. These papers have developed dynamic competitive models to focus on firms' behaviour of seeking temporal monopolistic benefits protected by patents. Basing their study on empirical data, Seaden *et al.* [5] analyse the strategic decision-making process in a construction firm. On the contrary, the structure for allocating potential benefit through the bidding process is more critical to firms' motivation in the construction market, rather than a dynamic motivator for technological development. Behavioural change due to the introduction of performance-based contracting has not been noted in previous literature. Extending the framework of Onishi *et al.* [6], this chapter aims to provide a theoretical framework for discussing the impact of PPP introduction from a performance-based contracting point of view, as well as relevant policy implications.

The Model

Settings

The government solicits competitive bidders according to the procedure shown in Figure. 5.1 for two work projects, each requiring a different technical competence. The condition to be considered for designing and selecting the method of work is called the 'design condition.' The types of design condition of the two works are represented by α and β, respectively, for descriptive purposes. For example, design condition α may indicate work requiring advanced methodological skill in hydro-technology, and β may indicate work requiring advanced methodological skill in geo-technology.

It is assumed that there are only two risk-neutral contractors, Contractor A and Contractor B, in the construction market. Under conventional procurement, the government solicits bids after specifying the method of work. The obligation of the contractors is to complete the work according to the method stipulated by the government; contractors cannot select methods of work at their own discretion. Then, both contractors spend cost $\bar{C}$ for each work, adopting the method of conventional technology determined by the government. On the other hand, under PPP, the government solicits bids specifying the required performance measures of service level for the work. This allows contractors to select a method of work that enables them to exploit their technical advantages, as long as the required performance level is reached.

Variation in design conditions, a specific feature of the construction industry, requires innovation. Contractors must make costly efforts to devise new technology in response to varying conditions. For the sake of simplicity in the model, contractor i ($i = A, B$) chooses an effort level denoted by $e_i = \{H, L\}$, wher $e_i = H$ represents a high level of effort and $e_i = L$ a low level of effort. Denote the effort cost for contractor i whose effort level $e_i = H$ as ξ_i. Assume that the cost of a high level of effort borne by both contractors is the same.

$$\xi_A = \xi_B = \xi, \tag{1}$$

and the cost of a low level of effort is 0.

Figure 15.1: Timeline of the basic model

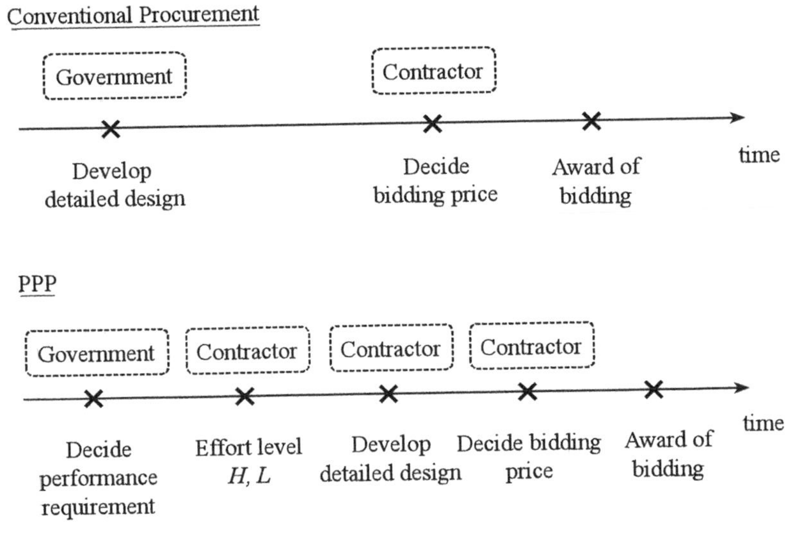

If contractor i chooses $e_i = H$, the construction cost for the work with design condition j is saved by ΔC_i^j compared to the cost when conventional technology is adopted. Assume that Contractor A has superior professional expertise for the construction work with type α design condition, compared to Contractor B and for the work with type β vice versa. This assumption implies that Contractor A, at the same effort cost, can develop more efficient technology for the work with design condition α, compared to Contractor B. For design condition type β, Contractor B is more efficient. Therefore, the following two conditions:

$$\Delta C_A^\alpha > \Delta C_B^\alpha, \Delta C_B^\beta > \Delta C_A^\beta \tag{2}$$

are assumed. The game where the contractors choose their effort level is formulated as the strategic-form game, which is played before the government solicits bids. After the contractors choose their effort level, information on the contractors' technologies, i.e. construction cost, is assumed to be common knowledge between them. By conducting competitive bidding, a contractor bidding the lowest price is successful and is awarded the project. If multiple bidders bid the same lowest price, those bidders are awarded stochastically with the same probability respectively.

Conventional procurement

As a benchmark to evaluate the effect of PPP with performance-based contracting, an equilibrium strategy under conventional procurement is derived. Under conventional procurement, because the government specifies the method of work, contractors cannot adopt new technologies. Therefore, both Contractors A and B bear construction cost $\bar{C}$. In this case, they both id the same price $\bar{C}$, which brings zero profit to them. If contractors bid the same price, one is awarded with the probability 1/2. The expected profit of Contractor i π_i^c is derived as

$$\pi_A^c = \pi_B^c = 0. \tag{3}$$

PPP

PPP allows contractors to select the method of work at their discretion. Key strategies under PPP are the bidding strategy and the effort level for technological development. Taking the expected payoff of bidding into consideration before bidding begins, contractors decide their effort level. The following analysis focuses on the game concerning the work with design condition. Given the contractors' effort levels, their expected payoffs through bidding are derived for the following four cases.

(a) Given $(e_A, e_B) = (L, L)$ both contractors bear the same amount of construction cost $\bar{C}$. In this case, the contractors bid the same bidding price $b_A^\alpha = b_B^\alpha = \bar{C}$. Each contractor is awarded with the probability 1/2. Therefore, Contractor i's expected payoffs are

$$\pi_A^\alpha = \pi_B^\alpha = 0. \tag{4}$$

Table 15.1: Payoff matrix (design condition α)

Design Cond. α		Contractor *B*	
		H	*L*
Contractor *A*	*H*	$(\Delta C_A^{\alpha} - \Delta C_B^{\alpha} - \varepsilon - \xi, -\xi)$	$(\Delta C_A^{\alpha} - \varepsilon - \xi, 0)$
	L	$(0, \Delta C_B^{\alpha} - \varepsilon - \xi)$	$(0, 0)$

Table 15.2: Payoff matrix (design condition β)

Design Cond. β		Contractor *B*	
		H	*L*
Contractor *A*	*H*	$(-\xi, \Delta C_B^{\beta} - \Delta C_A^{\beta} - \varepsilon - \xi)$	$(\Delta C_B^{\beta} - \varepsilon - \xi)$
	L	$(\Delta C_A^{\beta} - \varepsilon - \xi, 0)$	$(0, 0)$

(b) Given $(e_A, e_B) = (H, L)$, Contractor *A* bears the construction cost $\bar{C} - \Delta C_A^{\alpha}$ and Contractor *B* bears $\bar{C}$. In this case, Contractor *A* is awarded by bidding price $b_A^{\alpha} = \bar{C} - \varepsilon$. Therefore, Contractor *i*'s expected payoffs are

$$\pi_A^{\alpha}(H, L) = b_A^{\alpha} - (C - \Delta C_A^{\alpha}) - \xi$$
$$= \Delta C_A^{\alpha} - \varepsilon - \xi \quad (5a)$$

$$\pi_B^{\alpha}(H, L) = 0 \quad (5b)$$

where ε is a variable with arbitrary small positive value.

(c) Given $(e_A, e_B) = (L, H)$, Contractor *A* bears construction cost $\bar{C}$ and Contractor *B* bears $\bar{C} - \Delta C_B^{\alpha}$. In this case, Contractor *B* is awarded by bidding price $b_B^{\alpha} = \bar{C} - \varepsilon$. Therefore, Contractor i's expected payoffs are

$$\pi_A^{\alpha}(L, H) = 0 \quad (6a)$$

$$\pi_B^{\alpha}(H, L) = b_B^{\alpha} - (\bar{C} - \Delta C_B^{\alpha}) - \xi.$$
$$= \Delta C_B^{\alpha} - \varepsilon - \xi \quad (6b)$$

(d) Given $(e_A, e_B) = (H, H)$, Contractor *A* bears the construction cost $\bar{C} - \Delta C_A^{\alpha}$ and Contractor *B* bears $\bar{C} - \Delta C_B^{\alpha}$. In this case, Contractor *A* is awarded by bidding price $b_A^{\alpha} = \bar{C} - \Delta C_A^{\alpha} - \varepsilon$. Contractor i's expected payoffs are

$$\pi_A^{\alpha}(H, L) = b_A^{\alpha} - (C - \Delta C_A^{\alpha}) - \xi.$$
$$= \Delta C_A^{\alpha} - \Delta C_B^{\alpha} - \varepsilon - \xi \quad (7a)$$
$$\pi_B^{\alpha}(H, L) = -\xi \quad (7b)$$

Table 15.1 and Table 15.2 indicate the payoff matrices of this strategic form game in deciding the technological development effort for the construction work with design condition α and β respectively. The Nash equilibrium of this game for the construction work of type α (e_A^*, e_B^*) is

$$(e_A^*, e_B^*) = (H, L). \quad (8)$$

We can obtain the following Nash equilibrium of the technological development effort for work of type β as well.

$$(e_A^*, e_B^*) = (L, H) \tag{9}$$

As a conclusion for this basic model, we obtain **Proposition 1**.

Proposition 1 Under PPP with performance-based contracting, Contractor A exerts a technological development effort for construction work with design condition α and Contractor B does not. To the contrary, for construction work with design condition β, Contractor B exerts an effort and Contractor A does not.

Conventional procurement does not entice contractors to adopt new technologies because the government makes all decisions related to materials and technologies. Contractors cannot make use of their professional expertise, a fact that may be reflected in higher construction costs. Finally, the bidding prices proposed by contractors are likely to come very close, with the success of each contractor's bid being uncertain. In contrast, PPP enables contractors to adopt new technologies to save on construction costs. A noteworthy corollary of Proposition 1 is that the contractor is surely rewarded for making technological development efforts specializing in the technical field in which he holds an advantage. In practice, design conditions vary depending on construction work sites and facilities, with different types of construction work requiring different kinds of professional expertise. The introduction of the performance specification scheme makes it more likely that the contractor with the most appropriate professional expertise will be awarded a construction work project with certain design conditions. In other words, PPP is blessed with a mechanism that matches contractors' technical competencies with the conditions of the work. If this technological matching mechanism functions properly, contractors are sure to be awarded bids for construction work projects at which they are most proficient. This technological matching also has the positive effect of creating incentive for technological development efforts.

Rationale of conventional procurement

We have pointed out that the PPP scheme is superior to conventional procurement in terms of economic efficiency. However, conventional procurement, whereby the government selects the work method, i.e. the technology, has heretofore been the standard for public procurement. Procurement has not taken place in a competitive environment. From the government's point of view, its most important obligation is to ensure accountability regarding the amount of money to be spent for each construction project. To ensure accountability, it is most reasonable for the government to estimate the construction cost of each work based on a predetermined rule. This predetermined rule is regarded as the current quantity survey rule in Japan, where adopted technologies have been regulated as standard. Because these adopted technologies have been used for a number of actual projects, their reliability is considered proven. As a result, if budgetary constraints were not as severe as they currently are, the government could implement all necessary work simply by demanding budget resources in the amount of the projects' estimated cost using conventional technologies. Of course, benevolent governments may make an effort to develop new technologies and decide to adopt them as well. However, it is not realistic to expect public government officials to be familiar with all the new technologies in the construction industry or to be able to choose the most appropriate from among several. Therefore, as long as there are demands on the government to achieve accountability on technological choice, it is reasonable to have predetermined standard technologies.

Role of the Performance Test

Settings

The analysis in the above section shows that PPP fosters the technological development effort of the contractors because they can focus their efforts on the technical fields in which they hold an advantage. There has been sufficient accumulation of knowledge and experience with regard to the conventional technologies adopted by the government, whereas the lack of knowledge of new technology can cause doubts about its potential success in practice. Such uncertainty is inherent in new technology. If new technologies do not work as expected, realized performance level cannot reach the required performance level specified in the contract. The economic problem of failure in meeting the required performance level is its social cost. Therefore, once we take such technological uncertainty into consideration, we must analyse both construction cost reduction and expected social costs due to performance failure. Moreover, if contractors have private information related to the reliability of new technologies, they could be motivated to adopt unreliable new technology for excess cost saving. Such moral hazard behaviour by contractors might be alleviated by internalising social costs into the payoff of contractors by claiming a damage remedy. However, considering that facilities play the role of infrastructure, the social costs of performance failure could expand in various ways. Therefore, it is practically difficult for private firms to cover social costs resulting from a project's performance failure. In this case, there is a functional limitation of economic incentive schemes such as damage remedies to alleviate the moral hazard behaviour. In this section, the uncertainty of technological reliability is introduced into the performance specification scheme model established in the previous section. Moreover, the assumption of a contractor's limited liability is introduced in order to analyse the contractor's moral hazard behaviour.

Contractor A can develop a new technology by spending ξ, which realises a decrease of $\Delta C_A^{\beta M}$ in construction cost. However, this new technology may fail to attain the performance requirements, with a probability $\lambda_A^{\beta M}$. Herein, we call such a contractor's behaviour moral hazard and represent it by $e_A = M$. Failure to fulfil the performance requirement brings the social cost of D. Next, the following inequalities are assumed:

$$\Delta C_B^{\beta} < \Delta C_A^{\beta M} \tag{10a}$$

$$\Delta C_B^{\beta} > \Delta C_A^{\beta M} - \lambda_A^{\beta M} D, \tag{10b}$$

where (10a) assumes the condition that Contractor A can construct the facility at a lower cost than Contractor B, and (10b) assumes that considering the expected social costs resulting from performance failure, the new technology developed by Contractor B is socially more efficient than that of Contractor A. The maximum amount that the contractor can remedy for damage is T, and

$$D > T. \tag{11}$$

In this case, the contractor cannot remedy all social costs resulting from performance failure.

Nash equilibrium

Here we derive the Nash equilibrium within the strategic form of the game where contractors choose technological development strategies, considering the bidding process for construction work with the type of design condition β. Contractor A's moral hazard strategy is added to the model. Therefore, we derive the contractors' expected payoffs for the following two cases, given the contractors' strategies combination.

(a) Given $(e_A, e_B) = (M, H)$, if Contractor A is successful in bidding, Contract A bears the total cost including the construction cost and expected remedy $\bar{C} - \Delta C_A^{\beta M} + \lambda_A^{\beta M} T$, and Contractor B bears $\bar{C} - \Delta C_B^{\beta}$. In the case that the following inequality

$$\bar{C} - \Delta C_A^{\beta M} + \lambda_A^{\beta M} T < \bar{C} - \Delta C_B^{\beta}$$

$$\Leftrightarrow \Delta C_B^{\beta} < \Delta C_A^{\beta M} - \lambda_A^{\beta M} T \tag{12}$$

is satisfied, Contractor A is awarded by bidding price $b_A^{\beta} = \bar{C} - \Delta C_B^{\beta} - \varepsilon$ and gains a positive payoff. Contractor i's expected payoffs are thus

$$\pi_A^{\beta}(M, H) = b_A^{\beta} - (\bar{C} - \Delta C_A^{\beta M} + \lambda_A^{\beta M} T) - \xi \tag{13a}$$

$$= \Delta C_A^{\beta M} - \Delta C_B^{\beta} - \varepsilon - \lambda_A^{\beta M} T$$

$$\pi_B^{\beta}(M, H) = -\xi. \tag{13b}$$

(b) Given $(e_A, e_B) = (M, L)$, if Contractor A bids successfully, Contractor A bears the total cost including the construction cost and expected remedy $\bar{C} - \Delta C_A^{\beta M} + \lambda_A^{\beta M} T$, and Contractor B bears $\bar{C}$. In the case that (12) is satisfied, Contractor A is awarded the contract at bidding price $b_A^{\beta} = \bar{C} - \varepsilon$ and gains a positive payoff. Contractor i's expected payoffs are

$$\pi_A^{\beta}(M, L) = b_A^{\beta} - (\bar{C} - \Delta C_A^{\beta M} + \lambda_A^{\beta M} T) - \xi$$

$$= \Delta C_A^{\beta M} - \varepsilon - \lambda_A^{\beta M} T \tag{14a}$$

$$\pi_B^{\beta}(M, L) = 0 \tag{14b}$$

Table 15.3 indicates the payoff matrix of this strategic form of the game in deciding the technological development strategy for the construction work with design condition type. The Nash equilibrium of this game is (e^*_A, e^*_B) is (M, L).

Table15.3: Payoff matrix (moral hazard model)

Design Cond. β		Contractor B	
		H	L
Contractor A	H	$(-\xi, \Delta C_B^{\beta} - \Delta C_A^{\beta} - \varepsilon - \xi)$	$(0, \Delta C_B^{\beta} - \varepsilon - \xi)$
	L	$(\Delta C_A^{\beta} - \varepsilon - \xi, 0)$	$(0, 0)$
	M	$(\Delta C_A^{\beta M} - \Delta C_B^{\beta} - \varepsilon - \lambda_A^{\beta M} T, -\xi)$	$(\Delta C_A^{\beta M} - \varepsilon - \lambda_A^{\beta M} T, 0)$

Proposition 2 If Contractor A can adopt unreliable new technology for the construction work with design condition β, PPP may not motivate Contractor B's socially efficient effort level for technological development.

Contractors know in detail the characteristics of the new technologies they develop. It is difficult for the government to acquire correct information on the reliability of these new technologies. Even though informational asymmetricity exists, contractors may not adopt unreliable technologies under a rule requiring them to bear all social costs. However, if we consider a case where the contractor cannot bear all the social costs incurred by performance failure due to limited liability protection, it is possible that in order to emerge as the successful bidder he would adopt new technology that could, with excessively high probability, result in performance failure. Such moral hazard behaviour of contractors may remove the motivation for technological development.

Role of performance test

Figure 15.2: Technological uncertainty and performance test

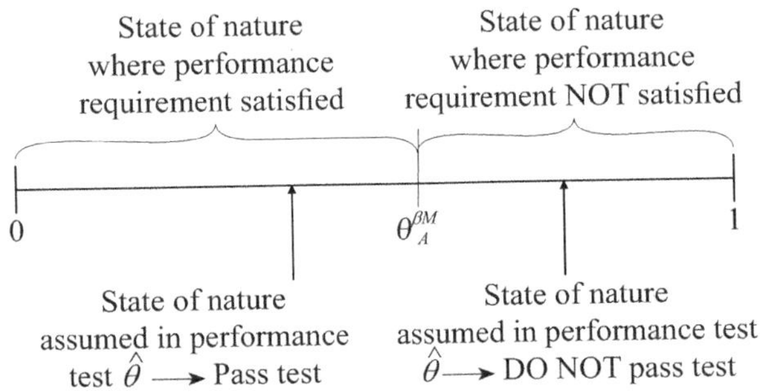

Table 15.4: Payoff matrix (Performance Test Model)

Design Cond. β		Contractor B	
		H	L
Contractor A	H	$(-\xi, \Delta C_B^{\beta} - \Delta C_A^{\beta} - \varepsilon - \xi)$	$(0, \Delta C_B^{\beta} - \varepsilon - \xi)$
	L	$(\Delta C_A^{\beta} - \varepsilon - \xi, 0)$	$(0, 0)$
	M	$(-\xi, \Delta C_B^{\beta} - \varepsilon - \xi)$	$(\Delta C_B^{\beta} - \varepsilon - \xi)$

Without any countermeasure to alleviate informational asymmetricity, the performance specification scheme could fail to function as shown in the basic model. Extending the moral hazard model, we analyse the role of performance testing conducted by the government or third-party agencies (certifier).

Final performance of a facility constructed adopting a new technology depends on the environmental condition (state of nature) realized during the in-service period, which cannot be forecast during the construction period. This state of nature is denoted by a stochastic variable θ, which is defined on the interval [0,1]. θ is assumed to be distributed on [0,1] according to a uniform distribution. θ close to 1 represents that the state of the environment is severe. New technologies developed by Contractor B satisfy the performance requirement for any state of nature on [0,1]. However, new technology developed by Contractor A satisfies the performance requirement for realized $\theta \in [0, \theta_A^{\beta M}]$, but not for $\theta \in [\theta_A^{\beta M}, 1]$. In this case, the probability that the required performance is not attained by new technology developed by Contractor A is represented by

$$\lambda_A^{\beta M} = 1 - \theta_A^{\beta M}. \tag{15}$$

The threshold $\theta_A^{\beta M}$ is known by both Contractor A and B, but not by the government. The certifier conducts performance tests in a certain state of environment $\hat{\theta}$. If, $\theta_A^{\beta M} \geq \hat{\theta}$ new technology developed by Contractor A passes the tests, and if $\theta_A^{\beta M} < \hat{\theta}$, it does not pass the tests. Contractors know the environment assumed in the performance test. Therefore, Contractor A, which has developed a new technology whose reliability is $1 - \lambda_A^{\beta M} \geq \hat{\theta}$, knows whether or not his technology will pass the performance test. Equation (10b) and (12) can be rewritten as

$$\frac{\Delta C_A^{\beta M} - \Delta C_B^{\beta}}{D} < \lambda_A^{\beta M} < \frac{\Delta C_A^{\beta M} - \Delta C_B^{\beta}}{L} \tag{16}$$

Therefore moral hazard can be alleviated by excluding Contractor A through the use of the performance test. Now let $\hat{\theta}$ be

$$\lambda_A^{\beta M} < \frac{\Delta C_A^{\beta M} - \Delta C_B^{\beta}}{D}$$

$$\Leftrightarrow \theta_A^{\beta M} > 1 - \frac{\Delta C_A^{\beta M} - \Delta C_B^{\beta}}{D} = \hat{\theta}. \tag{17}$$

In this case, a new technology developed by Contractor A cannot pass the tests. Performance testing can exclude from competitive bidding contractors who have developed unreliable technologies. However, let us assume that Contractor A can participate in competitive bidding if he adopts a conventional technology whose reliability has already been proven. The payoff matrix of this strategic form game with performance tests in deciding the technological development strategy is shown in Table 15.4. The Nash equilibrium of this game is (H, L).

Proposition 3 If the certifier conducts performance testing under some appropriate state of nature $\hat{\theta}$, the Contractor with unreliable technology can be excluded. This exclusion motivates the Contractor with professional expertise to develop new technologies that satisfy the certifier's criteria.

Performance testing plays a role not only in alleviating the risk of performance failure but also in motivating contractors to develop new technologies in their technical field. According to (17), the appropriate threshold $\hat{\theta}$ depends on 1) the social cost due to performance failure, 2) the amount of cost savings when unreliable technology is adopted and 3) the amount of cost savings when reliable technology is adopted. However, it is impossible for the certifier to know the construction costs of contractors. Therefore, in practice, the certifier determines the standard level of performance test based on the certifier's subjective judgment of the probability of moral hazard risk and social cost.

Policy implications

When the government stipulates the method of work, as in conventional procurement, it is difficult for the government to ensure accountability or to justify the adoption of new technology for a project. Therefore, it is reasonable for the government to make use of conventional standard technologies whose reliability has been guaranteed for any type of project. The opportunity for contractors to adopt new technologies has been constrained due to this need for accountability of the government. Even if a contractor develops a new technology, it cannot be adopted according to the conventional procurement system, under which contractors are not motivated to develop new technologies.

On the other hand, PPP is nothing more than a rule requiring the government to determine the technology to be adopted for each individual project based on the information revealed through the bidding or market mechanism. The performance-based contracting of PPP relieves the government of accountability regarding the selection of technology. According to the results analysed in this chapter, this scheme guarantees that a contractor who deals with a specific type of design condition most efficiently will be the successful bidder. However, technologies of contractors are evaluated only in terms of cost information under the general competitive bidding scheme. There is a possibility for moral hazard when contractors bid low prices and employ unreliable technology. Performance tests based on the appropriate state of environment should play an important role in excluding such contractors from competitive bidding.

An important implication of this chapter is that the economic value of PPP arises from a matching process between the technical needs that vary with individual projects and the contractors with the technical expertise to address those needs. The function of this matching mechanism will lead to technological differentiation among contractors and increase technological variety in the construction market.

To assure this matching between projects and contractors, the government should develop an information sharing system which identifies types of required professional expertise in each project so that contractors know which projects they should participate in among the vast number of construction projects in the market. To develop such an information system, a classification system of construction technologies is necessary. In Japan, for example, since 2006, the Ministry of Land, Infrastructure, and Transportation has operated NETIS (New Technology Information System), where new construction technologies are publicly introduced on a website, sorted according to a classification system. This information system could help improve the economic efficiency of the performance specification scheme by fostering matching between projects and contractors. However, in practice, it is difficult to confirm the reliability of a new technology by a one-shot performance test. The usefulness of a new technology can be confirmed only by applying it in practical work. Of course, the above discussion concerns a hypothetical scenario. The impact of PPP on technological development and differentiation must be empirically tested, though to do so is beyond the scope of this chapter.

Conclusion

One feature of PPP is its performance-based contracting scheme. This chapter provided a mechanism to determine why the performance-based contracting of PPP brings economic benefits through strategic interactions in the bidding process. PPP performance-based contracting endows contractors with the discretion to decide the method of work, enabling them to adopt new technologies. In contrast to conventional procurement, PPP is interpreted as a rule whereby the government selects technology through bidding. Bidding with performance-based contracting induces appropriate incentive for technological development and differentiation. Because site conditions for construction work differ from place to place, developing a tailor-made technology and technological differentiation is a key strategy for improving the economic efficiency of construction work. In sum, the performance-based contracting feature of PPP has a mechanism to realise matching between technical needs that vary with individual projects and those contractors with the technical expertise to address such needs. However, if there is a non-negligible possibility of moral hazard occurring when contractors bid an excessively low price and then employ unreliable technology, a performance test should play an important role in excluding such contractors from competitive bidding.

It should be noted that the above conclusions are derived based on the settings assumed in this chapter. Practically, the setting assumed in the model that in order to alleviate moral hazard the contractor should bear the entire social cost of performance failure may be unrealistic. In addition, the result of the performance test is not perfect. Excess risk bearing by contractors may not motivate them to adopt new technologies, because even if the reliability of technical uncertainty is quite small, they cannot eliminate the risk of bankruptcy. Appropriate risk sharing between the government and the contractor with regard to the uncertainty of adopting a new technology is a critical issue in PPP which is beyond the scope of this chapter. Nevertheless, information on the reliability of new technologies is critical for performance-based contracting of PPP, as the performance test is not always perfect. Therefore, the effective usage of PPP requires accumulating information on the reliability of new technologies through experience in practical work.

References

[1] Scotchmer, S. (2004) *Innovation and Incentives*. MIT Press.

[2] Dasgupta, P. and J. Stiglitz (1980) Uncertainty, industrial structure and innovation and the speed of R&D. *Bell Journal of Economics* **11**: 1–28.

[3] Loury, G. C. (1979) Market structure and innovation. *Quarter Journal Econonomic* **93**: 395-410.

[4] Reinganam, J.F. (1981) Dynamic games of innovation. *Journal of Economic Theory*, **25**: 21 – 41.

[5] Seaden, G., Guolla, M.,Doutriaux, J. and J. Nash (2003) Strategic decisions and innovation in construction firms. *Construction Management and Economics*. **21**: 603-612.

[6] Onishi, M., Shoji, Y. and K. Kobayashi (2008) Technological matching and economic value of performance specification in public procurement. *Proceedings of the IEEE Systems, Man, and Cybernetics Conference*. pp: 2881-2886.

Joint Ventures in Construction 2
ISBN 978-0-7277-5783-8

doi: 10.1680/jvc2.57838.213

Chapter 16

Private Finance Initiative Project Payment Deductions

Katsuji Ishihara, Kyoto University

Introduction

In Japan, the magnitude of public budgets available for public works is decreasing. It is expected that this trend will continue as Japan's population decreases and grows older. To avoid a shortage in public funds, we need public services to be delivered more efficiently and effectively. The Private Finance Initiatives (PFI) is a form of public-private partnership used to deliver capital investment projects efficiently and effectively that include capital projects such as schools, hospitals, roads, and water facilities. Instead of the public sector procuring capital assets and subsequently owning, operating, and regulating them, the private sector owns and operates them.

In 1999, the PFI was introduced in Japan as a method for procuring public services and since then, the PFI has been used in both the central government and the local government, as shown in Figure 16.1. As of July 31, 2011, implementation plans have been released for 391 PFI projects, of which 307 projects are in the operation and management stage. Today, the PFI is widely recognized as a procurement scheme for public works. However, in such a situation, a variety of problems have been actualized about the payment mechanism.

In Japan, the PFI Act specifies that "PFI is a project relevant to maintenance of a public facility etc., and is a project undertaken and effective by utilizing the private sector's fund and management and technical capabilities." The public sector transfers project risks to the consortium company that is owned by one or more private sector parties. The payment mechanism is one element that comprises the core of the PFI contract, and places the financial impacts on the risk allocation between the public sector and the private sector. A PFI project is not realized without cooperation among those private sector parties. Conversely, the project is realized by the close cooperation between the public and the private sectors.

Figure 16.1: The number of PFI projects in Japan

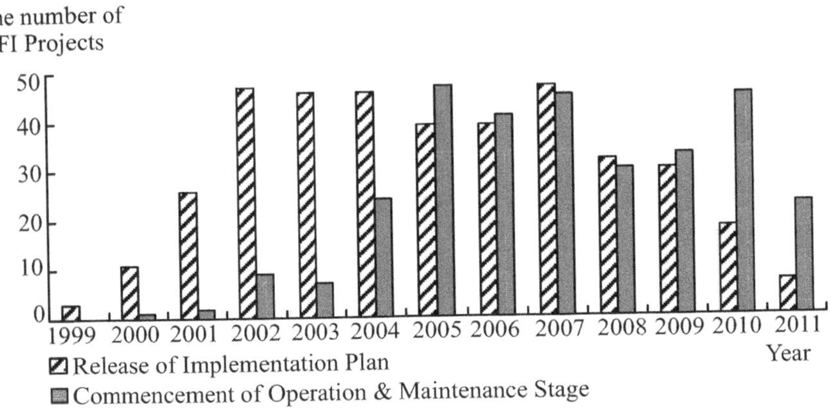

The monitoring process confirms whether the implementation is appropriately performed according to the requirements of the PFI contract. When a default on an obligation is detected by the monitoring process, the payment is deducted according to the importance and the degree of influence and of seriousness. Most PFI projects in Japan adopt this type of payment mechanism. It is intended that the payment mechanism urges a PFI consortium to satisfy the requirements postulated appropriately. However, in Japan, the accumulation of business know-how concerning contract management in the operation and management stage of PFI projects is insufficient. It has been illustrated that neither the monitoring process, nor the payment mechanism, necessarily functions to satisfy the requirements. We will now address the issue of constructing the good governance of PFI projects. The major focus of this chapter will be the payment deduction, which is a central part of the payment mechanism for PFI projects. The author will clarify the problems of payment deduction by reviewing some PFI projects in Japan, and will propose a concept of payment deduction that is appropriate for the PFI projects.

Figure 16.2: Schematic view of governance in PFI project [1]

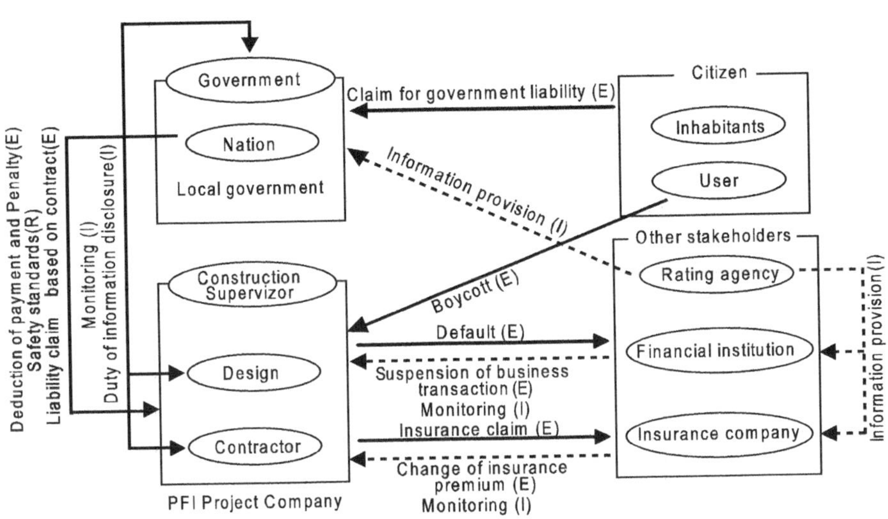

Note (1) Arrows denote the direction of influence.
(2) Dashed lines denote the factors to be considered further in the next phase.
(3) Type of governance, (R):Regulation, (E):Economic motive, (I):Information

Accident Occurred in the Nomura School Lunch Center PFI Project

Project profile

Sendai City, located in the Northeastern district of Japan, built a school lunch center that supplies approximately 11,000 meals to students within 26 schools every day (7,000 meals in the elementary schools and 4,000 meals in the junior high schools). The PFI consortium conducts, designs, constructs, operates (i.e., cooking and delivery), and maintains the Nomura school lunch center in Sendai City, using the PFI method. In addition, the facilities include an advanced school lunch cooking room for the children who have an allergy to certain food ingredients. The outline of the project is illustrated in Table 16.1.

Table 16.1: Project profile of the Nomura School Lunch Center [2]

Authority	School Education Department, Educational Bureau,
Subject matter of the contract	Design, construction, operation (cooking and delivery) and maintenance of Sendai City Nomura School Lunch Center.
Project scheme	BOT (Build, Operate and Transfer) Service purchase type
Project period	16.3 years (December 2006 – March 2023) Operation begins in April 2008.
Approximate project cost	9.5 billion yen
Successful bidder	Consortium of local major contractors and a cooking company

Outline of the payment mechanism of the Nomura School Lunch Center project

Payments to the consortium are made quarterly. The features of the payment mechanism are as follows:

(a) A unitary payment is adopted based on the features of Build-Operate-Transfer (BOT). The deduction point concerning operations is applied to aggregate payments for construction, maintenance, and operations.

(b) The payment mechanism decreases the reduction point to the excellent service provided by the Special Purpose Company (SPC).

(c) The ratchet system is adopted. Particularly, the deduction point is increased by 20% when there is a similar content of deduction that took place during the previous period.

Here, the payment is deducted based on the mechanism in Table 16.2 and Table 16.3. The point indicated in the table is added whenever the problem occurs within three months. Then, the amount of payment deduction is calculated according to the ratio shown in Table 16.3.

Table 16.2: Classification of the event related to the payment deduction for the Nomura PFI project [2]

Situation	Deduction Points
When especially a serious problem occurs with the school lunch supply (e.g., a person that died or incurred a serious illness due to food poisoning or an allergic reaction)	150 for each
When a serious problem occurs with the school lunch supply (e.g., a person that incurs a mild illness due to food poisoning or an allergic reaction)	120 for each

Situation	Deduction Points
When an immense problem occurs with the school lunch supply (e.g., no school lunch supply for more than half of the schools)	50 for each
(omit the description concerning the matter that has a medium influence)	
When there is a possibility that a significant influence on the school lunch supply will occur if the problem is not corrected (e.g., inconsistency with the sanitary supervision)	5 for each
When there is a possibility that a significant influence on the school lunch supply will occur if the problem is not corrected (e.g., many or important errors in writing, such as record keeping, book keeping, and documentation)	2 for each

Table 16.3: Ratio of payment deduction applied to the Nomura PFI project [2]

Situation	Ratio of Deduction
150 <	100 %
50 < Deduction point <150	(accumulated deduction point – 50)^2 /100 %
< 50	0 %

Outline of the accident

The accident occurred at the Nomura school lunch center on November 12, 2008. "Spaghetti meat sauce" was distributed to two children who had a wheat allergy. The school lunch center should have offered "Potato meat sauce," which did not contain wheat, to these two children. As a result, one child had temporary breathing difficulty caused by their acute allergic sensitivity, "Anaphylaxis shock." This child was hospitalized but recovered after a one night and was released on the second day. The other child's allergic reaction to the food was stomach aches. The inspection work that multiple people had to do was neglected. The city officer (dietitian) who had to attend the confirmation work was absent and did not probe for it.

The accident was considered as a serious problem and 120 points were incurred. As a result, they received approximately 50%, which was a deduction of 80 million yen, of their quarterly overall payment. In addition to the payment deduction, the project company should pay indemnity for the victims.

There were some arguments about this payment deduction. The city officer was attaching importance to the safety of food, including the prevention of food poisoning accidents and allergy mistakes. On the other hand, the PFI consortium believed that the large deduction executed in addition to the compensation for damages to the victims was too severe. In addition, Sendai City had previously experienced a similar accident with another PFI project. When the Miyagi offshore earthquake (M7.2) occurred in 2005, most of the ceiling of a indoor swimming pool, which was built and maintained in a PFI project, collapsed due to poor earthquake resistance strength and a few people were injured. In this case, the payment was deducted by only around 2 million yen, which corresponds with the operation and maintenance costs of the duration the facilities were closed due to the accident. The amount of the deduction for this case seems minor, judging from the fact that the public sector had to conduct numerous additional administrative projects.

Table 16.4 compares the payment deductions adopted in 12 PFI school lunch center projects in Japan. In the table, Project A describes the project concerned. There is a significant difference in both the ratio and the amount of payment deduction as shown in the Table 16.4. A considerable difference is recognized in the payment deduction to similar events occurred in similar projects. The difference is too large to explain reasonably even if the difference among the policy of each public sector is admitted.

Table 16.4: Classification of event related to deduction

Project	Delivery stop at all the schools by one day		One deceased by food poisoning	
	Ratio of deduction	Amount of deduction	Ratio of deduction	Amount of deduction
A	0%	0	100 %	102 million yen
B	1.5 day payment	2 million yen	7.5 day payment	12 million yen
C	0%	0	100 %	107 million yen
D	3%	6 million yen	40 %	21 million yen
E	20%	15 million yen	20 %	15 million yen
F	0%	0	100 %	94 million yen
G	20%	15 million yen	20 %	15 million yen
H	20%	23 million yen	100 %	45 million yen
I	0%	0	100 %	96 million yen
J	0%	0	100 %	28 million yen
K	20%	21 million yen	100 %	107 million yen
L	1.5 day	0.5 million yen	4.5 day payment	1.4 million yen

(1) Ratio of deduction is applied to quarterly payment.
(2) The running cost that became basis of the deduction calculation was assumed as about 4% of the project cost for H through L.
(3) It was assumed the suspension of business of three days at the death by food poisoning because the deduction was calculated based on the suspension for the projects I through L.
(4) Unitary payment is adopted in the project A and C.

Ideas and Situation Relating to the Deduction Mechanism in Japan

Guidelines issued by the Cabinet Office

In Japan, the payment deduction mechanisms in most of the PFI projects follow the description in the "Monitoring Guidelines [3]." The section relating to payment deductions is quoted as follows:

"Deductions and other measures should be balanced so that the service, itself, may not be damaged. It is necessary to decide on a deduction mechanism in consideration of the concern to rapid aggravation of a PFI consortium's financial situation. Furthermore, the importance of requirement for each service should be taken into account. It is required to deduct the payment based on whether the index set to a requirement is satisfied. For example, each service should be weighted based on the degrees of seriousness and of influence in case the service is not available. There is an idea of deducting payment when the accumulated value of the point corresponding to this weight becomes more than a prescribed value. There is also an idea of imposing the deduction corresponding to the period when the service is not available."

Problems concerning the payment deduction mechanism in the PFI projects of Japan

The number of PFI projects that are operating has been increasing. The case where payments have been deducted has occurred in many projects. The author has learned of a variety of dissatisfactions and opinions concerning payment deductions from the affected parties, such as the public sector, private companies, and financial institutions involved in the PFI projects. Some typical arguments are described as follows:

(a) The payment mechanism should be closely related to the requirements and the monitoring. However, we can find many cases where the criteria for payment deduction do not correspond with the items and priority of the requirements and the monitoring.

(b) Few PFI projects set the objective standard for the availability and performance of services at the tender stage of the project.

(c) Payment deductions are decided upon based on the subjective judgment of a public sector; therefore, the deductions can be excessive.

(d) Alternatively, payments are hardly deducted as long as the availability of the core services is satisfied. Additionally, there have been cases where public sectors hesitate to impose deductions because they are concerned that it will damage the ability of sub-contractors to rectify problems or because they fear threatening their relationship with the PFI consortium.

(e) The structure of "imposing a large amount of penalties when trouble is brought into the public sector" is a unilateral contract.

All of the above mentioned views and opinions originate from poor communications and shortage of examinations in the tender preparation stage. Both the public sectors and the PFI consortiums perform inadequate examinations about the monitoring process and payment deduction mechanisms compared with service requirements. Financial institutions have not conducted sufficient monitoring process concerning deductions before the tenders. Furthermore, after a successful bidder is selected, time for the contract negotiation is usually limited because of a tight schedule. Consequently, the payment deduction mechanism is not conferred in detail before the contract is finalized.

Considerations Concerning the Ideal Method of Payment Deductions

Scale of payment deductions

Private companies and financial institutions always make criticisms, stating that the scale of the deductions is too large. If the scale of likely payment deductions is overweighed, bidders will claim excessive risk premiums. However, if the scale is underweighted, bidders will not be sufficiently motivated to rectify the failure of the services. The scale of the deductions must be well designed. Some important matters at the time of considering the scale of the deductions include:

(1) The scale of the deductions should be examined on the basis of its importance to a public sector. Large amounts should be deducted for the unavailability of important parts of the service.
(2) Incentives on the PFI consortium are also important. Large deductions clearly strengthen the incentive on the PFI consortium to increase the value for money. If the reduction scale is too small, the PFI consortium obtains good revenues and expenditures by not conducting the improvement effort and by receiving deducted payments during the remaining project period.
(3) If the scale of the deductions is too severe, negative incentives may arise. For instance, when there is a service failure that causes a 100% (capped) deduction at an early stage of the payment period, the PFI consortium might not have an incentive to provide any remaining services during that period.
(4) If the scale of the deductions is too severe, the PFI consortium may feel the increase in the risk and may encourage higher pricing.
(5) The deterrent effect by the deductions is sometimes expected with the failure of the services, which could be expensive to remedy.
(6) In general, the role of a punitive forfeit is not expected in the deductions.
(7) In many payment mechanisms, although the maximum notional deduction exceeds 100% of the payment, the public sectors cap the deductions at 100% of the payment.
(8) The author believes that the role of the punitive compensation for damages may be given to the deductions, although the author admits that there is a difference in opinion.

It is necessary to clarify what functions we expect in the deductions. However, in Japan, when the scale of the deductions is considered in practice, the components of the deductions are not usually examined in detail.

In most PFI projects in Japan, a base amount payment corresponds to the costs of all of the operations and maintenance works for each payment period (in several cases, a quarter or half a year) or the cost of the corresponding work (for instance, the payment for guards). There are only a few examples where the base amount includes payments corresponding to the construction cost. Therefore, we can say that the scale of the deduction in the PFI projects in Japan is smaller than that of the impression received from the ratio.

Although the scale of the deductions is not the single most important factor of a payment deduction mechanism, it is a major factor. Large deductions for poor service may not be a severe mechanism if a public sector allows for long periods in which the PFI consortium can respond and if the service requirements are not excessively demanding. Small deductions may provide strong incentives to perform provided that the service requirements consist of tightly defined high level standards and short time periods are allowed for remediation.

A payment deduction mechanism is closely related to the adoption of the unitary payment in which the payment corresponding to the cost of construction may be deducted due to the failure of the operation services.

In a BTO project, it is, in general, thought that the payment corresponding to the construction cost is a definite claim and should not be deducted. Here, the BTO is the project scheme in which the SPC transfers a project property title to a government immediately after the completion of the construction. Conversely, in a BOT project, it is preferable to adopt the unitary payment for motivating the PFI consortium to perform properly. Here, the BOT is a project scheme in which

the SPC transfers a project property title to a government after a project period. In the case of the Nomura school center project, the BOT is adopted with the viewpoint of advancing the risk transfer to private companies and urging the use of their practical know-how.
In Japan, because the tax burden on PFI projects using the BTO scheme is small as compared with projects using the BOT scheme, most of the PFI projects adopt the BTO scheme, as illustrated in Figure 16.3. Furthermore, most of the BOT projects in Japan do not adopt the unitary payment in consideration of the opinions of the private companies. Accordingly, only a few PFI projects use the unitary payment system.

Many companies and financial institutions oppose the unitary payment system on basis of the following views and opinions:

(1) A construction sub-contractor claims compensation to an operation and maintenance sub-contractor, when the payment deduction due to the failure of the operation services influences the payment corresponding to the construction cost. The operation and maintenance sub-contractor may then weave such a risk and propose the higher tender price; consequently, the Value for Money (VFM) decreases.
(2) The risk rises for the shareholder. As a result, the shareholder comes to demand a high return and the Value for Money (VFM) decreases.
(3) The payment that corresponds to the construction cost cannot be treated as a definite claim; therefore, the scheme of the finances may become different from the ordinary scheme.

Figure 16.3: PFI project type in terms of ownership

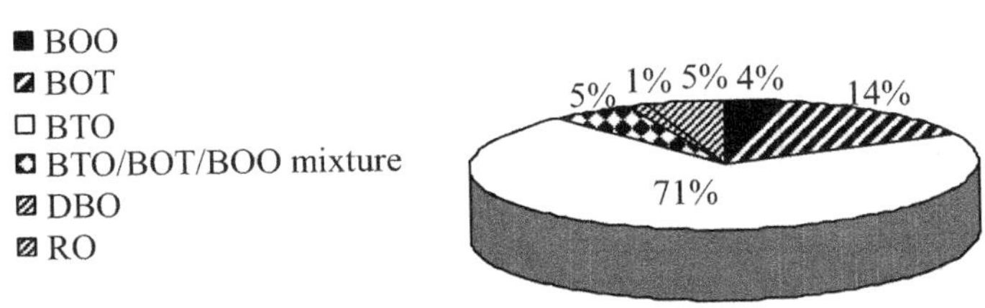

Payment mechanism with incentives and deductions

The service requirements and payment mechanisms are designed to deliver the required services. Therefore, the performance of an SPC beyond a required level should not ordinarily warrant further payments in Japan. However, for some projects, the public sectors may consider that there is a value for money created by the mechanism for payment increases and deductions.

Payments may be determined based on the quantity of services provided by the PFI consortium, as illustrated in Figure 16.4 (a). This is the so-called usage-based system used for the highway projects in the early stage after introduction of PFI in the UK. In Japan, one of the PFI library projects adopted a kind of meter rating system in which the payment to a project company is determined according to the amount of services measured using the numbers of users, and the number of lent books. The Nomura school cooking center project adopts the similar mechanism in which the payment to a project company is determined according to the number of supplied meals.

The PFI consortium expenditures are dependent on the amount of services provided. Therefore, it might be reasonable that a public sector determines payments based on the amount of services provided. There should be a minimum take-or-pay level to avoid the excessive transfer of usage risks. Then the payment deduction will be limited to the take-or-pay level, even if the amount of the services in use is below the level. In Figure 16.4(a), the payment for the amount of services beyond the contract base (Point A) should be set above the line AB, in case the increase in the service amounts result in value for money; otherwise, the PFI consortium would have an opposite motivation to prevent the service amounts from decreasing its profit.

Payment may be determined on the basis of the quality of services provided by the PFI consortium, as illustrated in Figure 16.4 (b). The author uses the term "service quality" instead of "service performance" because the latter may be related to the amount of services in many cases. The author considers the fact that the deduction and the increase are a series of mechanisms. The payment curve (JIFH in Figure 16.4(b)) can be appropriately determined, regardless of the position of the contract base (Point F). Assuming the same services, whether the payment is deducted or increased depends on the level of the contract base; however, it is just a technical matter. A public sector severely demands a high level of service quality and may make the series of mechanism that only uses the deduction as the extreme case. On the other hand, a public sector may set requirements at a low level and make the mechanism that only uses the increases in the payments. However, the payment mechanism with both the deductions and the increases should be incorporated in the payment curve in order to prepare a probable budget.

Notably payment should not be increased for excessive service. The public sector should determine what level of performance is required and should cap payments for out-performance, as shown in Figure 16.4 (b). However, most of the PFI projects in Japan set the cap level at the contract base (use the payment curve JIFG) because of the budget system. The budget systems of local governments require the approval of the assembly for each increase in payments. Consequently, financial institutions and private companies recognize that the payment increase is uncertain and the public sectors do not adopt the payment increase mechanism so as to evade the procedure of the assembly approval.

Figure 16.4: Schematic view of payment increases and deductions

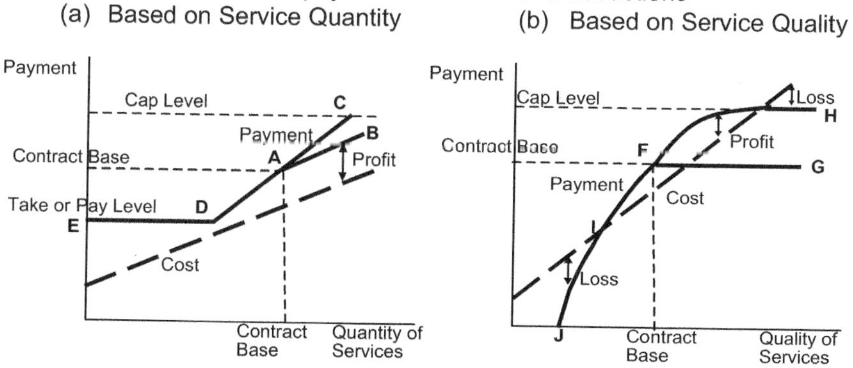

The author proposes the following measures to adopt the payment increase mechanism under the situation described above:

(1) Awarding "recovery points" that can only serve to offset the "negative" points for the deductions due to poor services. The Nomura project described in the previous section adopts the recovery point system, discussed later in this section,
(2) Using a contingency in a project budget, which is not currently used in Japan, and
(3) Saving deducted payments for the increase in the following period.

If a self-supporting scheme is adopted, the effect that is similar to the payment increase mechanism is achieved. However, this scheme can be applied to the project with the change in income from users. Additionally, private companies and financial institutions are reluctant to take income change risks and request applying the service purchase type, even if the self-supporting accounting type can be used. As a result, only 5% of the PFI projects follow the self-supporting accounting type, as shown in Figure 16.5.

Figure 16.5: PFI project type in terms of project cash flow

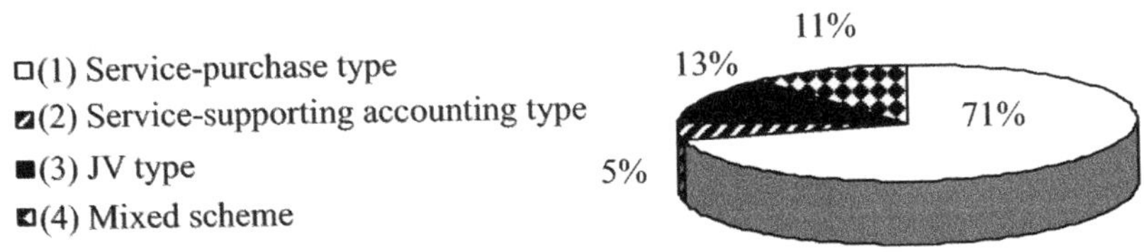

(1) Service purchase type: A project company's earning relies solely on the revenue from a public sector.
(2) Self-supporting accounting type: A project company's earning relies solely on the revenue from users.
(3) JV type: A project company's revenue comes from both a public sector and users..
(4) Mixed scheme: A project consists of service purchase type services and self-supporting type services.

Conclusion

Monitoring process is considered a means of confirming whether the implementation is appropriately performed according to the requirements of a PFI contract. When the implementation is not appropriately performed, the payment is deducted based on the payment deduction mechanism. As a result, securing the implementation according to the requirements of the contract is assumed. Therefore, the requirements, the monitoring process, and the payment mechanism should synchronize.

Today, more than 200 PFI projects entered into the operation and management stage, indicating that the PFI is widely recognized as a procurement scheme for public works. However, in such a situation, various problems have been actualized about the payment mechanism. This chapter discusses the immaturity of the payment deduction mechanism of the PFI projects in Japan.

In the Nomura PFI project referred in this chapter, significant amount of money was deducted from their quarterly payment, due to a child's hospitalization by allergic shock. This chapter mentioned the relevant factors and listed various problems concerning the payment deductions in the PFI projects. It is considered that most of the problems are recognized as an issue of the lack of examinations and negotiations in a tender process.

This chapter proposed the ideal method of payment deductions for PFI projects. It is considered that the scale of the deductions should be based on the functions expected of the deductions. The unitary payment may be effectively used with the payment deduction mechanism. The author discusses the importance of the payment increase mechanism and suggests a few feasible schemes for applying the mechanism.

The payment deduction mechanism must be properly applied in practice so that the PFI consortium is expected to have an incentive to offer good services. Public sectors should consider the payment deduction mechanism as an important part of the contract.

References

[1] Ishihara, K. and Onishi, M. (2006) Governance of PFI Projects - Lesson from the Accident of PFI Project-. *International Conference on Multi-National Joint Venture for Construction Works.*

[2] Sendai City (2006) *Draft of Project Contract of Nomura School Lunch Center PFI Project*, http://www.city.sendai.jp/kyouiku/kenko-k/nomura-kyusyoku/nyuusatu/index.html, [Accessed: 2009, August 15].

[3] The Cabinet Office of Japan (2003) *Monitoring Guidelines.*

Joint Ventures in Construction 2
ISBN 978-0-7277-5783-8

doi: 10.1680/jvc2.57838.225

Chapter 17

Moral Hazard Issues in PPP Projects

Lei Shi, Dalian University of Technology
Masamitsu Onishi, Kyoto University
Kiyoshi Kobayashi, Kyoto University

Introduction

Public-Private-Partnership (hereafter PPP) is an innovative scheme for the procurement of public services with private participation. The main players in a PPP project are the SPC (Special Purpose Company), the government, and the bank. In most PPP projects, a concession contract between the government and the SPC is signed to establish the performance levels and service fees that will rule the concession over its time framework. After the concession contract is signed, the SPC may close a debt contract with the bank for the provision of capital. The debt contract specifies the initial loan amount and the schedule of repayments over a contractual term.

Debt contracts are characterized by limited liability. If the SPC falls into default, the entire debt contract will default as well. Furthermore, the debt contract must be an incomplete contract, given the indescribability of all possible future phenomena. Moral hazard might arise from limited liability and contract incompleteness. In this chapter, we classify moral hazard into two types: that of the SPC and that of the bank. In a PPP project, the SPC might misuse operation funds raised from the bank in return for private benefits. This situation is referred to as the SPC's moral hazard. In a PPP project, the SPC ultimately selected in a tender is the one offering the minimum price. However, the SPC's moral hazard cannot be deterred in a regular tender.

In this chapter, a reserve tender system is introduced in order to overcome the SPC's moral hazard problem. In the reserve tender, an additional fee (quasi-rent) is paid to the SPC by the government. The problem with this system, however, is that the bank might expropriate the quasi-rent paid to the SPC. We refer to this situation as the bank's moral hazard. In addition, the bank's moral hazard may trigger the occurrence of the SPC's moral hazard.

Figure 17.1: Timing of the model

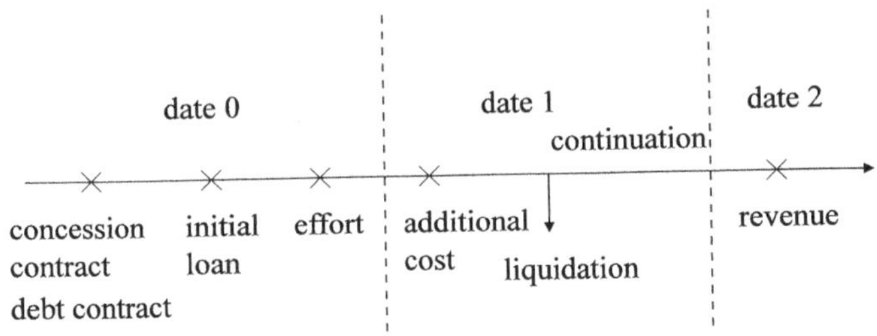

Other papers have addressed efficiency issues in privatization and PPP[1] - [3]. These studies analyze the impact of allocation control rights on the efficiency of projects by focusing on the underinvestment problem in relationship-specific investments, as well as on other inefficiencies in public services arising from contract incompleteness. However, there are currently only a limited number of studies on the moral hazard problem occurring in PPP projects.

Nevertheless, there is extensive literature [4] analyzing the moral hazard problem, which is apparently difficult to deter because it arises from asymmetric information between the principal (the government) and agents (the SPC and the bank). In order to resolve this problem, it is necessary to eliminate the incentives that trigger moral hazard. Hence, an agency cost arises, creating inefficiency. However, studies about moral hazard focus only on the problem as related to the private company. The possibility that the bank's moral hazard occurs in PPP projects is yet to be analyzed.

In this chapter, we investigate the possibility of the SPC's and the bank's moral hazard. We conclude that the bank's monitoring is valid for the deterrence of the moral hazard of both the SPC and the bank.

Basic Model

Assumptions

The PPP project is interpreted as a 3-period game, as shown in Figure 17.1. After signing the concession contract, the SPC enters into a debt contract. The amount of loan I is assumed to be equal to the total project cost, accounting for construction costs V and operation costs B, i.e. $I = B + V$. The SPC begins construction by allotting V. After construction is finished, the SPC chooses a level of effort in $e \epsilon \{e_H, e_L\}$. The cost related to effort $F(e)$ is defined as

$$F(e) = \begin{cases} B \text{ (if } e = e_H) \\ 0 \text{ (if } e = e_L) \end{cases} \tag{1}$$

That is, at date 0, the SPC after determining the level of effort e_i $(i = H, L)$incurred the expenditure $F(e)$. If the SPC chooses e_L, operation cost B will be accounted for as a private benefit. In the basic model, the level of effort chosen by the SPC is observable but unverifiable. Furthermore, information about private benefits is assumed to be common knowledge to the government, the bank, and the SPC. In this chapter, we focus on the role of the debt contract by assuming that the SPC has no equity in the investment and raises all funds by means of the debt contract. The initial loan I and repayment D are detailed in the debt contract. The repayment D includes not only the principal but also the risk premium accruing to the project, $D \geq I$. The repayment takes place at date 2 and the interest rate (discount rate) is assumed to be zero.

The SPC begins operation at date 1, at which default risks may occur. If the project falls into default, an additional loan C will be necessary to continue the project. The continuation cost C is assumed to be smaller than the total cost of the project I; therefore

$$I \geq C. \tag{2}$$

The project will be liquidated if the additional loan is unavailable. The probability of default risks is denoted by $1 - p_i (i = H, L)$, and depends on the level of effort e_i $(i = H, L)$ chosen by the SPC. Assume $p_H > p_L$.

The PPP project is completed at date 2. Accordingly, the concession and debt contracts are both liquidated. The government will pay the SPC the service fee R if the project is continued until date 2. We assume

$$1 + C > R \geq I. \tag{3}$$

If $R \geq I + C$, the SPC's payoffs become nonnegative and independent of the level of effort chosen by the SPC. Thus, only income is transferred from the government to the SPC. Assumption (3) implies that the government intends to maximize the Value for Money (VFM) of the project. Furthermore, the service fee will be high enough to cover the project total cost whenever $R \geq I$.

If the SPC falls into default, the government might cancel the concession contract. In this case, the bank has the right to decide whether to step in to avoid project termination or continuation. If the bank accepts the termination of the concession contract, the government must pay compensation A to the bank in return for assets belonging to the project. Assume compensation price A satisfies

$$0 \leq A \leq I. \tag{4}$$

That is, an additional cost C shall be borne by the government. Finally, the government will determine whether to find an alternative SPC or directly execute the project. This chapter considers the case in which the project is continued by the initial SPC even if the initial concession contract is terminated by the government.

Nonetheless, the bank can also choose to avoid cancellation of the concession contract. In this instance, the bank has no other choice but to agree to modify the initial debt contract, and closes a new debt contract with the SPC. In this chapter, we assume that the externalities of the project are huge enough always to make it socially efficient to continue the project even when the SPC falls into default.

The last assumption states that the level of effort e_H is socially optimal.

$$p_H R + (1 - p_H)(R - C) - I \geq p_L R + (1 - p_L)(R - C) - V \tag{5}$$

(5) can be simplified as

$$\Delta = \Delta p C - B \geq 0. \tag{6}$$

In (6), Δ represents the social cost arising from moral hazard, and $\Delta p = p_H - p_L$.

The bank's decision making

Suppose now that we arrive at a situation in which the additional cost C becomes necessary to continue the project at date 2; i.e. the SPC falls into default. Next, consider the bank's decision regarding whether to accept the termination of the concession contract or not. If not, the bank would have to decide whether to continue the project or to liquidate it. Given (2), (3),

$$R \geq C. \tag{7}$$

Furthermore, given (7), the bank will always choose to continue the project since it can assure a partial payback of the debt – that is, if an equilibrium leading to the liquidation of the project does not exist. Assuming the bank has all the bargaining power, the new repayment D' will be determined as $D' = R$. The bank's payoff can be written as

$$\pi_i^{bank}(C) = R - C - I < 0. \tag{8}$$

On the other hand, if the government's decision of canceling the concession contract is accepted, the bank will earn a payoff

$$\pi_i^{bank}(C) = A - I. \tag{9}$$

Accordingly, the condition that makes the bank accept the termination of the initial concession contract is written as

$$A - I \geq R - C - I. \tag{10}$$

The LHS of (10) represents the bank's payoff arising from the bank's acceptance of the termination of the concession contract. The RHS of (10) represents the bank's payoff when the bank does not accept the termination.

The equilibriums of the basic game

(a) If $A < R - C$ (case A)

In this case, the bank rejects the cancellation of the concession contract. First, consider the case where default does not occur; i.e. the SPC can complete the project without an additional loan. Suppose the SPC chooses e_H. The initial loan I is used to cover construction costs V and operation costs B. However, if the SPC chooses e_L, operation costs B may be misused for private benefit. The payoffs earned by the SPC become

$$\pi_H^{spc}(0) = R - D, \qquad \pi_L^{spc}(0) = R + B - D. \tag{11}$$

Then consider the case where the SPC falls into default at date 1; i.e. an additional loan C becomes necessary. Given (7), the bank agrees to provide the additional loan to assure the continuation of the project. However, given (3), the initial debt contract should be modified. Under the assumption that the bank has all the bargaining power, the bank ends up earning the whole service fee R. Therefore, the SPC's payoffs are written as

$$\pi_H^{spc}(C) = 0, \qquad \pi_L^{spc}(C) = B. \tag{12}$$

On the other hand, the bank's payoffs become

$$\pi_H^{bank}(0) = D - I, \qquad \pi_H^{bank}(C) = R - I - C < 0. \tag{13}$$

And the bank's expected payoff can be written as

$$\prod_i^{bank}(D) = p_i D + (1 - p_i)(R - C) - I, \ (i = H, L). \tag{14}$$

Next, consider the problem of choosing the level of effort facing the SPC at date 0. The SPC's expected payoffs are written as

$$\prod_H^{spc}(D) = p_H(R - D), \tag{15a}$$

$$\prod_L^{spc}(D) = p_L(R - D) + B. \tag{15b}$$

And the incentive constraint imposed on the SPC to choose e_H is written as

$$\prod_H^{spc}(D) - \prod_L^{spc}(D) = \Delta p(R - D) - B \geq 0. \tag{16}$$

Before the SPC chooses its level of effort, the bank must determine the amount of repayment D at date 0. Assume the financial market is perfectly competitive. If the level of effort chosen by the SPC is known, the repayments for the bank to break even must be $D_i = \{(I - (1 - p_i)(R - C)\}p_i$, $(i = H, L)$. The condition ensuring that the SPC will choose e_H is written as

$$I + C > R \geq v_1. \tag{17}$$

Define $v_1 = I + (1 - p_H)\ C + p_H B/\Delta p$ (See **Appendix I**).The implementation condition with respect to the level of effort e_L is written as

$$v_1 > R \geq v_2. \tag{18}$$

Define $v_2 = I + (1 - p_L)\ C$ (See **Appendix I**). Therefore, the equilibriums obtained in this case are:

$$\begin{cases} \text{Equilibrium } A & D^* = D_H,\ e^* = e_H, \\ \text{Equilibrium } B & D^* = D_L,\ e^* = e_L. \end{cases}$$

b) If $A \geq R - C$ (case B)

In this case, the initial concession contract is cancelled by the government when the project falls into default. The government will then pay compensation A to the bank in return for the project's assets. After receiving the assets, the government may enter into a new concession contract with the initial SPC to continue the project. Under this setting the SPC's payoffs can also be defined as in (12)-(13). The bank's payoffs can therefore be written as

$$\pi_i^{bank}(0) = D - I\ (i = H, L), \qquad \pi_i^{bank}(C) = A - I\ (i = H, L) \tag{19}$$

And the bank's expected payoff is written as

$$\pi_i^{bank}(0) = D = p_i D + (1 - p_i)\ A - I\ (i = H, L) \tag{20}$$

Finally, the government's payments become

$$\pi_i^{pub}(0) = R\ (i = H, L), \qquad \pi_i^{pub}(C) = A + C\ (i = H, L) \tag{21}$$

Now consider the problem that the SPC chooses the level of effort at date 0. Given the amount of dept D, the SPC's expected payoff are written as (15a) and (15b). The incentive constraint imposed on the SPC to choose e_H is alsi written as (16). Given the level of effort chosen by the SPC, the repayments for the bank to break even must be $D_i' = \{I - (1 - p_i)A\}/p_i$, $(i = H, L)$. The implementation condition with respect to e_H is written as

$$I + C > R \geq v_3. \tag{22}$$

Define $v_3 = \{I - (1 - p_H)A\}/p_H + B/\Delta p$ (See **Appendix I**). On the other hand, the SPC would choose e_L under the following condition

$$v_3 > R \geq v_4 \tag{23}$$

Define $v_4 = \{I - (1 - p_H)A\}/p_L$ (See **Appendix I**). The equilibriums obtained in this case are

$$\begin{cases} \text{Equilibrium } C & D^* = D', e^* = e_H \\ \text{Equilibrium } D & D^* = D_L', e^* = e_L \end{cases}$$

Tender System and Equilibriums

Tender system

Consider a tender process for a concession contract. The bidder offering the minimum price will ultimately be selected to constitute the SPC in the PPP project. Suppose now that perfect competition is realized during the tender process. There is extensive literature [5], [6] investigating the tender mechanism. We first analyze the case in which the bid price is determined under the assumption of perfect competition and under the condition that no constraints are imposed on the bid prices. The equilibriums in this case refer to regular tender equilibriums. In the regular tender, bidders deciding to make an effort might not be selected as a result of excessive competition. To avoid excessive competition, we introduce an instrument called the reserve tender, in which a constraint is imposed on the minimum price. We refer to the equilibriums obtained under the reserve tender as reserve tender equilibriums.

Regular tender equilibriums

Suppose the government discloses the compensation price A at the beginning of the project. A regular tender is implemented to determine the service fee R, and subsequently, the government signs the concession contract with the bidder offering the minimum price. The regular tender equilibriums are obtained in the following 3 cases.

Referring to the equilibrium prices obtained in case *A*

In this case, $A < \min\{v_1 - C, v_2 - C\}$. Denote the minimum service fee (hereafter marginal service fee) in equilibrium A by $R^A_{\min}$ written as $R^B_{\min} = v_1$. The marginal service fee $R^B_{\min}$ corresponding to equilibrium B is written as $R^B_{\min} = v_2$. When the service fee is determined by regular tender, the bidder offering the minimum price obtains the concession rights. Thus, the regular tender equilibrium price denoted by v^* for this case is written as

$$\begin{cases} \text{Equilibrium } A' & D^* = D^*_L,\ e^* = e_L,\ v^* = v_2, \\ \text{Equilibrium } B' & D^* = D^*_H,\ e^* = e_H,\ v^* = v_1, \end{cases}$$

$$D^*_H = I + (1 - p_H)C - (1 - p_H)B/\Delta p,\ D^*_L = I + (1 - p_L)C.$$

Referring to the equilibrium prices obtained in case *B*

In this case, $A \geq \max\{v_1 - C, v_2 - C\}$. The marginal service fee $R^C_{\min}$ and $R^D_{\min}$ corresponding to equilibriums C and D are written as $R^C_{\min} = v_3$, $R^D_{\min} = v_4$ respectively. The regular tender equilibriums for this case are written as

$$\begin{cases} \text{Equilibrium } C' & D^* = D'_L,\ e^* = e_L,\ v^* = v_4, \\ \text{Equilibrium } D' & D^* = D'_H,\ e^* = e_H,\ v^* = v_3, \end{cases}$$

Referring to the equilibrium prices obtained in case *A* or in case *B*

Consider now that the compensation price A satisfies $\max\{v_1 - C, v_2 - C\} > A \geq \min\{v_1 - C, v_2 - C\}$. First, assume $v_1 - C > A \geq v_2 - C$. The marginal service fees are v_1 with respect to the level of effort e_H and v_4 with respect to the level of effort e_L. The regular tender price v^* is equal to v_4, since $v_1 > v_2 \geq v_4$ (See Appendix II). Then, assume $v_2 - C > A \geq v_1 - C$. The marginal service fee is equal to v_3 with respect to the level of effort e_H, or v_2 with respect to the level of effort e_L. The regular tender price v^* is thus equal to v_3, since $v_2 > v_1 \geq v_3$ (See **Appendix II**). The regular tender equilibriums in this case are therefore written as

$$\begin{cases} \text{Equilibrium } E' & D^* = D'_L,\ e^* = e_L,\ v^* = v_4, \\ \text{Equilibrium } F' & D^* = D'_H,\ e^* = e_H,\ v^* = v_3, \end{cases}$$

The regular tender equilibriums analyzed above are summarized in Table 17.1 (See **Appendix II**).

Table 17.1: Regular tender equilibriums in the basic game

Regular Tender Equilibriums	Equilibrium Conditions	D^*	e^*	v^*	Π^{spc}	Π^{pub}
A'	$v_1 > v_2, A < v_2 - C$	D^*_L	e_L	v_2	B	v_2
B'	$v_1 \leq v_2, A < v_1 - C$	D^*_H	e_H	v_1	$p_H B/\Delta p$	v_1
C'	$v_1 > v_2, A \geq v_1 - C$, or $v_1 \leq v_2, A > I - \frac{p_H p_L B}{(\Delta p)^2} \geq v_2 - C$	D'_L	e_L	v_4	B	v_2
D'	$v_1 \leq v_2, I - \frac{p_H p_L B}{(\Delta p)^2} \geq A \geq v_2 - C$	D'_H	e_H	v_3	$p_H B/\Delta p$	v_1
E'	$v_1 - C > A \geq v_2 - C$	D'_L	e_L	v_4	B	v_2
F'	$v_2 - C > A \geq v_1 - C$	D'_H	e_H	v_3	$p_H B/\Delta p$	v_1

Note) Π^{spc}, Π^{pub} represents the expected payoff earned by the SPC and the bank respectively.

Reserve tender equilibriums

In equilibriums A', C', E', moral hazard occurs (the SPC chooses e_L). To resolve this problem, assume a reserve tender is implemented; i.e. a lower limit is imposed on the bid prices. Under the reserve tender, the bidder offering the minimum price still higher than the reserve price is selected as the SPC.

Suppose now the reserve price is determined as $\tilde{R} = v_1$ with respect to regular tender equilibriums A^t, C^t, and $\tilde{R} = v_3$ with respect to equilibrium E^t. The reserve tender equilibriums are written as

$$\begin{cases} \text{Reserve tender equilibrium } A^r, E^r & \tilde{D} = D^*_H, \tilde{e} = e_H, \tilde{R} = v_1, \\ \text{Reserve tender equilibrium } C^r & \tilde{D} = D^*_H, \tilde{e} = e_H, \tilde{R} = v_3, \end{cases}$$

In reserve tender equilibriums A^r, C^r and E^r, moral hazard does not occur, and the socially optimal contract is realized. The SPC's expected payoffs then are written as

$$\tilde{\prod}^{spc}_{A^r} = \tilde{\prod}^{spc}_{C^r} = \tilde{\prod}^{spc}_{E^r} = R = \{I + (1 - p_H)C\} = p_H B/\Delta p. \tag{24}$$

That is, the government has to pay $p_H B/\Delta p$ in addition to the expected cost of the project. The SPC's expected payments are written as

$$\tilde{\prod}^{gov}_{A^r} = \tilde{\prod}^{spc}_{C^r} = \tilde{\prod}^{spc}_{E^r} = v_1 = I + (1 - p_H)C + p_H B/\Delta p. \tag{25}$$

On the other hand, in regular tender equilibriums B^t, D^t, and F^t, the SPC's moral hazard does not occur. The regular tender equilibrium prices, therefore, can be considered as the reserve equilibrium prices. In all reserve tender equilibriums, the SPC's expected payoffs are $p_H B/\Delta p$, and social efficiencies are accomplished.

The bank's moral hazard problem

With the reserve tender equilibriums, the government pays the quasi-rent to the SPC to deter moral hazard. However, if the bank has all the bargaining power in its relationship with the SPC, the quasi-rent may be expropriated by the bank. The situation whereby the bank's strategic behavior is not verifiable to the government is referred to as the bank's moral hazard.

Assume now that the bank has all the bargaining power. In the reserve tender equilibriums, the service fee $\tilde{R}$ is determined as

$$\begin{cases} \tilde{R} = v_3, A \geq v_1 - C \\ \tilde{R} = v_1, v_1 - C > A \end{cases} \tag{26}$$

First, consider the case where $A < v_1 - C$. In this case, the equilibrium bid price is determined as v_1, and the repayment as D^*_H. The SPC earns the quasi-rent $p_H B/\Delta p$ by choosing e_H since the incentive constraint (16) is satisfied. However, the quasi-rent involved in the equilibrium bid price is known to both the SPC and the bank. Since the bank has all the bargaining power, the quasi-rent might be expropriated by the bank. Now suppose that the bank demands the repayment $D^\circ = v_1 = D^*_H + p_H B/\Delta p$ instead of D^*_H. Consequently, the incentive constraint (16) does not hold; i.e $\Delta p(v_1 - D^\circ) - B < 0$ The SPC then chooses e_L, and its expected payoff is written as $\tilde{\prod}^{spc}_L(D^\circ) = p_L(v_1 - D^\circ) + B = B$. Conditions in which the bank's moral hazard occurs are

$$v_1 > v_2, A < v_2 - \mathrm{C} \text{ or } v_1 \geq v_4, v_1 - C > A \geq v_2 - C. \tag{27}$$

If (27) does not hold, repayment v_1 is smaller than marginal repayment v_2 or v_4, which are determined in the case where the SPC chooses e_L. This means that the bank's expected payoff should be negative if the repayment was determined as v_1; i.e. the bank's moral hazard does not occur.

Then, consider the case where $I \geq A > v_1 - C$. The equilibrium bid price and the repayment in the reserve tender are determined as v_3 and D^t_H respectively. Now, suppose that the bank demands the repayment $D^{\circ\circ} = v_3 = D_H^t + B/\Delta p$ instead of D_H^t. Given the bid price v_3, the incentive constraint (16) does not hold. The SPC thus chooses e_L, and earns the expected payoff $\tilde{\Pi}_L^{spc}(D^{\circ\circ}) = p_L(v_3 - D^{\circ\circ}) + B = B$. The condition in which the bank's moral hazard occurs is

$$v_3 > v_4, A > \max\{v_1 - C, v_2 - C\}. \tag{28}$$

The results analyzed above are summarized in Table 17.2 (See **Appendix III**).

Table 17.2: The possibility of the bank's moral hazard

Compensation Price		Bid Prices	The Order Relation of the Marginal service Fee	Repay-ment	SPC's moral hazard	Bank's moral hazard
$A < v_1 - C$	$A < \min\{v_1 - C, v_2 - C\}$	v_1	$v_1 > v_2$	v_1	×	×
			$v_1 \leq v_2$	D_H^*	O	O
	$v_1 - C > A \geq v_2 - C$		$v_1 > v_4$	v_1	×	×
$A \geq v_1 - C$	$A \geq \max\{v_1 - C, v_2 - C\}$	v_3	$v_3 > v_4$	v_3	×	×
			$v_3 \leq v_4$	D_H^t	O	O
	$v_2 - C > A \geq v_1 - C$		$v_3 < v_2$	D_H^t	O	O

Note: Mark O means moral hazard does not occur; mark × means moral hazard occurs.

Monitoring

Assumptions

Let us now formalize the idea that monitoring reduces the extent of the SPC's moral hazard. In PPP projects, both the government and the bank play a disciplinary role with regard to the SPC's behavior. However, if default risks are all undertaken by the government, the bank will have no incentive to monitor the SPC's behavior. We investigate the conditions for monitoring, and the impacts of monitoring on the equilibriums. The bank can choose whether to implement monitoring or to shirk. Denote the bank's behaviors by $b = b_j$ $(j = M, N)$. b_M signifies that the bank implements monitoring, and b_N signifies that the bank shirks. Assume that the bank can reduce private benefit from B to 0 by choosing b_M. Monitoring, however, bears a cost c. Assume that $c \leq B$.

Equilibriums

Assume that information related to monitoring is commonly held between the bank and the SPC. Consider the condition that the bank chooses b_M. The payoffs the SPC obtains when choosing e_H are written as

$$\tilde{\pi}_H^{spc}(0) = p_H(R - D), \tilde{\pi}_H^{spc}(C) = 0 \tag{29}$$

On the other hand, the bank's payoffs are written as

$$\tilde{\pi}_H^{bank}(0) = D - I - c, \tag{30a}$$

$$\tilde{\pi}_H^{bank}(C) = \begin{cases} R - I - C - c, (A < R - C) \\ A - I - c, (A \geq R - C) \end{cases} \quad (30b)$$

The bank's expected payoffs thus are

$$\tilde{\pi}_H^{bank}(D) = \begin{cases} p_H D + (1 - p_H)(R - C) - I - c, (A < R - C) \\ p_H D + (1 - p_H)A - I - c, (A \geq R - C) \end{cases} \quad (31)$$

Given (30), the marginal repayments are obtained as $D_m = \{I + c - (1 - p_H)(R - C)\}/p_H$ corresponding to $A < R - C$, and $D_{m'} = \{I + c - (1 - p_H)A\}/p_H$ corresponding to $A \geq R - C$. Given D_m and $D_{m'}$, the expected payoffs obtained by the SPC when choosing e_H are written as

$$\tilde{\pi}_H^{spc}(D_m) = p_H(R - D_m),\ \tilde{\pi}_H^{spc}(D_{m'}) = p_H(R - D_{m'}) \quad (32)$$

When the bank implements monitoring, the implementation condition is written as

$$R \geq v_{m'}, (A < v_m - C), R \geq v'_{m'}, (A \geq v' - C). \quad (33)$$

Define $v_m = I + (1 - p_H)C + c$, $v' = \{I + c - (1 - p_H)A\}p_H$. So, the equilibriums obtained in this case are

$$\begin{cases} \text{Equilibrium } M & \breve{D}^* = D_m, \breve{e}^* = e_H, \breve{b}^* = b_M \\ \text{Equilibrium } M' & \breve{D}^* = D_{m'}, \breve{e}^* = e_H, \breve{b}^* = b_M \end{cases}$$

Now given the equilibriums M and M', we get the tender equilibriums regarding the bank's monitoring behavior. First, consider the case that $A < v_m - C$. The marginal service fee corresponding to equilibrium M is written as $R^M_{\min} = v_m$ Since $v_m < \min\{v_1, v_2\}$,the incentive constraint ensuring that the SPC chooses e_H is not satisfied. Furthermore, the implementation condition with respect to e_L does not hold. As a result, the project will not be implemented if the bank chooses b_N. Assume that the financial market is completely competitive. The bank chooses b_M under the marginal service fee v_m. The tender equilibrium thus is obtained as

$$\{\text{Equilibrium } M'' \quad \breve{D}^* = D_m, \breve{e}^* = e_H, \breve{b}^* = b_M, \breve{v}^* = v_m.$$

In tender equilibrium M'', the bank implements monitoring, and the SPC chooses e_H. Then, consider the case that $A \geq v_m - C$. The marginal service fee corresponding to equilibrium M' is written as $R^{M'}_{\min} = v_{m'}$. $v_{m'} < \min\{v_1, v_2, v_3\}$is obtained. First assume $v_2 - C > A \geq v_m - C$. If monitoring is not implemented, both the incentive constraint (16) and (22) are not satisfied. Furthermore, since $v_{m'} < v_2$ the implementation condition corresponding to e_L does not hold. Thus, the bank chooses b_M. Then, assume $A \geq v_2 - C$. The incentive constraints (16), (22) will not be satisfied if the bank chooses b_N. The condition for the bank's monitoring thus is written as

$$v_{m'} < v_4. \quad (34)$$

The necessary and sufficient condition for (34) is written as

$$I - p_L c/\Delta p > A \geq v_m - C \quad (35)$$

However, if

$$I \geq A \geq I - p_L c/\Delta p \quad (36)$$

$v_{m'} \geq v_4$ holds. The bank then does not monitor the SPC's behaviors under service fee v_4. Thus, we obtain the following equilibriums.

$$\begin{cases} \text{Equilibrium } M^{m'} \quad \breve{D}^* = D_{m''}, \breve{e}^* = e_{H}, \breve{b}^* = b_{M}, \\ \breve{v}^* = v_{m'} \text{ (if (35) holds)}, \\ \text{Equilibrium } M^{m''} \quad \breve{D}^* = D'_{L}, \breve{e}^* = e_{H}, \breve{b}^* = b_{N}, \\ \breve{v}^* = v_{4}, \text{ (if (36) holds)}, \end{cases}$$

We subsequently obtain the following proposition.

Proposition If the compensation price A satisfies $A < I - p_L c/\Delta p$ the bank implements monitoring which can deter the SPC's moral hazard.

Conclusion

In this chapter, we have pointed out the possibility of the SPC's moral hazard and the bank's moral hazard in PPP projects. We have left certain issues for future research. First, in this chapter, we assume that the bank can perfectly observe the SPC's behaviors. However, the moral hazard, especially that caused by the SPC's technology, cannot be deterred by the bank's monitoring. Second, since monitoring is costly, attention should be paid in the future to conducting empirical studies about the costs of monitoring.

Appendix

Appendix I

Proof of equilibriums in the basic game. In case A, the conditions that the SPC chooses the level of effort e_H with regard to the marginal repayment D_H is written as

$$\Delta p(R - D_H) - B \geq 0. \tag{37a}$$

$$R \geq D_H. \tag{37b}$$

(37a) is the incentive constraint assuring that the SPC chooses e_H, (37b) represents the implementation condition. Given (37a) and (37b), the condition that the SPC chooses e_H is written as $I + C > R \geq v_1$. Define $v_1 = I + (1 - p_H)C + B/\Delta p$. On the other hand, the condition that the SPC chooses e_L is written as $v_1 > R \geq D_L$, which can be simplified as $v_1 > R \geq v_2$. Define $v_2 = I + (1 - p_L)C$. In a similar way, we get the equilibriums in case B (omitting proof).

Appendix II

Proof of the regular tender equilibriums. Given (7), $v_1 - v_2 = p_L B/\Delta p - \Delta$. The necessary and sufficient condition is $\Delta < p_L B/\Delta p$, $A < v_2 - C$ with respect to equilibrium A', and $\Delta \geq p_L B/\Delta p$, $A < v_1 - C$ with respect to equilibrium B'. On the other hand,

$$v_3 - v_4 = B/\Delta p - \Delta p(I - A)/p_H p_L \tag{38}$$

Since $v_3 - v_4$ is an increasing function with respect to $A \in [\max\{v_1, v_2\} - C, I]$, the maximum of $v_3 - v_4$ is obtained when $A = I$, and the minimum when $A = \max\{v_1, v_2\} - C$. We thus obtain

$$v_3 - v_4 = \begin{cases} B/\Delta p & A = I \\ (p_L B/\Delta p - \Delta)/P_L, & A = v_1 - C \\ (p_L B/\Delta p - \Delta)/P_H, & A = v_2 - C \end{cases}$$

$v_3 > v_4$ holds when $A > v_1 - C > v_2 - C$. On the other hand, assume $A > v_2 - C \geq v_1 - C$. Given (38), $v_3 > v_4$ holds when $I \geq A > I - (p_H p_L B)/(\Delta p)^2$, and $v_3 \leq v_4$ holds when $v_2 - C < A \leq I - (p_H p_L B)/(\Delta p)^2$. (The necessary and sufficient conditions with regard to equilibrium C' and D' are written as

$$\text{equilibrium } C' \quad \begin{cases} \Delta < p_L B/\Delta p,\ A \geq v_1 - C \\ \Delta \geq p_L B/\Delta p,\ A > I - (p_H p_L B)/(\Delta p)^2 \end{cases}$$
$$\text{equilibrium } D' \quad \Delta \geq p_L B/\Delta p,\ I - (p_H p_L B)/(\Delta p)^2 \geq A > v_2 - C.$$

Appendix III

The proof of the order relation of the marginal service fee. (Case $A < v_1 - C$) Suppose $A < \min\{v_1 - C, v_2 - C\}$. The marginal repayment is D_H^* with respect to the level of effort e_H, and D_L^* with respect to e_L. The necessary and sufficient condition for $v_1 > v_2 = D_L^*$ is obtained in appendix II. Given $D^\circ = v_1$, the bank's expected payoff is written as $p_L D^\circ + (1 - p_L)(v_1 - C) - I$, which is larger than $p_L D_L^* + (1 - p_L)(v_2 - C) - I$, $= 0$ when the marginal repayment is D_L^* and the marginal service fee is v_2. Then, suppose $v_1 - C > A \geq v_2 - C$. The marginal debt repayment is v_1 with respect to e_H, and $D_L' = v_4$ with respect to e_L. We can obtain that $v_1 > v_2 \geq v_4$. Under repayment $D^\circ = v_1$, the bank's expected payoff $p_L D^\circ + (1 - p_L)(v_1 - C) - I$ is larger than $p_L D_L' + (1 - p_L)A - I = 0$ (Case $A \geq v_1 - C$) Assume $A \geq \max\{v_1 - C, v_2 - C\}$. The marginal debt repayment is D_H' with respect to e_H, D_L' with respect to e_L. The necessary and sufficient condition assuring $v_3 > v_4 = D_L'$ is written in appendix II. Given $D^{\circ\circ} = v_3$ the bank's expected payoff is written as $p_L D^{\circ\circ} + (1 - p_L)A - I$, which is larger than $p_L D_L' + (1 - p_L)A - I = 0$. Next, assume $v_2 - C > A \geq v_1 - C$. The marginal debt repayment is D_H' with respect to e_H, $D_L^* = v_2$ with respect to e_L. $v_2 > v_1 \geq v_3$ is finally obtained.

Acknowledgements

This work is supported by Science Foundation of Ministry of Education of China (08JC790012); Scientific Research Foundation for Returned Scholars, Ministry of Education of China; the Fundamental Research Funds for the Central Universities (DUT11RW415).

References

[1] Hart, O., Shleifer, A. and Vishny, W. (1997) The proper scope of government: theory and an implication to prisons. *Quarterly Journal of Economics* **112**: 1127-1161.

[2] Bennett, J. and Iossa, E. *(2005) Public-Private Parnership: An Incomplete Contract Approach*. Research Report, Department of Economics and Finance, Brunel University.

[3] Schmidt, K.M.(1996) The costs and benefits of privatization: an incomplete contracts approach. *Journal of Law, Economics and Organization* **12**: 1-24.

[4] Laffont, J. and Martimort, D. (2002) *The Theory of Incentives*. Princeton University Press.

[5] Klemperer, P. (2004) *Auctions Theory and Practice*. Princeton University Press.

[6] Milgrom, P. (2004) *Putting Auction Theory to Work*. Cambridge University Press.

Joint Ventures in Construction 2
ISBN 978-0-7277-5783-8

doi: 10.1680/jvc2.57838.237

Chapter 18

International Labor Movement and Welfare Effects: Malaysia as a Case Study

Sharina Farihah Hasan, International Islamic University Malaysia
Kakuya Matsushima, Kyoto University
Kiyoshi Kobayashi, Kyoto University

Introduction

It is important to keep in mind three basic principles of international labor movement [1]. First, international labor movement is a diverse phenomenon, and its economic impact in one location or another depends heavily on the particular circumstances involved. Second, basic data on international labor movement and remittances are lacking, so predicting the impact of policy changes can be problematic. This underlines the need for better data and more research. Third, international labor movement has social and political implications that may be just as important as the economic implications that are provided in this chapter. Despite their importance, internal labor movement, skilled labor movement, and the political and social impacts of international labor movement are beyond the scope of this chapter.

This chapter focuses on the international labor movement scenario in Malaysia. It aims to report on a study of the inward movement of unskilled foreign labor into Malaysia due to the country's high reliance on this type of labor. There have been many studies conducted on the impact of unskilled labors on the social and political aspects of the country. This chapter will offer another view by looking at the topic from an economic perspective.

With Malaysia as a case study, this chapter investigates the quantitative effects of the inflow of unskilled foreign labors on a host country, particularly on the welfare of local unskilled labors. A computable general equilibrium (CGE) modeling framework based on a GTAP 7 data base (GTAP, or Global Trade Analysis Project, is an alternative data source established in 1992, with the objective of lowering the cost of entry for those seeking to conduct quantitative analyses of international economic issues in an economy wide framework) [i] and a GMig2 model (Global Migration model version 2 was developed based on significant changes made to the standard GTAP model and database to incorporate the movement of labors) [ii] are used to measure the effects of the inflow of unskilled foreign labor on the welfare of local labors in the country.

International Labor Movement in Malaysia

Malaysia is an exporter as well as an importer of labor. The country began relying on foreign labor due to the structural changes and labor market segmentation that first emerged in the early 1970s, when industry faced severe labor and skill shortages [2]. Local labors began emigrating to the rapidly growing Newly Industrializing Economies, where wages were much higher.

According to NEAC [2], the first upsurge of foreign labor inflow into Peninsular Malaysia took place in the early 1970s and coincided with structural changes arising from the implementation of the New Economic Policy aimed at reducing poverty and equity imbalances. The inflow involved Indonesian labors engaged mainly in the plantation and construction sectors. Based upon the rapid industrialization and growth of the late 1980s, the employment of foreign labor became a complementing factor supporting production activity amidst a tight labor market. The entry of foreign workers was generally allowed as an interim measure to meet the excess demand for unskilled labor in selected sectors. Instead of diminishing as originally planned, however, reliance on this cheap labor has apparently risen.

In Malaysia, the inflow of international labor are divided into two categories; (1) foreign labor, and (2) expatriates [2].

According to the Malaysian Economic Report 2008/2009 [3], at the end of March 2008, the number of registered international labor inflow totaled 2.1 million (including expatriates) in Malaysia, accounting for 18% of total employment. Foreign labors are mainly engaged in the manufacturing sector (36%), agriculture (16.1%), and construction (14.9%). In 2007, Indonesians comprised 75.6% of total foreign labor, followed by Nepalese, 14.5%, and Indians, 9.1% [4].

Table 18.1 shows statistics of foreign labors in Malaysia, according to proportion of skill. The high figure for unskilled foreign labors illustrates the country's high reliance on this type of labor.

Table 18.1: Statistics of foreign labors in Malaysia (Proportion of skilled and unskilled labor)

Year	TOTAL		Foreign[1]		Local	
	skilled 25%[2]	unskilled 75%[2]	skilled	unskilled	Skilled[3]	Unskilled[3]
2004	2,636,400	7,909,200	34,519	1,324,981	2,601,881	6,584,219
2005	2,723,225	8,169,675	33,448	1,586,552	2,689,777	6,583,123
2006	2,789,775	8,369,325	35,052	1,804,948	2,754,723	6,564,377
2007	2,852,400	8,557,200	38,090	1,871,910	2,814,310	6,685,290
2008	2,901,475	8,704,425	33,062	2,066,938	2,868,413	6,637,487
2009[4]	2,950,800	8,852,400	n.a	2,124,576[5]	n.a	6,727,824
2010[4]	3,000,964	9,002,891	n.a	2,160,694[5]	n.a	6,842,197

Sources:

1 Data are gained from Economic Report (various issues [3]-[6]), Ministry of Finance, Malaysia

2 The 25% and 75% proportions of skilled and unskilled labor, respectively (total), are derived from the 4-year trend in employment. The 4-year trend (as reported in the Labor report survey 2006 and 2007, Department of Statistics, Malaysia) indicated that skilled and unskilled labors in the labor market have been employed in such proportions.

3 The 30% and 70% proportions of skilled and unskilled local labor are based upon the 5-year trend in employment (balance of skilled and unskilled labor after deduction of number of foreign labors from total employment). The 5-year trend (as reported in the Labor report survey [7], [8], Department of Statistics, Malaysia) indicated that skilled and unskilled local labors in the labor market have been employed in such proportions.

4 Estimates – based on 1.7% labor force growth

5 Estimates – based on data in 2008 (24% of total unskilled labor is foreign labor)

n.a. Not available

Foreign labor has become a vital component of the construction workforce in Malaysia. An estimated 60% of all manual labors in the construction sector were foreign nationals, a figure revised upwards to 70% in 1991, and then to 80% in 1995 [9]. Legal and illegal foreign labors dominated the sector's employment, accounting for nearly 70% of its workforce, according to Narayanan & Lai [10]. Table 18.2 shows the number (in thousands) of legal unskilled foreign labors in the Malaysian construction sector. The figures were obtained from Economic Report (various issues), Ministry of Finance, Malaysia. In contrast to this estimate, the number of unrecorded illegal foreign labors is much higher. The International Labor Organization in its Tripartite Meeting on Social and Labor Issues Concerning Migrant Workers in the Construction Industry, Geneva [11] reported that at least two-thirds of all foreign labors in Malaysia are illegal.

Table 18.2: Number (in thousands) of legal unskilled foreign labors in the Malaysian construction sector

Year	Total Employment in the Construction Sector	Unskilled Foreign Labor	Percentage
2004	798,200	269,100	34%
2005	759,600	245,916	32%
2006	755,200	272,547	36%
2007	757,300	280,787	37%
2008	760,000	307,974	41%
Source: Economic Reports (various issues [3] - [6])			

In Malaysia, local youth would rather be unemployed than work in the construction industry. This is attributed to archaic employment practices, outdoor work, and the prevalence of temporary and casual labor. The industry has been forced to rely on foreign labor, mostly from neighboring Indonesia, for the past two decades. Official estimates put foreign labor at 80% of the entire construction workforce in 1992 [9]. There is consensus among employers in the industry that it will continue to depend on imported labor (regularized or otherwise) in the foreseeable future. As work in construction has come to be associated with foreign labors (many of them illegal), the industry's status has deteriorated accordingly. A similar situation is found in some developed countries where construction work has for many years been undertaken largely by foreigners, or by their children [11]. Also, recently the local population in many more countries has begun to shun construction work.

The substantial rise in the population of foreign labors, both legal and illegal, therefore calls for the development of better strategy and policy. In efforts to sustain rapid economic growth in a tight labor situation, effective labor management poses an immense challenge to the authorities. Various problems such as illegal entry, overstaying, and abscondence require persistent attention by the government.

There were plans to reduce the number of foreign labors to about 1.5 million by 2010, in particular unskilled labors. To reduce reliance on unskilled foreign labor, employers are encouraged to move up the value chain by adopting new technology and introducing mechanization and automation in the production process. To achieve the target, the government has implemented various measures including introducing and encouraging the use of the Industrialized Building System (IBS), especially in government projects, and increasing automation and mechanization in the manufacturing and agriculture sectors. Other measures under consideration relate to improving the skills of local workers. The government had allocated RM100 million under the Ninth Malaysia Plan's Mid-Term Review to train 15,000 local labors in various fields [12].

Government's standpoint toward regulating foreign labor

The purpose of foreign labor management policy is to prevent excessive reliance on the 'cheap labor' offered by foreign labors and to serve as a measure for adjusting the labor market in response to economic trends [13]. On the one hand, the government is trying to ensure that there are not too many foreign labors in the country, but on the other hand the government wants to help industries avoid labor shortages detrimental to the nation's economy.

Policies include the control and regulation of the inflow of foreign labors, the elimination of irregular labor movement, and the protection of labors' rights. Employers are obligated by the laws of Malaysia to give equal treatment to foreign labors, with the rights and benefits accrued to local labors accruing equally to foreign labors. Such laws, however, are not always applied equally in practice; evidently, not all employers comply with the requirement that foreign labors receive benefits equal to those of local labors [10].

A study done by Abdul Rashid [9] suggested that there exists a wage hierarchy, especially in the construction sector, with Malaysians occupying the apex. Nonetheless, it is not high wages but abundant job opportunities that stimulate migration to Malaysia. Furthermore, in order to control the inflow of foreign labors, application approval is based on the merits of each case and is subject to conditions that will be determined periodically. Employers' applications to hire foreign labors will be considered only when efforts to find qualified local citizens and permanent residents have failed [14].

Although their contributions to the country's economic growth are undeniable, the unprecedented influx of unskilled foreign labors, following the unabated high growth beginning in the mid-1980s, has raised several social, political, and economic concerns. The popular perception has been that foreign labors might be negatively affecting local unskilled labors, that is, by depressing their welfare in terms of both wages and living standards. This study was undertaken to address the question of how foreign unskilled labor affects local unskilled labor. Answers will be validated through the results of the quantitative analysis employed in the next section.

Application of GMig2 Model and Data Base for Malaysia

The model and data necessarily make a number of simplifying assumptions, which should be noted at the outset. Although every effort was made to collect data on the flow of unskilled foreign labor, data are scarce and of questionable quality, and assumptions had to be made to fill in the gaps.

Assumptions

The model assumes a two-country model: Malaysia as a labor importing country and Rest of World (ROW) as a labor exporting country with only one-way labor movement between the countries in the model. The focus of the study is the impact of legal unskilled foreign labors on the host country. Malaysia imports only unskilled foreign labors, and therefore unskilled foreign labors are internationally mobile from ROW. The model does not consider the movement of skilled labor or the effects of brain drain. Thus, in this case, skilled labor is immobile.

Neither does the model consider illegal foreign labors. It is undeniably true that illegal immigration has an impact on labor importing countries; however, due to the scarcity of data, illegal labors are not included in the study. 5-sector aggregation (Agriculture, Manufacturing, Construction, Services and Transportation) is generated for both countries using GTAP7Agg. The GMig2 model is used, and most of the same assumptions are applied here.

There are 5 factors of production (skilled labor, unskilled labor, capital, land, and natural resources) and 5 sectors (agriculture, manufacturing, construction, services and transportation). Labors are divided into 3 categories: unskilled local labor, skilled local labor, and unskilled foreign labor. The wage rate between the two countries differs and unskilled foreign labors are assumed to be of different levels of productivity, with unskilled local labor possessing a higher level of productivity. This is used to measure the magnitude of unskilled labor inflow into the host country. The nominal number of unskilled foreign labors is adjusted into a real number by multiplying with the productivity level.

Further assumptions are made about the labor market in the labor importing country. Different labor markets are based on types of labor: skilled and unskilled. Foreign labor is employed only in the unskilled labor market. Capital is internationally immobile. It is, however, mobile within individual countries. Meanwhile, other factors such as land and natural resources are immobile.

The model considers a situation whereby remittance is allowed into the home country. In the GMig2 Model and Data Base, remittances flowing out of the host country back to the home country are assumed to be a constant portion of income. Consequently, as the number of foreign labors or their wages increase, so do remittances. Remittances then flow back to the permanent residents of the home country, therefore reducing the income of the foreign labors in the host country but increasing the income of permanent residents back home. This also affects a country's balance of payments. Foreign labors spend the rest of their earnings on consumption in the host country. Other factors related to production, land, and natural resources are immobile, with income assumed to accrue to permanent residents of the respective countries. Malaysia is assumed to be a small open economy under the condition of full employment, with a market environment of perfect competition.

The Equations

The equations listed in this section describe the model in the study. Refer to **Appendix** for the relationships of the variables presented here.

Equation 1

Utility maximization of regional household

$$\max_{QP_i,\, OG_i,\, QCGDS_i} UU = QP_i^{\alpha}\, OG_i^{\beta},\, QCGDS_i^{\gamma}$$

Subject to: $Y = (PP_i * QP_i) + (PG_i * QG_i) + (PCGDS_i * QCGDS_i)$

$$QP_i = Y\frac{\alpha}{\alpha+\beta+\gamma} \qquad QG_i = Y\frac{\beta}{\alpha+\beta+\gamma} \qquad QCGDS_i = Y\frac{\gamma}{\alpha+\beta+\gamma}$$

Notation:

QP_i Quantity of composite tradable commodity i demanded by Private Household
QG_i Quantity of composite tradable commodity i demanded by Government Household
$QCGDS_i$ Quantity of capital goods sector supplied
α, β, γ Share parameter, $\alpha+\beta+\gamma = 1$
Y Total income of the regional household
PP_i Demand price of composite tradable commodity i for Private Household
PG_i Demand price of composite tradable commodity i for Government Household
$PCGDS_i$ Price of investment goods

Equation 2
Composite goods for domestic consumption by Private and Government Household

$$\left[(PIM_i * QIM_i)^{\alpha} + (PM_i * QDC_i)^{\alpha}\right]^{\frac{1}{\alpha}} = (PP_i * QP_i) + (PG_i * QG_i)$$

Notation:

PIM_i	Market price of aggregate imports of tradable commodity *i* demanded by Private and Government Household
QIM_i	Quantity of demanded aggregate imports of commodity *i* demanded by Private and Government Household
QDC_i	Quantity of composite tradable commodity *i* demanded by Private and Government Household
PM_i	Market price of composite tradable commodity *i* demanded by Private and Government Household

Equation 3
Distribution of domestic production for domestic consumption and exports

$$\underset{QO,\,QXS,\,QD}{Max} \quad \pi = (PFOB_i * QXS_i + PM_i * QD_i) - (\tau + PS_i)QO_i$$

Subject to: $QD_i = \theta(QXS_i^{\alpha} + QD_i^{\alpha})^{\frac{1}{\alpha}}$

$$QXS_i = \left(\frac{\theta^{\sigma}(\tau_i + PS_i)}{PFOB_i}\right)^{\frac{1}{1-\sigma}} QO_i \qquad QD_i = \left(\frac{\theta^{\sigma}(\tau_i + PS_i)}{PM_i}\right)^{\frac{1}{1-\sigma}} QO_i$$

Notation:

PS_i	Supply price of non-savings commodity *i*
QO_i	Quantity of non-savings commodity *i* output or supplied
$PFOB_i$	World (fob) price of tradable commodity *i* exported
PM_i	Market price of composite tradable commodity *i* demanded by Regional Household
QXS_i	Quantity of exports of tradable commodity *i*
QD_i	Quantity of composite tradable commodity *i* demanded by Regional Household
τ	Tax rate on production of commodity *i* per unit
π	Profit of firm
θ	Output augmenting technical change
σ	Share parameter

Equation 4
Domestic production

$$QO_i = \left[QVA_j^{\alpha} + \sum_j QF_{i,j}{}^{\alpha}\right]^{\frac{1}{\alpha}}$$

Notation

QVA_j	Quantity of value added (composite endowments) in firms of sector *j*
$QF_{i,j}$	Quantity of intermediate input *i* demanded by firms in sector *j*

Equation 5
Composite intermediate goods

$$VFA_{i,j} = \left[(PFM_{i,j} * QFM_{i,j})^{\alpha} + (PFD_{i,j} * QFD_{i,j})^{\alpha}\right]^{\frac{1}{\alpha}}$$

Notation:

$VFA_{i,j}$	Value of composite intermediate goods *i* demanded by firms in sector *j*
$PFM_{i,j}$	Demand price of imported intermediate goods *i* for firms in sector *j*
$QFM_{i,j}$	Quantity of imported intermediate goods *i* demanded by firms in sector *j*
$PFD_{i,j}$	Demand price of domestically produced intermediate goods *i* for firms in sector *j*
$QFD_{i,j}$	Quantity of demanded domestically produced intermediate goods *i* by firms in sector *j*

Equation 6

Behavior of value added

$$QVA_j = \delta(S_j^{\alpha} + U_j^{\alpha} + K_j^{\alpha} + LD_j^{\alpha} + NR_j^{\alpha}]^{\frac{1}{\alpha}}$$

Notation:

QVA_j	Quantity of value added in sector *j*
δ	Primary factor augmenting technical change in sector *j*
S_j	Quantity of skilled labor demanded by firms in sector *j*
U_j	Quantity of unskilled labor demanded by firms in sector *j*
K_j	Quantity of capital endowment demanded by firms in sector *j*
LD_j	Quantity of land endowment demanded by firms in sector *j*
NR_j	Quantity of natural resources endowment demanded by firms in sector *j*

Equation 7

Market clearing of endowments

$$S_j = \sum_j S_j \qquad U_j = D_U + M_U \frac{PR_M}{PR_D} = \sum_j U_j$$

$$K_j = \sum_j K_j \qquad LD_j = \sum_j LD_j \qquad NR_j = \sum_j NR_j$$

Notation:

$\sum_j S_j$	Total skilled labor in all sectors
$\sum_j U_j$	Total unskilled labor demanded by firms in sector *j*
D_U	Number of unskilled local labor demanded by firms in sector *j*
M_U	Number of unskilled migrant labor demanded by firms in sector *j*
PR_M	Productivity of unskilled migrant labor demanded by firms in sector *j*
PR_D	Productivity rate of unskilled local labor demanded by firms in sector *j*
$\sum_j K_j$	Total of Capital endowment demanded by firms in sector *j*
$\sum_j LD_j$	Total of Land endowment demanded by firms in sector *j*
$\sum_j NR_j$	Total of Natural resources demanded by firms in sector *j*

Equation 8
Market closure

$$EX - IM - REM = SAVE \qquad VKB = \sum_j K_j$$

Notations:

EX	Value of Export
SAVE	Value of Savings
REM	Remittance
IM	Value of Imports
VKB	Value of capital stock

The equilibrium solution is realized by solving the above equations. In the general equilibrium model discussed earlier, the equations derived 22 endogenous variables: *PP, PG, PCGDS, PIM, PM, PS, PFOB, PFM, PFD, Y, QP, QG, QCGDS, QIM, QDC, QO, QXS, QD, QVA, QF, QFM,* and *VFA* if τ, θ, δ, *S, U, K, LD, NR*, PR_M and PR_D are given exogenously. And α, β, γ, σ are the parameters.

Data sets for analysis

Data aggregation for empirical analysis is generated from the GTAP 7 Data Base. However, in order to simulate the GMig2 model, the model necessitates further data to complement the GTAP 7 Data Base. In addition, a downloadable database for GMig2 model is not sufficient for the Malaysia-ROW model. Additional ad hoc data was formulated based on various sources, mainly the Malaysian Economic Report (various issues), the Malaysian Labor Force Survey Report, and other publications by the Department of Statistics Malaysia.

The GMig2 Data Base is used as the main reference in formulating the aforementioned additional data. Consequently, all the data are compiled in Base Data.har file for simulation purposes.

Policy Analysis

According to economic reports [5], there are plans to reduce the number of foreign labors to 1.5 million by 2010. This policy is tested to evaluate its effect on the economy. Based on Table 18.1, as of end of March 2008, there were about 2.1 million legal unskilled foreign labors in the country. However, it is important to note that this number does not reflect economic conditions then, whereby many labors have been retrenched and lost their job due to the global economic crisis that has affected the world economy since the end of 2008. Given the unavailability of current data on the number of unskilled foreign labors in the country, data from year 2004 to 2008 are used in the simulation.

Based on Table 18.1, it is obvious that the inflow of unskilled foreign labor is increasing every year. Therefore, the study reported in this chapter aims to investigate the quantitative effects of the inflow of unskilled foreign labor into Malaysia, particularly to analyze the effects on the wages and welfare of local labors in the country.

Changes in the labor movement are modeled by 'shocking' the allocation of unskilled foreign labors in the model. The software used for this study is the current Source-code version (Release 8.0, October 2002) of GEMPACK. GEMPACK (General Equilibrium Modeling PACKage) is a suite of general-purpose economic modeling especially suitable for general and partial equilibrium models (refer Harrison & Pearson [15] for details). For simulation purposes, WinGem, the Windows interface of GEMPACK, is utilized.

The Results

The outcomes of the experiments are being discussed in two sections. The first section investigates the 5-year effects of increased inflow of unskilled foreign labor (2005 to 2010). Meanwhile, the second section examines the effects of policy intervention should it be implemented.

The effects of increasing inflow of foreign labor

The increase in inflow of unskilled foreign labor from labor exporting country (ROW) is found to have an overall positive impact on the welfare of the labor force (local and foreign) in the host country (Malaysia). Details of the effects are presented in the following tables and figures.

Table 18.3 depicts some of the outcomes from the increased inflow of unskilled foreign labor. Percentage change shown in Table 18.3 is relative to base year 2004. Malaysia experiences increases in real GDP as a consequence of an increased supply of labor to be utilized by firms in production. Due to the country's high reliance on unskilled labor, the increased supply of this labor causes an increase in output.

Table 18.3: Results of the simulation (Percentage change is relative to base year 2004)

	Real GDP (%)	Real wages of unskilled (%)	Real return to capital (%)	Household income (%)	Household income of local population (%)	Household income of foreign labor in Malaysia (%)	Remittances (%)	EV (USD Millions)	Change in Current Account Balance (USD Millions)
2005	0.73	-1.64	0.59	0.36	-0.08	17.65	17.65	463.28	-150.01
2006	1.33	-2.96	1.08	0.66	-0.13	31.93	31.93	850.09	-267.62
2007	1.51	-3.36	1.23	0.75	-0.14	36.23	36.23	968.65	-302.36
2008	2.03	-4.48	1.65	1.03	-0.18	48.56	48.56	1313.82	-400.15
2009	2.18	-4.81	1.78	1.11	-0.19	52.15	52.15	1415.78	-428.11
2010	2.28	-5.01	1.86	1.16	-0.19	54.38	54.38	1479.65	-445.42

The wages of unskilled labors fall with the increase in supply due to the inflow of foreign labors, while the wage of local skilled labor increases. With the assumption that unskilled local and foreign labors are a composite function linearly substitutable with the difference level of productivity, the wages of existing foreign labors are affected in the same way as those of local labors. However, foreign labors do not get the benefits of increased returns to capital, which are assumed to be owned by permanent/local residents. Returns to capital increase as greater labor supply and demand for goods increase the demand for capital. The rise in returns to capital will in the long run lead to an increased level of investment that would result in higher capital stocks and production (not within the scope of the model).

With the increase in foreign unskilled labor in Malaysia, the country's total income rises. However, if incomes of the local labor force and foreign labor are to be considered separately, the result shows that the total income of local labor decreases, although rental of capital increases (the percentage is very small since the increase in rental of capital is lower than the decrease in real wage). On the other hand, the total income of unskilled foreign labor increases as their number increases.

The decrease in prices of goods consequently increases demand in regional households and exports. Even though prices for imports increase, imports also rise in Malaysia due to increases in total income and the number of consumers. The demand for imports moves in the same direction as the inflow of unskilled foreign labors. The current account, which takes into account remittance flows, tends to decline as more remittances leave the country.

In the standard GTAP model, the measure of welfare change used is the Equivalent Variation (EV), which is obtained from the income and utility function of each regional household. In GMig, however, the EV of the host region has to be divided into two components, as with income: the EVs for local and foreign labors. The welfare of local labors is a function of the utility derived from their income. Meanwhile, the utility of these labors is a function of their income, number, and the prices of goods they purchase with this income. Given this utility, their EV can be calculated.

The welfare of foreign labors is similarly assumed. Their welfare is a function of utility derived from their income, from which remittances paid have already been subtracted.Their utility is a function of their income, number, and the price level in the host region, and thereby the EVs can be calculated. Consequently, the welfare change for a host region can therefore be found by summing the parts for local and foreign labor. Figure 18.1 indicates that the increased inflow of unskilled foreign labor results in an increase in total welfare in Malaysia.

All residents in Malaysia and ROW gain in terms of real incomes as a result of increased foreign labor in Malaysia. The Malaysian economy gains from increases in returns to capital and increased tax income, while ROW economies gain from increased remittances and wages.

Figure 18.1: The effect of the inflow of unskilled foreign labor on the welfare of the Malaysian population

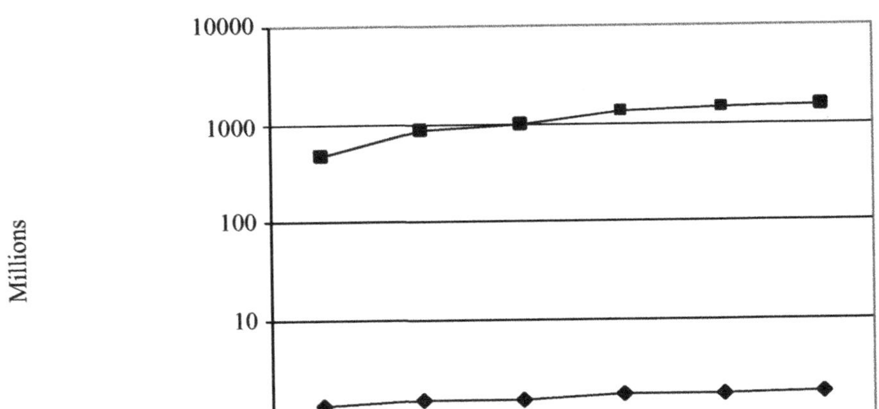

	2005	2006	2007	2008	2009	2010
Real Number (Millions)	1,269242	1,443958	1,497528	1,65355	1,699661	1,728555
EV (USD Millions)	463.28	850.09	968.65	1313.82	1415.78	1479.65

When particular attention is given to the construction sector, the result shows that firm price of unskilled labor in the sector decreases with the increased inflow of unskilled foreign labor into the sector. Hence, firms increase their demand for labors in the sector. The decrease in the price of labor reduces the cost of production and therefore reduces the price of output. This situation consequently increases the demand for construction output in Malaysia. Table 18.4 below illustrates the aforementioned situations.

Table 18.4: Results for construction sector

	Firm price for unskilled labor in construction (%)	Demand for unskilled labor in construction (%)	Change of construction output (%)
2005	-1.75	2.82	1.48
2006	-3.15	5.16	2.69
2007	-3.57	5.88	3.05
2008	-4.77	7.95	4.11
2009	-5.11	8.56	4.42
2010	-5.33	8.95	4.61

The effects of policy implementation

However, an experiment was conducted on the policy intervention intended to reduce the inflow of unskilled foreign labors to 1.5 million by 2010. Table 18.5 illustrates the outcome should the policy be implemented. The difference between 2010 without the policy and 2010 with the policy suggested reverse outcomes. Obverse to the discussion in the previous subsection, a reduction in unskilled foreign labor would cause real GDP to fall due to the subsequent reduction in output.

Table 18.5: Results of the simulation (with and without policy implementation)

	2010 (without policy)	2010 (with policy)	Difference
Real GDP (%)	2.28	0.49	-1.79
Terms of Trade (%)	-0.45	-0.10	0.35
Change in Trade Balance (% of GDP)	1071.19	229.62	-841.57
Change in Current Account Balance (USD Millions)	-445.42	-101.51	343.91
Real wages of unskilled (%)	-5.01	-1.11	3.90
Real return to capital (%)	1.86	0.40	-1.46
Household income (%)	1.16	0.24	-0.91
Household income of local population (%)	-0.19	-0.05	0.14
Household income of foreign labor in Malaysia (%)	54.38	11.87	-42.51
Remittances (%)	54.38	11.87	-42.51
EV (USD Millions)	1479.65	309.97	-1169.67
Firm price for unskilled labor in construction (%)	-5.33	-1.18	4.15
Demand for unskilled labor in construction (%)	8.95	1.89	-7.06
Change of construction output (%)	4.61	0.99	-3.62

The wage of unskilled labor still decreases, albeit at a lower rate, with the smaller supply of unskilled foreign labors entering the country. With policy implementation, however, the real wage of unskilled labor increases by 3.90%. Returns to capital would be lower by 1.46%, consequently reducing the total income. Nonetheless, the total income of the local population would eventually increase by 0.14%.

The decreased number of unskilled foreign labors would result in decreases in exports and imports. The current account balance will fare better with fewer remittances leaving the country. Total welfare will nevertheless be reduced by 79% (refer Table 18.5).

In the construction sector, the firm price of unskilled local labor still decreases by -1.88 percent with policy implementation, as compared to -5.33 percent in its estimated condition. The decrease in price ultimately instigates demand for more unskilled labors in the sector. However, the difference between the two scenarios is obvious; i.e., 8.95% without the policy and 1.89% with the policy, the lower decreases in the price thus creating lower construction output.

Assuming that the number of unskilled foreign labors is reduced to 1.5 million by 2010, the Malaysian economy would incur losses from decreased returns to capital and decreased tax income, while ROW economies would incur losses in the form of decreases in remittances and wages.

Conclusion

This chapter discusses general equilibrium modeling to measure the quantitative effects of unskilled foreign labors on a host country, with Malaysia as a case study. The model presented in this chapter is based on the GMig2 model, developed using the GTAP 7 database. In addition, the scenario of international labor movement into Malaysia is discussed as the main focus of the study.

A two-country, five-sector, and five-factor model is formulated to signify theoretically the scenario of international labor movement into Malaysia. Assumptions are made to simplify the analysis. Considering Malaysia's high reliance on unskilled foreign labor, the model assumes that inflows of foreign labor involve only this type of labor; therefore, the international movement of skilled labor and the issue of brain drain are ignored.

Results of the simulation are discussed in two parts. Part 1 looks at the effects of the increasing inflow of unskilled foreign labor into Malaysia from year 2005 to 2010. The result indicates that with the rising numbers of unskilled foreign labors, the Malaysian economy gains in terms of increasing returns to capital and tax incomes. Although unskilled local labor experiences decreases in real wages, the total welfare of the Malaysian population eventually increases, possibly due to the fact that permanent residents begin trading with foreign labors, thereby increasing the aggregate welfare.

Part 1 signifies that although it is theoretically proven that the perception of the increasing inflow of foreign labors would depress the wages of local unskilled labors, the situation would in fact increase their total welfare due to the reason mentioned above.

On the other hand, Part 2 of the results discussing implementation of the policy to reduce the number of unskilled foreign labors to 1.5 million by 2010 shows reverse effects. Such a reduction would cause GDP to fall and would reduce total income, thus decreasing the total welfare of the country's population.

Extremely labor-intensive, the construction sector is inevitably affected by the inflow of foreign labor and is highly dependent upon foreign labors willing to do work generally shunned by locals. Results show that the firm price of unskilled labor decreases with an increase in the supply of labor, hence creating more demand for the labor. With policy implementation reducing worker numbers, however, the lower decreases in firm price eventually reduce construction output.

Assuming that the number of unskilled foreign labors is reduced to 1.5 million by 2010, the Malaysian economy would incur losses from decreased returns to capital and decreased tax incomes while ROW economies would suffer losses in the form of decreases in remittances and wages.

References

[1] The World Bank (2006) *Global Economic Prospects: Overview and Global Outlooks 2006*. The International Bank for Reconstruction and Development, Washington D.C.

[2] National Economic Action Council (NEAC), (2003). *Impact of Foreign Workers on the Malaysian Economy*. NEAC, Kuala Lumpur, http://www.neac.gov.my (Accessed: 2005, June 23).

[3] Ministry of Finance Malaysia (2008) *Economic Report 2008/2009*. Percetakan Nasional Berhad, Kuala Lumpur.

[4] Ministry of Finance Malaysia (2007) *Economic Report 2007/2008*. Percetakan Nasional Berhad, Kuala Lumpur.

[5] Ministry of Finance Malaysia (2006) *Economic Report 2006/2007*. Percetakan Nasional Berhad, Kuala Lumpur.

[6] Ministry of Finance Malaysia (2005) *Economic Report 2005/2006*. Percetakan Nasional Berhad, Kuala Lumpur.

[7] Department of Statistics, Malaysia (2007) *Labor Force Survey Report 2006*. Putrajaya, Malaysia.

[8] Department of Statistics, Malaysia (2008) *Labor Force Survey Report 2007*. Putrajaya, Malaysia.

[9] Abdul Rashid Abdul Aziz (2001) Foreign workers and labour segmentation in Malaysia's construction industry. *Construction Management and Economics* **19**: 789-798.

[10] Narayanan, S. and Lai, Y.W. (2005) The Causes and Consequences of Immigrant Labour in the Construction Sector in Malaysia. *International Migration* **43(5)**: 31-57.

[11] International Labour Office (1995) *Social and labor issues concerning migrant workers in the construction industry/Triparte Meeting on Social and Labour Issues Concerning Migrant Workers in the Construction Industry*. Geneva.

[12] NST Online (21/07/2008), *Recruitment of Foreign Workers: Subra: Let my ministry handle it*. http://www.hst.com.my (Accessed: 2008, July 21).

[13] Watanabe, M. (2003) Economic Development and International Labour Migration in Malaysia, in Hayase, Y. (Editor) *International Migration In APEC Member Economies: Its Relations With Trade, Investment And Economic Development*. APEC Study Center publications in FY 2002/2003. APEC Study Center, Institute of Developing Economies.JETRO, Japan.

[14] MIDA (2009) *Employment of Foreign Workers*. http://www.mida.gov.my/en_v2/index.php?page=employment-of-foreign-workers.

[15] Harrison, W.J and Pearson, K.R (2002) *GEMPACK Document No. GPD-1*. Centre of Policy Studies and Impact Project, Monash University, Australia.

Endnotes

(i) Futher readings on GTAP - Hertel, Tom (ed.) (1997). Global Trade Analysis: Modeling and Applications, New York: Cambridge University Press and refer www.gtap.org/.

(ii) Further reading on GMig2 Model - refer https://www.gtap.agecon.purdue.edu/models/ date of sites labor_migration.asp.

Appendix

Structure of the model to show the flow of variables

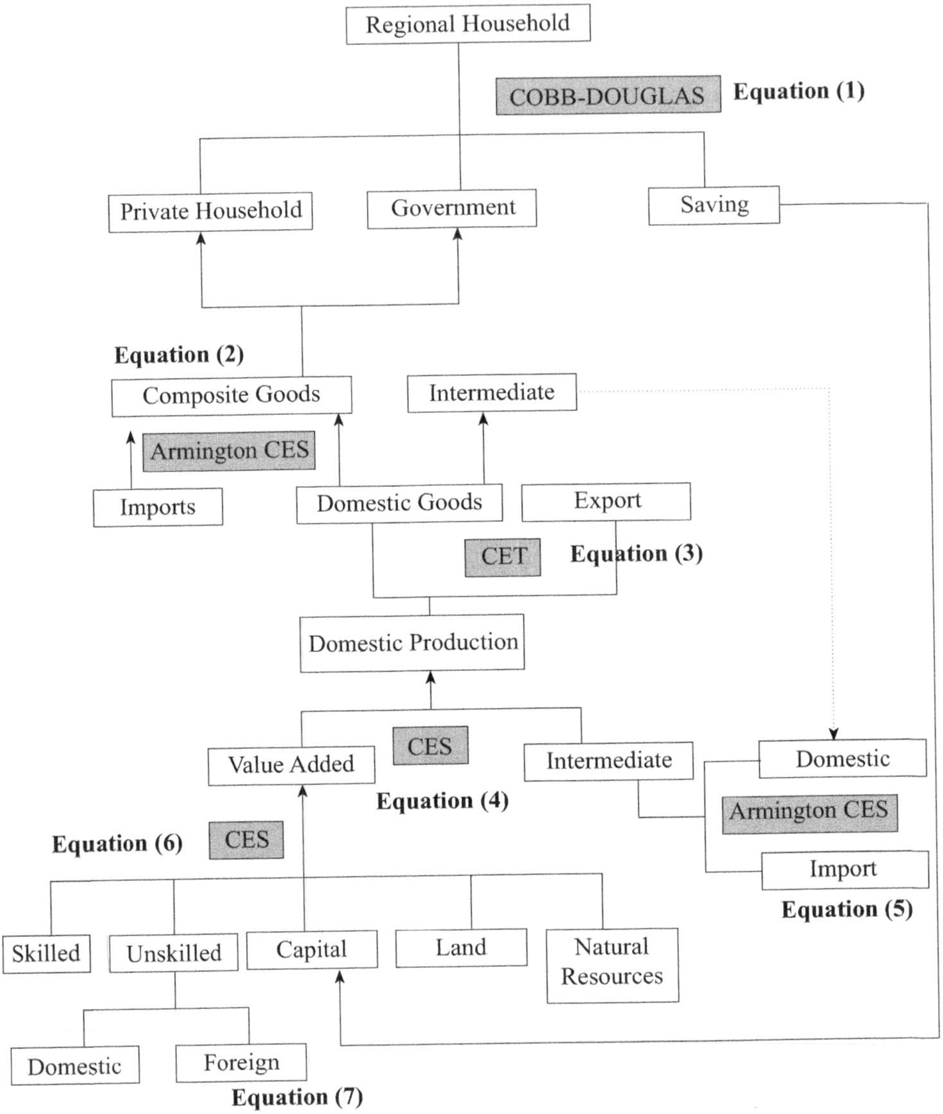

Joint Ventures in Construction 2
ISBN 978-0-7277-5783-8

doi: 10.1680/jvc2.57838.251

INDEX

Printed and bound by CPI Group (UK) Ltd, Croydon, CR0 4YY
16/09/2026
14963383-0002